Sustainable Practices for Agriculture and Marketing Convergence

Jabulani Garwi
University of the Free State, South Africa

Reason Masengu
Middle East College, Oman

Option Takunda Chiwaridzo
University of Science and Technology, Beijing, China

A volume in the Practice, Progress, and Proficiency in Sustainability (PPPS) Book Series

IGI Global
PUBLISHER of TIMELY KNOWLEDGE

Published in the United States of America by
IGI Global
Engineering Science Reference (an imprint of IGI Global)
701 E. Chocolate Avenue
Hershey PA, USA 17033
Tel: 717-533-8845
Fax: 717-533-8661
E-mail: cust@igi-global.com
Web site: http://www.igi-global.com

Library of Congress Cataloging-in-Publication Data

Names: Garwi, Jabulani, 1984- editor. | Masengu, Reason, 1982- editor. | Chiwaridzo, Option Takunda, 1990- editor.
Title: Sustainable practices for agriculture and marketing convergence / edited by Jabulani Garwi, Reason Masengu, Option Chiwaridzo
Description: Hershey, PA : Engineering Science Reference, 2024 | Includes bibliographical references and index. | Summary: "The book has a primary objective of highlighting the significance of sustainable agriculture in addressing the environmental impacts commonly associated with conventional farming methods"-- Provided by publisher
Identifiers: LCCN 2023059495 (print) | LCCN 2023059496 (ebook) | ISBN 9798369320112 (hardcover) | ISBN 9798369320129 (ebook)
Subjects: LCSH: Sustainable agriculture--Economic aspects. | Alternative agriculture. | Agricultural industries.
Classification: LCC S494.5.S86 S893 2024 (print) | LCC S494.5.S86 (ebook) | DDC 338.1068--dc23/eng/20240207
LC record available at https://lccn.loc.gov/2023059495
LC ebook record available at https://lccn.loc.gov/2023059496

This book is published in the IGI Global book series Practice, Progress, and Proficiency in Sustainability (PPPS) (ISSN: 2330-3271; eISSN: 2330-328X)

British Cataloguing in Publication Data
A Cataloguing in Publication record for this book is available from the British Library.

Practice, Progress, and Proficiency in Sustainability (PPPS) Book Series

ISSN:2330-3271
EISSN:2330-328X

Editor-in-Chief: Ayman Batisha, International Sustainability Institute, Egypt

MISSION

In a world where traditional business practices are reconsidered and economic activity is performed in a global context, new areas of economic developments are recognized as the key enablers of wealth and income production. This knowledge of information technologies provides infrastructures, systems, and services towards sustainable development.

The **Practices, Progress, and Proficiency in Sustainability (PPPS) Book Series** focuses on the local and global challenges, business opportunities, and societal needs surrounding international collaboration and sustainable development of technology. This series brings together academics, researchers, entrepreneurs, policy makers and government officers aiming to contribute to the progress and proficiency in sustainability.

COVERAGE

- Socio-Economic
- Knowledge clusters
- Green Technology
- E-Development
- Innovation Networks
- Strategic Management of IT
- ICT and knowledge for development
- Technological learning
- Outsourcing
- Environmental informatics

IGI Global is currently accepting manuscripts for publication within this series. To submit a proposal for a volume in this series, please contact our Acquisition Editors at Acquisitions@igi-global.com or visit: http://www.igi-global.com/publish/.

The Practice, Progress, and Proficiency in Sustainability (PPPS) Book Series (ISSN 2330-3271) is published by IGI Global, 701 E. Chocolate Avenue, Hershey, PA 17033-1240, USA, www.igi-global.com. This series is composed of titles available for purchase individually; each title is edited to be contextually exclusive from any other title within the series. For pricing and ordering information please visit http://www.igi-global.com/book-series/practice-progress-proficiency-sustainability/73810. Postmaster: Send all address changes to above address.

Titles in this Series

For a list of additional titles in this series, please visit:
http://www.igi-global.com/book-series/practice-progress-proficiency-sustainability/73810

Sustainable Development in AI, Blockchain, and E-Governance Applications
Rajeev Kumar (Moradabad Institute of Technology, India) Abu Bakar Abdul Hamid (Infrastructure University, Kuala Lumpur, Malaysia) Noor Inayah Binti Ya'akub (Infrastructure University, Kuala Lumpur, Malaysia) Hari Om Sharan (Amity University, Kolkata, India) and Sandeep Kumar (Rightzone Technologies Pvt. Ltd., India)
Engineering Science Reference • copyright 2024 • 282pp • H/C (ISBN: 9798369317228) • US $285.00 (our price)

Operational Research for Renewable Energy and Sustainable Environments
Joshua Thomas (University of Wollongong Malaysia, Malaysia) Gerhard Wilhelm Weber (Poznan University of Technology, Poland) Roman Rodríguez Aguilar (Universidad Panamericana, Mexico) Elias Munapo (North West University, South Africa) and Pandian Vasant (Ton Duc Thang University, Vietnam)
Engineering Science Reference • copyright 2024 • 350pp • H/C (ISBN: 9781668491300) • US $240.00 (our price)

Sustainable Solutions for E-Waste and Development
Rajesh Kumar K. V. (Woxsen University, India) Hemachandran Kannan (Woxsen University, India) Dmitry Spodarets (Data Phoenix, USA) Parvez Alam Khan (Universiti Teknologi PETRONAS, Malaysia) and Bikash Kumar Pradhan (Woxsen University, India)
Engineering Science Reference • copyright 2024 • 396pp • H/C (ISBN: 9798369310182) • US $265.00 (our price)

Teaching and Learning for a Sustainable Future Innovative Strategies and Best Practices
Chai Lee Goi (Curtin University, Malaysia)
Information Science Reference • copyright 2024 • 341pp • H/C (ISBN: 9781668498590) • US $230.00 (our price)

For an entire list of titles in this series, please visit:
http://www.igi-global.com/book-series/practice-progress-proficiency-sustainability/73810

IGI Global
PUBLISHER of TIMELY KNOWLEDGE

701 East Chocolate Avenue, Hershey, PA 17033, USA
Tel: 717-533-8845 x100 • Fax: 717-533-8661
E-Mail: cust@igi-global.com • www.igi-global.com

Table of Contents

Detailed Table of Contents

This chapter explores the connection between sustainable agriculture and the Sustainable Development Goals (SDGs). It discusses various practices like conservation agriculture, organic farming, agroforestry, and precision agriculture, and how they contribute to various SDGs. It focuses on SDG 2 (Zero Hunger), SDG 6 (Clean Water and Sanitation), SDG 15 (Biodiversity Preservation), and SDG 1 and 8 (Rural Development). The chapter also discusses barriers to widespread adoption, including economic, technological, and sociocultural factors. It uses case studies to illustrate successful models and offers policy recommendations, emphasizing national policies aligning with sustainable agriculture, fostering international cooperation, and investing in education and capacity building. The chapter provides valuable insights for policymakers, researchers, and practitioners in agriculture, sustainability, and development.

The agricultural sector faces considerable challenges as a result of climate change requiring the implementation of new and sustainable methods. This chapter delves into climate-smart agriculture (CSA), an all-encompassing strategy for boosting agricultural resilience and profitability that incorporates environmental, economic, and social considerations. Various agricultural settings are taken into account with topics including soil, water, crop, livestock, forest, and fisheries management. Precision farming, reforestation, and animal husbandry are all examples of CSA methods. Increased productivity, stable food supplies, lessened vulnerability, and lower expenses are just some of the ways it boosts economic benefits. The results of successful CSA programs can be learned from case studies. Challenges and future prospects are discussed in the latter section of the chapter, along with calls for further study, technical advancements, and policy changes.

This chapter explores the emerging issues in the agriculture supply chain. The chapters also aims to address the following objectives: (1) to examine sustainable agriculture practices that can reduce the environmental footprint, (2) to examine the environmental, logistical, and market-related challenges faced by agricultural supply chains, (3) to determine the role of digitalization and data analytics, including IoT applications and precision agriculture, and blockchain technology within agriculture supply chains, (4) to explore the future trends and opportunities in agriculture

The purpose of the study was to explore how Indigenous knowledge systems (IKS) could be integrated with modern farming techniques in order to enhance sustainable agriculture in selected rural areas. The sustainable livelihoods approach and post-colonial theory were used as the theoretical framework for the study. The interpretive approach was adopted for data generation by the researchers. Data were generated using individual interviews, focus group discussions (FGD), and field observations. Data were presented and analyzed using thematic approach. The study found that sustainable indigenous farming strategies including indigenous land management strategies, indigenous water conservation methods, and biodiversity management strategies could be integrated with modern farming techniques to reduce land degradation, loss of biodiversity, and lessen the impacts of climatic change in rural communities. The study recommends that farmers incorporate indigenous strategies with modern forms of farming in enhancing sustainable agriculture.

This chapter explores the transformative role of ICTs in revolutionizing agricultural sector of developing nations like Zimbabwe. By scrutinizing the current state of traditional and inefficient agricultural marketing in Zimbabwe, the study advocates for the potential of ICTs to enhance competitiveness and profitability. The benefits of digitizing marketing include minimizing information asymmetry, facilitating price discovery, expanding market linkages, improving financial access, and enabling traceability. Simultaneously, the paper delves into key challenges such as infrastructure limitations, skills gaps, affordability issues, gender disparities, and cultural factors that must be addressed. To create an enabling environment for digitizing agricultural marketing, the study recommends targeted policy and regulatory interventions. It concludes that strategic investments in ICT infrastructure and the capacity-building of farmers are crucial to fully unlock the potentials of ICTs, leading to improved incomes and enhanced rural livelihoods in Zimbabwe.

Contract farming is increasingly becoming a major source of funding for most agricultural commodities on the basis of its ability to ensure a consistent supply of raw materials to agro processing firms. Drawing from the market imperfection theory and the information asymmetry theory, extant literature has documented a number of factors that contribute to side marketing, but is very limited on profiling the common demographic factors attributed to that practice. Therefore, this study sought to develop a comprehensive profile of contract farmers that are likely to side market. A sample of 400 tobacco contract farmers drawn from Bindura District was surveyed. The data was analysed using logistic regression analysis, and the results indicated that the men, youth, single, primary level educated, and communal farmers were more likely to side market than their counterparts in the same demographic groups. The study recommended that agro-processing firms should be cautious when they enter into farming contracts with the flagged demographic groups.

For millions of people in developing nations, the agriculture industry is essential to their livelihoods, economic growth, and food security. It does, however, confront several difficulties, such as constrained resource availability, inefficient markets, and susceptibility to environmental fluctuations. Within this framework, integrating emerging technology offers a strong chance to transform marketing for agriculture and promote sustainable growth. This chapter explores the revolutionary prospective of using cutting-edge technologies for sustainable agriculture marketing in underdeveloped nations using a case study in Zimbabwe. Through an analysis of the convergence of technology, agriculture,

participation and securing policy support, and prioritising climate-resilient and gender-inclusive practises. Future research directions include investigations into the program's adaptability, explorations of its social and economic implications scrutinising its climate change resilience, and fostering global collaboration for knowledge exchange.

Preface

In an era marked by unparalleled challenges facing the agricultural sector, the call for innovative and sustainable solutions has become imperative. The confluence of environmental degradation, resource limitations, and the specter of climate change necessitates a departure from traditional agricultural paradigms toward a more sustainable trajectory. Concurrently, the landscape of marketing has undergone profound transformations propelled by technological advancements, evolving consumer preferences, and the ascendance of online platforms. It is within this dynamic milieu that this meticulously curated volume, aptly titled "Sustainable Practices for Agriculture and Marketing Convergence," embarks on its mission to elucidate the intersection between sustainable agriculture and marketing strategies, providing a nuanced understanding of this critical junction.

This compendium serves as an indispensable resource for a diverse array of stakeholders, spanning researchers, academics, policymakers, and industry professionals, all eager to delve into the intricate interplay between sustainable agriculture and marketing dynamics. By dissecting the convergence of these two domains, the volume sheds light on the seamless integration of sustainable practices across the agricultural value chain, encompassing production, distribution, and consumption. The narrative unfolds with an examination of the pivotal role played by marketing in advocating for sustainable agricultural practices while simultaneously fostering market opportunities for environmentally friendly products.

At its essence, the book endeavors to underscore the pivotal role of sustainable agriculture in mitigating the adverse environmental impacts associated with conventional farming methodologies. Historically, industrialized agriculture has been intrinsically linked to deforestation, soil erosion, water contamination, and greenhouse gas emissions. In contrast, sustainable agriculture prioritizes the alleviation of these deleterious effects through the adoption of eco-friendly techniques such as organic farming, precision agriculture, agroforestry, and integrated pest management. Through the presentation of compelling case studies and exemplary practices, the volume seeks to inspire and catalyze widespread adoption of sustainable agricultural methodologies.

Beyond its exploration of sustainable agriculture, this volume serves as a vibrant forum for researchers, scholars, practitioners, policymakers, and students to delve into the nexus of sustainable agriculture and marketing. By contributing cutting-edge research findings, innovative methodologies, and real-world case studies, the book endeavors to enrich the existing knowledge base. It integrates diverse perspectives from across disciplinary boundaries, fostering a deeper appreciation of the challenges and opportunities inherent in the convergence of sustainable agriculture and marketing realms. Furthermore, the volume explores potential solutions and provides insights into future trajectories within these domains.

Crafted to appeal to a broad readership, this volume caters to the interests of researchers and scholars in agriculture, environmental studies, marketing, economics, and sustainability. Likewise, practitioners, policymakers, and professionals in agriculture, agribusiness, marketing, and sustainable development will find practical applications and novel approaches to enhance their endeavors. Students embarking on studies in fields related to agriculture, marketing, and sustainability will benefit from the comprehensive coverage of topics and the synthesis of theoretical frameworks with practical insights. In essence, "Sustainable Practices for Agriculture and Marketing Convergence" aspires to empower a diverse audience, inviting them to deepen their understanding and actively contribute to the transformative journey of sustainable agriculture and marketing practices.

ORGANIZATION OF THE BOOK

Chapter 1: Sustainable Agriculture and the SDGs: A Convergence Approach

This chapter explores the connection between sustainable agriculture and the Sustainable Development Goals (SDGs). It discusses various practices contributing to SDGs, barriers to adoption, successful models, and policy recommendations. The chapter aims to provide valuable insights for policymakers, researchers, and practitioners in agriculture, sustainability, and development.

Chapter 2: Climate-Smart Agriculture for Resilience and Profitability

This chapter provides a comprehensive strategy for boosting agricultural resilience and profitability through climate-smart agriculture (CSA). It covers various agricultural settings, discussing soil, water, crop, livestock, forest, and fisheries management. The economic benefits, challenges, and future prospects of CSA are explored through case studies and discussions on the role of precision farming.

Chapter 3: Agriculture Supply Chains: Agriculture Supply Chains

This chapter delves into emerging issues in the agriculture supply chain, aiming to examine sustainable practices, environmental challenges, and the role of digitalization and data analytics. It explores future trends and opportunities, utilizing a critical review methodology, providing stakeholders with a comprehensive resource for informed decision-making and sustainability promotion.

Chapter 4: Fair Trade and Ethical Sourcing in Agriculture: Paving the Way for Sustainable Supply Chains

Examining the practices and implications of fair trade and ethical sourcing in the agricultural supply chain, this chapter identifies principles, benefits, and challenges associated with these practices. Through an extensive literature review, it aims to present case studies and recommend best practices for implementing fair trade and ethical sourcing.

Chapter 5: Cultural Ecology and Agriculture in Rural Areas in Zimbabwe

Concentrating on indigenous wisdom in agriculture, this chapter showcases the resilience of traditional practices in rural Zimbabwe. Highlighting Indigenous Knowledge Systems (IKS), the study explores how these practices contribute to sustainable farming, incorporating qualitative methods and interviews with subsistence farmers.

Chapter 6: Incorporating Indigenous Knowledge Systems (IKS) in Enhancing Sustainable Agriculture in Selected Rural Areas of Zimbabwe

Focused on integrating Indigenous Knowledge Systems (IKS) with modern farming techniques, this chapter explores sustainable indigenous farming strategies. It emphasizes the reduction of land degradation, biodiversity loss, and climate change impacts in rural communities by incorporating indigenous land management and water conservation methods.

Chapter 7: Digitizing Marketing in Agriculture: Leveraging Information Communication Technologies for Success in Zimbabwe

This chapter explores the transformative role of information communication technologies (ICTs) in revolutionizing the agricultural sector in developing nations like Zimbabwe. It examines the current state of traditional marketing, discusses the benefits of digitizing marketing, and addresses challenges, recommending policy interventions for creating an enabling environment.

Chapter 8: Side Marketing, a Threat to Contract Farming Viability in Zimbabwe: Demographic Profiling of Contract Farmers Proclivity to Side Marketing

Focusing on contract farming, this chapter profiles farmers likely to engage in side marketing. Through logistic regression analysis, it identifies demographic factors associated with side marketing, offering insights for agro-processing firms to cautiously enter contracts with specific demographic groups.

Chapter 9: Contemporary Agriculture Marketing Strategies for Smallholder Farmers in a Developing Context-Echoes From Zimbabwe: From the Farm to the Table

Exploring the transformative potential of cutting-edge technologies in underdeveloped nations, this chapter focuses on sustainable agriculture marketing. Using a case study in Zimbabwe, it analyzes the convergence of technology, agriculture, and marketing, offering stakeholders significant perspectives and practical recommendations for leveraging innovation in smallholder agricultural marketing practices.

Chapter 10: Alternative Dispute Resolution Techniques for Intertwined Artisanal Mining-Communal Farming Conflict

Focusing on artisanal mining and its impact on sustainable agriculture, this chapter reviews alternative dispute resolution (ADR) techniques. It discusses facilitated mediation, negotiation, and arbitration, emphasizing the importance of land-use zoning agreements to achieve synergy and market convergence, offering recommendations to prevent conflicts and promote sustainability.

Chapter 11: Enhancing Food Security through Sustainable Agriculture. A Case Study of the Pfumvudza/Intwasa Program in Zimbabwe

Focusing on the Pfumvudza program, this chapter exemplifies how sustainable agriculture can empower smallholder farmers and contribute to global food security. It recommends continuous monitoring, technology integration, community participation, and policy support, providing valuable lessons for nations grappling with food insecurity.

IN CONCLUSION

As we bring this meticulously curated collection to a close, the resonance of its collective wisdom echoes the commitment of our contributors to fostering a sustainable future in agriculture and marketing convergence. The diverse chapters presented in this edited reference book, *Sustainable Practices for Agriculture and Marketing Convergence*, have illuminated critical intersections, from supply chains and climate-smart agriculture to indigenous knowledge systems, digitization, and fair trade.

Each chapter is a testament to the evolving landscape of agriculture, marked by challenges that demand innovative solutions and opportunities that beckon transformative change. The exploration of sustainable practices and their integration into every facet of the agricultural value chain underscores a shared commitment to environmental stewardship, economic viability, and social responsibility.

From the examination of emerging issues in agriculture supply chains to the intricacies of precision irrigation, the contributors have delved into the complexities of our global food systems. The chapters addressing artisanal mining, indigenous knowledge, and the Pfumvudza program in Zimbabwe demonstrate a nuanced understanding of the diverse challenges and contexts in which sustainable agriculture must thrive.

Our journey has taken us through the realms of data-driven precision agriculture, the transformative potential of information communication technologies, and the intricate dynamics of fair trade and ethical sourcing. The demographic profiling of contract farmers and the exploration of the Sustainable Development Goals as they relate to agriculture further enrich our understanding.

In closing, this edited volume seeks to serve as a compass for researchers, scholars, practitioners, policymakers, and students navigating the dynamic landscape of sustainable agriculture and marketing. By integrating diverse perspectives, addressing challenges, and presenting innovative solutions, our contributors have not only expanded the existing body of knowledge but have also sparked a dialogue that we hope will reverberate across disciplines and borders.

As editors, we extend our deepest gratitude to the esteemed contributors who have shared their expertise, insights, and passion for sustainable practices. We invite readers to embrace this collection as a source of inspiration and knowledge, fostering a collective commitment to the sustainable transformation of agriculture and marketing practices for the betterment of our planet and future generations.

Jabulani Garwi
University of the Free State, South Africa

Reason Masengu
Middle East College, Oman

Option Takunda Chiwaridzo
University of Science and Technology Beijing, China

Chapter 1
Sustainable Agriculture and the SDGs:
A Convergence Approach

Muhammad Asim
University of Agriculture, Faisalabad, Pakistan

Aamir Raza
https://orcid.org/0009-0001-1867-2660
University of Agriculture, Faisalabad, Pakistan

Muhammad Safdar
https://orcid.org/0009-0006-1779-6967
University of Agriculture, Faisalabad, Pakistan

Mian Muhammad Ahmed
University of Agriculture, Faisalabad, Pakistan

Amman Khokhar
University of Agriculture, Faisalabad, Pakistan

Mohd Aarif
https://orcid.org/0000-0001-9986-4818
Global Research Network, Noida, India

Mohammed Saleh Al Ansari
https://orcid.org/0000-0001-9425-0294
University of Bahrain, Bahrain

Jaffar Sattar
https://orcid.org/0009-0006-6107-9156
Khwaja Fareed University of Engineering and Information Technology, Rahim Yar Khan, Pakistan

Ishtiaq Uz Zaman Chowdhury
North South University, Bangladesh

ABSTRACT

This chapter explores the connection between sustainable agriculture and the Sustainable Development Goals (SDGs). It discusses various practices like conservation agriculture, organic farming, agroforestry, and precision agriculture, and how they contribute to various SDGs. It focuses on SDG 2 (Zero Hunger), SDG 6 (Clean Water and Sanitation), SDG 15 (Biodiversity Preservation), and SDG 1 and

DOI: 10.4018/979-8-3693-2011-2.ch001

8 (Rural Development). The chapter also discusses barriers to widespread adoption, including economic, technological, and sociocultural factors. It uses case studies to illustrate successful models and offers policy recommendations, emphasizing national policies aligning with sustainable agriculture, fostering international cooperation, and investing in education and capacity building. The chapter provides valuable insights for policymakers, researchers, and practitioners in agriculture, sustainability, and development.

1. INTRODUCTION

1.1. Background and Significance

The Sustainable Development Goals (SDGs) provide a comprehensive and inclusive appeal for collective efforts to eradicate poverty, safeguard the environment, and secure universal peace and prosperity by the year 2030. The agricultural sector plays a crucial role in the attainment of these objectives, with a particular emphasis on SDG2, which seeks to "End hunger, achieve food security and improved nutrition, and promote sustainable agriculture". Sustainable agriculture is an agricultural approach that prioritizes the cultivation of crops and livestock over an extended period, with the aim of minimizing environmental impact. The concept entails the effective administration of agricultural resources to meet evolving human demands, while simultaneously preserving or improving environmental quality and conserving natural resources (Paroda and Joshi, 2019). In this context, the convergence method pertains to the amalgamation of sustainable farming practices with the goals and objectives outlined in the SDGs. This approach acknowledges the interconnectedness between several objectives, such as promoting sustainable agriculture, empowering small-scale farmers, advancing gender equality, eradicating rural poverty, promoting healthy lifestyles, addressing climate change, and other topics encompassed by the 17 Sustainable Development Goals.

1.2. Objectives of the Chapter

The main objectives of this chapter are:

- To understand the concept of sustainable agriculture and its significance in the context of the SDGs.
- To explore various sustainable agricultural practices and their potential to contribute towards achieving the SDGs.

- To discuss case studies highlighting how sustainable agriculture is being practiced in countries around the world.
- To provide practical guidelines for implementing sustainable agriculture in line with the 2030 Agenda.

2. SUSTAINABLE AGRICULTURE AND ITS RELEVANCE TO THE SDGS

2.1. Defining Sustainable Agriculture

Sustainable agriculture is an agricultural system that seeks to meet the needs of current human populations while also preserving the Earth's capacity to support future generations. The integration of three primary objectives is observed, namely environmental health, economic profitability, and social and economic equity. The agricultural approach under consideration prioritizes the cultivation of sustainable crops and livestock with the aim of minimizing environmental impact over an extended period of time. The concept encompasses the effective administration of agricultural resources in order to meet evolving human demands, while simultaneously preserving or improving environmental quality and conserving natural resources (Czyzewski and Kryszak, 2022).

Sustainable agriculture places significant emphasis on the cultivation of a wide range of crops, including heritage plants, which are frequently selected based on their compatibility with the specific climatic conditions prevalent in a given region. Sustainable agriculture proponents advocate for the implementation of polyculture as an alternative to industrial monoculture, wherein numerous crops are cultivated simultaneously instead of relying on a single crop. Sustainable agriculture can be defined as a farming approach that prioritizes the preservation of the environment, the enhancement and conservation of natural resources, and the optimal utilization of nonrenewable resources.

2.2. The Role of Sustainable Agriculture in Achieving the SDGs

The SDGs, also known as the Global Goals, are a collection of 17 interlinked objectives designed to serve as a "shared blueprint for peace and prosperity for people and the planet, now and into the future as shown in figure 2. They address the global challenges we face, including poverty, inequality, climate change, environmental degradation, peace, and justice. There are many ways in which sustainable agriculture and the SDGs are intertwined. The implementation of sustainable agricultural practices plays a pivotal role in the attainment of SDG2, which encompasses the eradication of hunger, the establishment of food security, and the enhancement of nutritional standards. By pushing for the implementation of sustainable agricultural methods, it is possible to enhance food production and provide universal access to an adequate, secure, and nourishing food supply. The promotion of sustainable agriculture is aimed at facilitating the efficient utilization of natural resources. This is consistent with multiple Sustainable Development Goals (SDGs), namely SDG6 (Clean Water and Sanitation), SDG7 (Affordable and Clean Energy), SDG12 (Responsible Consumption and Production), and SDG15 (Life on Land). The mitigation of climate change is a significant responsibility assigned to the agricultural sector, as emphasized by the Sustainable Development Goal 13. The use of sustainable agricultural techniques has the potential to mitigate greenhouse gas emissions and enhance the adaptability of farming systems in the face of climate change consequences.

To effectively address the issue of poverty eradication, it is imperative to prioritize rural development and allocate resources towards sustainable agriculture. These strategies have proven to be influential in achieving the first Sustainable Development Goal (SDG1) of ending poverty, as well as promoting economic growth, as outlined in SDG8. Sustainable agriculture has the potential to alleviate poverty within communities by enhancing the production and revenue of small-scale farmers. The promotion of gender equality can also be facilitated by the implementation of sustainable agriculture practices, so contributing to the achievement of Sustainable Development Goal 5 (SDG5). The participation of women in the agricultural sector is of paramount importance in numerous regions across the globe. The enhancement of agricultural production and sustainability can be achieved through the promotion of gender equality in access to agricultural resources, services, and opportunities. In summary, it can be said that agriculture serves as the primary connecting element among the 17 Sustainable Development Goals (SDGs), surpassing other sectors in its significance. The act of allocating resources towards sustainable agriculture has the potential to effectively tackle a range of issues, such as hunger, malnutrition, poverty, water and energy usage, climate change, as well as unsustainable production and consumption patterns (Omilola and Robele, 2017).

3. SUSTAINABLE AGRICULTURE PRACTICES

3.1. Conservation Agriculture

Conservation Agriculture (CA) is an agricultural strategy that endeavors to mitigate the loss of arable land and restore degraded lands. The practice encourages the perpetuation of a continuous soil covering, little disruption of soil, and the introduction of a variety of plant species. This strategy effectively promotes biodiversity and natural biological processes in both the above ground and below ground environments, hence leading to enhanced water and nutrient utilization efficiency and ultimately resulting in improved and sustained agricultural productivity. Conservation agriculture encompasses three fundamental principles as shown in figure 2: (i) the practice of direct seeding or planting with minimal soil disturbance, (ii) the maintenance of permanent or semi-permanent soil cover by minimal soil disturbance or direct seeding, and (iii) the implementation of varied crop rotation.

Soil interventions, such as mechanical soil disturbance, are minimized or avoided completely, while external inputs, such as agrochemicals and plant nutrients derived from minerals or organic sources, are applied in an optimal manner. These inputs are carefully managed to ensure they do not impede or disrupt the natural biological processes. Conservation Agriculture promotes effective agronomic practices, including timely operations, and enhances land management for both rainfed and irrigated agricultural production. In conjunction with established best practices such as the utilization of high-quality seeds, as well as the implementation of integrated approaches to pest control, nutrient management, weed management, and water management, Conservation Agriculture serves as a fundamental framework for the intensification of sustainable agricultural production (Corsi and Muminjanov 2019).

3.2. Organic Farming

Organic farming is an environmentally conscious agricultural practice that places a strong emphasis on sustainability, biodiversity, and the well-being of both the environment and the animals involved. It adheres to a set of principles that exclude the use of synthetic chemicals, such as pesticides and herbicides, favoring natural alternatives like compost and cover crops for crop protection and nourishment. Soil health is paramount in organic farming, with practices like crop rotation and reduced tillage enhancing fertility and reducing erosion. Biodiversity is encouraged, allowing natural predators to manage pests and fostering diverse cropping systems. In organic livestock farming, animal welfare is a top priority, providing animals with outdoor access and organic feed. Non-GMO cultivation is a common practice, and organic products are typically certified, meeting specific organic farming standards. Despite the potential for lower yields and higher labor costs, the demand for organic products continues to rise, driven by consumer preferences for healthier, environmentally responsible food options.

3.3. Agroforestry

Agroforestry can be defined as the deliberate incorporation of trees and shrubs into agricultural practices involving crops and livestock, with the aim of generating advantageous outcomes in terms of the environment, economy, and society (Umrani, 2010). Agroforestry, alternatively referred to as agro-sylviculture, is a method of land management wherein the cultivation of trees or shrubs is integrated with crops or pastureland. The integration of agricultural and forestry technology facilitates the development of land-use systems that are characterized by enhanced diversity, productivity, profitability, health, and sustainability. The agricultural approach being discussed here involves maximizing the advantages derived from intentional combinations of trees and/or shrubs with crops and/or livestock, so harnessing the potential of biological interactions. The practice encompasses the development and utilization of trees and shrubs in conjunction with crops and livestock within agricultural systems.

Agroforestry can be described as a form of agriculture that incorporates the presence of trees inside farming systems and agricultural environments. This includes the integration of trees on farms and in agricultural landscapes, cultivation practices within forested areas and along forest edges, as well as the creation of tree-crop combinations. The practice encompasses a diverse array of trees that are conserved, restored, cultivated, or supervised within agricultural environments, where they interact with annual crops, livestock, and wildlife. The sustainable integration of trees into agriculturally productive landscapes is crucial for maintaining the benefits they offer, especially when natural forests are being removed for agricultural and other developmental purposes (Haq *et al.,* 2023). This phenomenon has a significant historical background that spans over thousands of years.

3.4. Precision Agriculture

Precision agriculture (PA) refers to a data-centric approach employed in agriculture, where information technology is utilized to enhance the effectiveness and ecological integrity of crop cultivation. Precision agriculture technology assist farmers in gathering and analyzing data pertaining to various aspects of their farms, including but not limited to soil moisture, nutrient concentrations, and crop vitality (Bucci *et al.,* 2018). The data can thereafter be utilized to make well-informed judgments on the management of various inputs, including but not limited to fertilizer, water, and pesticides.

Precision agriculture (PA) has the potential to provide numerous advantages to farmers, encompassing:

- Enhanced crop productivity: Precision agriculture (PA) can facilitate the augmentation of crop yields by ensuring the timely provision of essential resources to crops.

- One potential benefit of precision agriculture (PA) is its ability to assist farmers in reducing costs through the more effective utilization of inputs. One illustrative instance involves the implementation of variable-rate fertilization, which has the potential to decrease fertilizer consumption without compromising crop productivity.
- Enhanced environmental sustainability: Precision agriculture (PA) can facilitate the mitigation of farmers' environmental footprint through the implementation of reduced pesticide and herbicide usage, as well as the conservation of water and other valuable resources.

The following is an illustrative instance of the practical application of performance appraisal (PA). The utilization of precision agriculture (PA) techniques is employed by a corn farmer in the state of Iowa to effectively manage his agricultural operations. The individual use sensors to quantify the moisture content of the soil, subsequently utilizing variable-rate irrigation techniques to administer water solely in areas where it is required. In addition, the utilization of unmanned aerial vehicles (UAVs) is employed by the individual to acquire visual data of the agricultural plot. Subsequently, specialized software is employed to discern and categorize certain plants exhibiting signs of distress or sickness. Subsequently, he proceeds to apply specific therapies to those particular plants.

3.5. Crop Rotation and Diversification

Crop rotation and diversification are essential practices in the realm of sustainable agriculture. Crop rotation encompasses the agricultural technique of cultivating diverse varieties of crops within a given geographical region, alternating them over various seasons. This approach offers numerous advantages. To begin with, the provision of diverse food sources for soil bacteria contributes to the improvement of soil health, hence enhancing its overall condition. Additionally, it contributes to the management of weeds and the control of pests. Various crops exhibit varying growth patterns and possess diverse root structures, hence potentially disrupting the life cycles of pests and weeds. This phenomenon can serve as a natural mechanism for controlling these unwanted organisms. Furthermore, the implementation of crop rotation and diversification practices contributes to the advancement of environmental sustainability by mitigating adverse environmental effects, all the while ensuring the economic viability of agricultural operations (Lemaire *et al.,* 2015). Finally, when viewed through an economic lens, these methods provide farmers with a variety of income streams and mitigate production risk by avoiding dependence on a certain crop. Consequently, the use of crop rotation and diversification strategies is gaining momentum as effective methods for ensuring the long-term sustainability of crop production.

4. SUSTAINABLE AGRICULTURE AND FOOD SECURITY (SDG 2)

4.1. Ensuring Access to Safe and Nutritious Food

The assurance of access to food that is both safe and nutritious is a fundamental component of food security. This encompasses the provision of sufficient, safe, and nutritious food to meet individuals' dietary requirements, while also addressing the aspects of physical, social, and economic accessibility. The matter at hand holds significant significance as it serves to address not only the issues of hunger, poverty, and malnutrition, but also plays a crucial role in shaping the national security and political and economic autonomy of nations. The primary objective of the Sustainable Development Goals (SDGs) is to eradicate hunger and guarantee universal access to safe, nutritious, and adequate food throughout the year, with a special focus on impoverished individuals and those in vulnerable circumstances, such as infants. Nevertheless, the majority of nations are not making sufficient progress towards achieving the nutrition targets set by the World Health Organization or the Sustainable Development Goal 2 targets. The COVID-19 pandemic has further intensified the issue of malnutrition and underscored the imperative for ensuring food safety.

Hence, it is imperative to undertake the necessary measures to overhaul our food systems in order to guarantee universal access to a diet that is both safe and nutritionally balanced. The adoption of a comprehensive approach encompassing policy, research, and monitoring and evaluation is necessary in this context. Additionally, this entails the resolution of supply chain deficiencies and the use of technology-based remedies tailored to particular circumstances (Hendriks *et al.,* 2023).

4.2. Sustainable Crop and Livestock Production

The establishment of sustainable crop and livestock production serves as a fundamental component within the framework of sustainable agriculture. The concept encompasses the effective and equitable allocation of resources to promote economic sustainability, environmental well-being, and social justice. The crop-livestock system is a land use approach that involves the integrated practice of livestock husbandry and crop cultivation. likewise, it is important to note that the production system does not have any adverse effects on the environment. The rapid growth of the global human population has led to an escalation in the prevalence of food insecurity and malnutrition, particularly in developing nations. Hence, there exists a pressing want for an enhanced, streamlined, and environmentally conscious production system that can adequately cater to the nutritional demands of the growing global populace, including both agricultural and livestock outputs. The integration of agricultural and livestock production systems offers a promising possibility to sustainably produce and deliver food from both plant and animal sources, while minimizing negative effects on the delicate environment. Within this integrated system, the interaction between crop cultivation and animal rearing results in a synergistic relationship that facilitates the achievement of ecosystem equilibrium and the promotion of sustainable intensification of production. One notable contribution is the utilization of manure as a vital source of natural fertilizer (Baiyeri *et al.,* 2019).

Additionally, the utilization of livestock as draft animals can enhance productivity in areas with limited automation. Livestock holds significant value as assets within vulnerable communities. Silvopastoral systems, encompassing the integration of shrubs and trees bearing food leaves or fruits alongside herbage, represent a potentially viable strategy for achieving sustainable livestock production. These systems have the potential to offer a well-rounded nutritional intake for livestock, boost the welfare of animals, and contribute to the preservation and enrichment of biodiversity. In a nutshell, the attainment of food security, the preservation of biodiversity, and the long-term viability of agricultural systems heavily rely on the implementation of sustainable practices in crop and livestock production (Lemaire *et al.,* 2014).

4.3. Reducing Food Loss and Waste

The issue of food shortage continues to be a significant concern in different regions across the globe, with approximately one-third of the overall food output being thrown out as food loss and waste (FLW). Approximately 25% of the land, water, and fertilizer utilized for agricultural cultivation is attributed to this phenomenon, despite the anticipation of resource scarcity and environmental limitations that are projected to impede global food production (Shafiee-Jood and Cai, 2016). The reduction of food loss and waste (FLW) has garnered significant attention in recent times due to its potential to improve food security and promote environmental sustainability. The reduction of food loss and waste is an imperative matter with substantial significance for the economy, ecology, and society at large. Approximately 24% of the total caloric output intended for human consumption is currently lost or wasted (Lipinski *et al.,* 2013). The loss and waste of agricultural produce signify an inefficient use of resources, leading to diminished returns for farmers and heightened costs for consumers. From an environmental perspective, the issue of food loss and waste has numerous consequences, such as the emission of greenhouse gases and the wasteful utilization of water and land resources.

There exist multiple solutions aimed at mitigating the issue of food loss and waste. Some of the strategies that can be employed to address these challenges include advancements in technology and infrastructure, optimization of packaging methods, improvement of food management practices, and alignment of supply with demand. For example, inadequate infrastructure or a deficiency in cold storage facilities might impede the transportation of food from its origin to the final consumer. The implementation of appropriate packaging techniques can contribute to the preservation of food quality and extend its shelf life, thereby mitigating the occurrence of spoiling. Minimizing food loss and waste has the potential to emerge as a prominent global approach in attaining a sustainable food future. If the existing rate of food loss and waste were reduced by 50%, the global demand for food per annum would decrease by around 22% compared to the projected scenario of maintaining current practices.

5. SUSTAINABLE AGRICULTURE AND WATER MANAGEMENT (SDG 6)

5.1. Efficient Water Use in Agriculture

The issue of optimizing water utilization in agricultural practices is of paramount importance, given that the agricultural sector ranks among the most water-intensive industries on a global scale. In order to tackle this matter, a range of strategies and practices have been devised to enhance the efficiency of water utilization in agricultural activities. Drip irrigation and micro-irrigation systems are prominent instances. These techniques facilitate the targeted delivery of water to the root zone of plants, hence reducing water loss due to evaporation and runoff. Through the implementation of precise irrigation control, farmers have the ability to customize these systems in order to effectively address the distinct water requirements of various crops. As a consequence, there is a notable reduction in water consumption without compromising the well-being of the crops.

Soil moisture monitoring is an additional crucial component of effective water utilization. Farmers have the ability to utilize soil moisture sensors in order to ascertain the precise water needs of their crops. This technological advancement enables precise irrigation scheduling, effectively mitigating the issues of excessive watering and inadequate watering. The use of this system guarantees the provision of appropriate water quantities to crops at the required times, hence promoting water resource conservation and enhancing crop productivity. The selection of crops also plays a significant role in enhancing water use efficiency. Farmers have the ability to decrease water requirements by selecting crop varieties that are well adapted to the specific climate and soil conditions of their region. Certain agricultural crops exhibit greater resistance to drought or possess lower water demands, and the deliberate choice of such crops can contribute to the alleviation of water resource strain.

Crop rotation and cover cropping are two sustainable agricultural strategies that have the potential to improve water efficiency. These agricultural approaches have been shown to enhance soil quality, minimize the requirement for excessive irrigation, and effectively manage pest populations. Crop rotation is a farming practice that entails the sequential cultivation of diverse crop varieties within a given field during consecutive growing seasons. This method serves to disrupt pest life cycles and mitigate the dependence on water-demanding crops. Crops are strategically sown in the intercropping spaces between cash crops with the purpose of safeguarding and enhancing the soil quality, hence contributing to a more efficient utilization of water resources. Mulching is an additional method employed to enhance water utilization efficiency within the agricultural sector. The application of mulch, such as straw or plastic film, over the surface of the soil serves to enhance soil moisture retention, inhibit the growth of weeds, and mitigate evaporation. The function of this layer is to provide protection, hence enhancing water conservation and promoting effective water management on a broader scale.

Furthermore, the implementation of conservation tillage measures also plays a significant role in enhancing water efficiency. The implementation of reduced or no-till farming techniques has been found to have several beneficial effects on agricultural practices. These strategies contribute to the preservation of soil moisture, the mitigation of water runoff, and the improvement of soil health. Through the implementation of reduced soil disturbance and the practice of retaining crop residue, these strategies effectively enhance the soil's water-holding capacity, hence increasing its efficiency in retaining moisture. The optimization of water utilization in agricultural practices can be further augmented by the integration of advanced technologies and precision agriculture techniques. The utilization of GPS-guided tractors, drones, and sophisticated data processing techniques empowers farmers to enhance the efficiency of resource allocation. These technologies offer instantaneous data on the health of crops and the conditions of soil, enabling accurate decision-making in the use of irrigation and fertilizers.

Efficient water utilization in the agricultural sector is necessary to foster sustainable and conscientious farming methodologies. The practice not only facilitates the preservation of water resources but also entails a reduction in energy consumption, mitigation of environmental consequences, and a positive contribution to the long-term sustainability of the agricultural industry. Collaboration among governments, agricultural organizations, and farmers is vital in order to advance and execute these practices, thereby guaranteeing a future for agriculture that is both water-efficient and sustainable.

5.2. Sustainable Irrigation Practices

The implementation of sustainable irrigation practices plays a crucial role in the responsible stewardship of water resources within the agricultural sector. The purpose of their design is to achieve a harmonious equilibrium between the demands of crop cultivation, environmental preservation, and the sustainable management of water resources over an extended period. Drip irrigation and micro-irrigation systems are widely recognized as highly efficient and sustainable ways of irrigation. These irrigation systems effectively transport water directly to the root zone of plants, resulting in a notable reduction in water loss due to evaporation and runoff. Through the implementation of precise irrigation control, farmers have the ability to customize these systems in order to align with the distinct water needs of various crops, hence promoting optimal water utilization and enhancing crop health. Rainwater harvesting is a sustainable practice that entails the collection and storage of rainwater for the specific purpose of irrigation. This methodology has the potential to decrease the dependence on groundwater and surface water resources, rendering it especially advantageous in areas characterized by periodic precipitation patterns.

The implementation of effective irrigation techniques is crucial in promoting the sustainable utilization of water resources. The implementation of sprinklers, pivots, or furrow irrigation systems has the potential to mitigate water wastage. The implementation, upkeep, and functioning of these systems are of paramount importance in ensuring sustainability Effective water management methods play a crucial role in mitigating water losses. These practices encompass various measures, including the regulation of water flow, the upkeep of appropriate furrow and basin designs, and the prevention of leaks within irrigation systems. In order to ensure the efficiency of water systems, it is imperative to carry out regular maintenance and implement appropriate modifications. The utilization of wastewater treatment and subsequent application in agricultural practices can additionally help to the promotion of sustainability. The implementation of wastewater treatment and reuse in agricultural operations serves to mitigate the need for freshwater resources and mitigate pollution, so exemplifying an environmentally conscientious approach.

Precision agriculture, propelled by advanced technologies such as GPS-guided tractors and unmanned aerial vehicles (UAVs), enhances resource utilization efficiency, particularly water, through the provision of up-to-date information on crop vitality and soil conditions. The utilization of data-driven methodologies promotes the sustainability of agricultural practices by empowering farmers to make well-informed decisions on the allocation of resources. The significance of sustainable irrigation scheduling lies in its ability to customize irrigation practices according to the growth phases of crops and prevailing meteorological conditions, hence minimizing the wasteful utilization of water resources.

Government restrictions and incentives are frequently employed as mechanisms to encourage the adoption of sustainable irrigation practices. Potential measures that can be implemented encompass water pricing mechanisms, incentives aimed at promoting the adoption of water-efficient technology, and mandates for the development and implementation of water management plans. Moreover, the implementation of sustainable irrigation practices heavily relies on the presence of teaching and training programs. It is imperative to provide farmers and agricultural experts with comprehensive education regarding the importance of sustainable irrigation, as well as provide them with the necessary training to effectively apply these techniques. The dissemination of knowledge and the promotion of awareness are essential factors in fostering a sustainable and water-efficient agriculture industry.

5.3. Water Quality and Agriculture

The relationship between water quality and agriculture is highly interconnected, and recognizing this interdependence is crucial in the pursuit of SDG 6, which seeks to ensure universal access to clean water and sanitation. Agriculture, an essential means of survival and the production of food, exerts a substantial influence on water quality through diverse ways. Significantly, the application of fertilizers and pesticides in agricultural practices has the potential to result in the discharge of nutrients and the pollution of aquatic ecosystems. The introduction of an excessive amount of nutrients, specifically nitrogen and phosphorus, into rivers, lakes, and seas can result in the contamination of water and the development of detrimental algal blooms, hence causing adverse effects on aquatic ecosystems. These challenges undermine the achievement of SDG 6, which aims to guarantee universal access to clean and safe water. Additionally, pesticides and herbicides, which are crucial instruments in contemporary agriculture for safeguarding crops against pests and weeds, have the potential to unintentionally contaminate water supplies due to inappropriate usage or the runoff originating from agricultural fields. The presence of this contamination has the potential to create hazards for both the natural ecosystem and the well-being of human beings. Furthermore, inadequate land management techniques, such as excessive tilling and the removal of forests, have a role in the process of soil erosion. This, in turn, leads to the transportation of sediment-laden runoff, which exacerbates the deterioration of water quality. In addition, the practice of cattle farming presents water quality concerns. Animal agriculture is responsible for the production of manure, a substance that contains both diseases and an abundance of nutrients. Insufficient management of manure can result in the contamination of water bodies and pose potential health hazards.

In order to address the potential negative effects of agriculture on water quality and contribute to the achievement of SDG 6, the implementation of various solutions is needed. An essential measure entails the use of sustainable farming methods aimed at mitigating the reliance on excessive fertilizers and chemicals. Precision agriculture, crop rotation, and cover cropping are agricultural practices that have been identified as effective strategies for mitigating nutrient runoff and pollution, while also ensuring the sustainability of agricultural productivity. The implementation of vegetative buffer zones between agricultural regions and water bodies can serve as a natural filtration mechanism, thereby mitigating the pollution of water sources. Effective waste management practices in agriculture, with a specific focus on the appropriate handling of manure and crop residues, are of utmost importance in order to mitigate the potential pollution of adjacent water sources. Governments possess the capacity to assume a central role in the realm of environmental stewardship through the establishment and enforcement of rules pertaining to the utilization of fertilizers, pesticides, and herbicides, so effectively mitigating their adverse effects on the environment. The regular monitoring of water quality is of paramount importance in order to promptly identify any potential issues and implement appropriate corrective actions. Furthermore, the implementation of soil conservation measures, including the reduction of tillage, the establishment of forests, and the creation of new forests, is necessary in order to mitigate soil erosion and protect the quality of water resources.

Education and outreach initiatives targeting farmers, with the objective of enhancing their understanding of the significance of water quality and imparting knowledge on sustainable agricultural methods, play a pivotal role in facilitating transformative outcomes. The pursuit of research and innovation within the agricultural domain holds the potential to provide novel methodologies and technologies that effectively mitigate the environmental repercussions associated with this sector. Consequently, such advancements offer enduring resolutions to the predicaments concerning water quality and agriculture. Examining the relationship between water quality and agriculture is crucial not just in the context of attaining Sustainable Development Goal 6, but also for guaranteeing the long-term viability and adaptability of the agricultural industry. Through the implementation of these practices and policies, it is possible to ensure the provision of sustenance and economic opportunities for the expanding global populace, while also protecting the well-being of our water resources and the ecosystems that depend on them.

6. SUSTAINABLE AGRICULTURE AND BIODIVERSITY (SDG 15)

6.1. Conserving and Restoring Ecosystems

The preservation and rehabilitation of ecosystems, within the framework of Sustainable Agriculture and Biodiversity (SDG 15), play a pivotal role in endeavors aimed at safeguarding the intricate equilibrium of terrestrial life. This argument emphasizes the need of conserving and restoring natural habitats and ecosystems, such as forests, marshes, grasslands, and other natural regions that support a diverse range of plant and animal species by providing them with shelter and resources. Ecosystems, characterized by their complex interconnections among organisms, make significant contributions to the preservation of biodiversity and serve as crucial agents in maintaining ecological balance. Conservation measures encompass the implementation of strategies aimed at safeguarding and preserving prevailing ecosystems, often accomplished through the creation and management of designated areas such as national parks and wildlife reserves. These regions impose restrictions on human activity in order to safeguard the life and welfare of many animals. In contrast, restoration efforts primarily aim to rehabilitate ecosystems that have undergone deterioration or destruction as a result of diverse human activities. The restoration programs encompass activities such as reforestation, wetland rehabilitation, and invasive species eradication, with the objective of reinstating ecosystems to a state that is more representative of their natural and robust condition.

The preservation of biodiversity is significantly influenced by the protection and restoration of ecosystems. By implementing measures to preserve these natural habitats, we not only ensure the protection of individual species, but also conserve the integrity of entire ecological communities comprising plants, animals, and microorganisms. This strategy aids in the preservation of biodiversity, a crucial factor for the resilience and stability of ecosystems. Furthermore, it guarantees the preservation of ecosystem services, which play a vital role in the overall welfare of humanity. Ecosystems provide essential services, like water purification, carbon sequestration, and pollination, which are critical for supporting agriculture, ensuring a clean water supply, regulating climate, and facilitating other key tasks. The objective of this goal is to safeguard terrestrial ecosystems, thereby making a significant contribution towards the overarching target of halting the decline in biodiversity. The preservation of these ecosystems is crucial for preserving ecological balance and promoting the sustainable use of terrestrial resources, as they are intricately interconnected with the overall health of the planet. The prioritization of ecosystem protection and restoration is crucial for sustainable agricultural practices, as it plays a fundamental role in the attainment of Sustainable Development Goal 15. By conserving habitats, maintaining biodiversity, and sustaining vital ecosystem services, these activities contribute significantly to the achievement of SDG 15.

6.2. Biodiversity in Agricultural Landscapes

Agricultural landscapes frequently exhibit high levels of biodiversity, serving as valuable habitats for a diverse array of plant and animal species. The presence of diverse species within an ecosystem plays a significant role in providing essential services to the environment. These services include but are not limited to pollination, insect management, maintenance of soil health, preservation of water quality, and regulation of climate. The provision of these services is crucial in ensuring the long-term viability of agricultural output and the maintenance of food security. Nevertheless, agricultural operations have the potential to endanger biodiversity due to factors such as habitat degradation, excessive resource extraction, environmental contamination, and the impacts of climate change. Hence, the implementation of sustainable farming practices that effectively protect and enhance biodiversity is of utmost significance.

Agroforestry is an example of a practice that has the potential to make a positive contribution towards the achievement of SDG 15. The practice encompasses the incorporation of trees inside agricultural systems, so augmenting biodiversity, enhancing soil quality, sequestering carbon, and yielding economic advantages. The user's text is too short to be rewritten academically. By actively preserving and augmenting biodiversity within agricultural landscapes, we may effectively contribute to the achievement of SDG 15, while also safeguarding the long-term viability and resilience of our agricultural systems (Kufeoglu, 2022).

6.3. Promoting Sustainable Land Use

The promotion of sustainable land use constitutes a fundamental component of SDG 15. The concept demands the implementation of land management strategies that are environmentally and socially sustainable, ensuring that they satisfy the requirements of the current generation while safeguarding the capacity of future generations to fulfill their own demands. This encompasses sustainable forest management, together with efforts to combat desertification and halt and reverse land deterioration. The preservation of terrestrial ecosystems is facilitated by sustainable land use practices, which encompass the provision of various goods such as raw materials for construction and energy, as well as food. Additionally, sustainable land use contributes to the provision of ecosystem services, including carbon capture, maintenance of soil quality, habitat provision for biodiversity, preservation of water quality, and regulation of water flow and erosion control (Rehman *et al.,* 2022).

These factors have a significant role in mitigating the potential hazards associated with natural disasters, including floods and landslides. Additionally, they contribute to the regulation of climate patterns and the preservation of agricultural systems' production.

The conservation of a wide range of land-based organisms requires targeted efforts to ensure the protection, restoration and promotion of the protection and sustainable management of different ecosystems. These initiatives encompass not only the preservation of the inherent value of these ecosystems but also the guarantee of their accessibility and utilization by future generations. The integration of sustainable lifestyles is a fundamental component of this method, as it aims to achieve a harmonious equilibrium between human need, agricultural efficiency, and the preservation of the environment. By implementing these specific strategies, our goal is to ensure the long-term welfare of land-based ecosystems. These measures aim to maintain the ecological equilibrium and biodiversity that these ecosystems offer, while also aligning with the wider objectives of environmental sustainability and SDG 15.

7. SUSTAINABLE AGRICULTURE AND RURAL DEVELOPMENT (SDG 1 AND 8)

The adoption of sustainable agriculture and rural development plays a crucial role in the attainment of SDGs 1 and 8. Sustainable agriculture plays an important role in improving the well-being of rural populations and aligning with global objectives such as poverty eradication, inclusive economic growth, and the promotion of decent work and economic progress. This is achieved through poverty reduction, livelihood enhancement, inclusive economic growth promotion, and the development of resilience in rural communities.

7.1. Reducing Poverty and Enhancing Livelihoods

The core components of Sustainable Development Goal 1, which is focused on the eradication of poverty, involve the reduction of poverty levels and the improvement of livelihoods. Agriculture continues to serve as a predominant means of generating income and employment in numerous rural regions, particularly those situated in developing nations. The implementation of sustainable agricultural methods assumes a pivotal role in reducing poverty within these regions. By advocating for the adoption of sustainable agricultural practices, rural communities have the potential to enhance agricultural output and augment their income, so improving their overall quality of life. These approaches not only result in increased agricultural productivity but also contribute to the preservation of the environment in the long run. By implementing strategies such as crop rotation, organic farming, and water management, small-scale farmers have the capacity to safeguard their means of subsistence and mitigate their susceptibility to economic volatility. The promotion of sustainable agriculture as a means to improve livelihoods is in line with the primary goal of SDG 1, which aims to eradicate poverty and enhance the welfare of individuals residing in rural regions.

7.2. Promoting Inclusive Economic Growth

The promotion of inclusive economic growth constitutes a major objective within the framework of Sustainable Development Goal 8 (SDG 8), which pertains to the attainment of decent work and sustainable economic progress. Sustainable agriculture plays a crucial role in advancing this objective by fostering economic prospects for a diverse range of individuals within society, encompassing small-scale farmers, women, and underprivileged populations residing in rural regions. In contrast to traditional agriculture, which tends to prioritize larger and financially secure companies, sustainable agriculture aims to foster equity by ensuring that economic growth is inclusive and benefits all parts of the population. The utilization of this approach provides small-scale farmers, who are frequently the fundamental support system of rural societies, with the means to engage in market activities, enhance their agricultural output, and attain stable and sustainable sources of income. This particular kind of economic development not only enhances the strength of rural economies, but also promotes social inclusivity and gender parity, thus strengthening the underlying principles of Sustainable Development Goal 8.

7.3. Building Resilience in Rural Communities

The cultivation of resilience within rural communities is an essential element in the promotion of sustainable agriculture and rural development. Community resilience entails providing these communities with the necessary resources and expertise to effectively endure and recuperate from a range of adversities, such as climate change, natural calamities, and economic disruptions. The promotion of sustainable agriculture is of utmost importance in bolstering the resilience of rural communities. This is achieved through the implementation of many methods, including agroecology, crop diversity, and soil conservation. These methodologies enhance the adaptability of agriculture to dynamic environmental conditions, enhance the resilience of crops, and mitigate susceptibility to unforeseen disruptions. As rural communities enhance their resilience, they acquire improved capabilities to effectively adjust to evolving circumstances and sustain their means of living, even when confronted with challenging conditions. This enhances their capacity to disrupt the cycle of poverty, in accordance with the goals of SDG 1. Simultaneously, it guarantees the economic welfare and advancement of rural communities, aligning with the principles of SDG 8.

8. INTEGRATION OF SDGS AND SUSTAINABLE AGRICULTURE

Sustainable agriculture encompasses more than food production; it involves fostering a future in which ecological integrity, financial success, and societal welfare harmoniously coexist. This is in complete accordance with the 17 Sustainable Development Goals (SDGs), which serve as a comprehensive worldwide plan for achieving a more improved world.

8.1. SDG 1: No Poverty

Intercropping food crops with trees generates revenue from the sale of wood and fruit, enhances soil fertility, and offers a variety of food options for communities. In Niger, this method resulted in a 50% increase in farmer revenue. Fair trade projects provide a direct connection between farmers and customers, guaranteeing equitable pricing and enhancing lives. Fair trade coffee cooperatives in Mexico had a significant 30% rise in their revenue.

8.2. SDG 2: Zero Hunger

Climate-smart agriculture refers to agricultural practices and techniques that are designed to mitigate the impacts of climate change and contribute to sustainable food production. Methods such as water collection and the cultivation of drought-resistant crop cultivars enhance food production even in susceptible areas. Rainwater collection initiatives in Ethiopia resulted in a 20% boost in agricultural production. Conservation agriculture involves reducing tillage and using cover crops to enhance soil health and productivity, resulting in increased crop yields. This method resulted in a 50% increase in maize yields in Kenya.

8.3. SDG 3: Good Health and Well-being

Organic farming reduces the risk of exposure to detrimental chemicals and encourages a wide range of crop production, resulting in more nourishing meals and enhanced nutrition. Organic farming in Bangladesh resulted in a 24% increase in vitamin A consumption among youngsters.

8.4. SDG 6: Clean Water and Sanitation

Precision irrigation involves the use of sensors and drip systems to reduce water wastage and maximize water efficiency, hence safeguarding valuable resources. California achieved a water save of more than 2 billion gallons in a single year with the use of precision irrigation techniques. Wetland restoration is the process of repairing damaged wetlands, which effectively purifies contaminants found in agricultural runoff, hence protecting the quality of water for both human populations and natural ecosystems.

8.5. SDG 13: Climate Action

Climate Action focuses on the process of carbon sequestration. Implementing afforestation and including cover crops in agricultural systems sequesters carbon dioxide and alleviates the impacts of climate change. Implementing cover crop cultivation on a just 10% of agricultural land in the United States has the potential to capture and store around 50 million tons of carbon dioxide annually.

8.6. SDG 15: Life on Land

Agroecological methods and sustainable land management practices help protect biodiversity and restore damaged ecosystems, in line with SDG 15. The Convention on Biological Diversity highlights the significance of sustainable agriculture in conserving biodiversity and maintaining the well-being of terrestrial ecosystems.

Incorporating sustainable agriculture within the wider Sustainable Development Goals (SDG) framework requires continuous study, cooperation, and innovation. By endorsing and expanding these practices, we may pave the way towards a future in which food security, environmental sustainability, and social equality are interconnected.

9. CHALLENGES AND BARRIERS TO SUSTAINABLE AGRICULTURE

9.1. Economic and Policy Challenges

The practice of sustainable agriculture is seeing continuous expansion as it endeavors to produce food and energy in a manner that is environmentally and socially sustainable, with the objective of meeting the needs of a growing global population while promoting inclusive growth and development (Barker and Chapman, 2020).

Nevertheless, the subject at hand encounters various economic and policy obstacles. The global population is projected to reach 9.1 billion by the year 2050, representing a 34 percent increase compared to the present. There is a pressing need to augment food production in accordance with the objectives of sustainability. The growing need for food creates notable prospects for the agricultural sector, although it also brings problems in terms of enhancing productivity growth.

The topic of environmental sustainability is of great concern when considering the global food system. It is worth noting that agriculture currently occupies a significant portion, approximately 40%, of the earth's surface, surpassing all other human activities in terms of land use. However, the expansion of agricultural land can have detrimental effects on the environment, such as deforestation, increased greenhouse gas emissions, and a decline in biodiversity. Therefore, it is crucial for policies to tackle these challenges and prioritize the enhancement of environmental sustainability, which includes the reduction of greenhouse gas emissions. The detrimental impacts of climate change are presenting unprecedented difficulties to agricultural production systems. Enhancing the capacity to adapt and exhibit resilience in response to climate change and other unanticipated disruptions represents a significant challenge (Diakosavvas and Frezal, 2019).

A significant proportion of those experiencing extreme poverty reside in rural regions, wherein the primary economic endeavor frequently revolves around food production (Brooks *et al.,* 2019). Currently, there exists a global count of over 570 million farms, with a substantial number of individuals engaged in various occupations within the food industry. In order to secure a sustainable living for farmers, it is imperative that policies effectively handle the various economic issues at play. In order to tackle these difficulties, it is imperative to adopt a comprehensive approach that takes into account all facets of sustainable agriculture. Policies aimed at addressing certain difficulties frequently give rise to synergistic or trade-off consequences in relation to other objectives.

9.2. Knowledge and Technology Gaps

The implementation of sustainable agricultural practices plays a pivotal role in the attainment of food security and the advancement of environmental sustainability. Nevertheless, it is imperative to acknowledge the existence of various gaps in knowledge and technology that must be effectively addressed in order to fully harness its potential. The comprehension of the paths leading to sustainable agriculture and the measurement thereof remains challenging due to the inadequate integration of both environmental and socioeconomic components. The interplay between the many Sustainable Development Goals (SDGs) pertaining to agriculture involves intricate trade-offs and synergistic relationships. The interactions also exhibit variation across different geographic locations. The integration of technology into the agricultural sector is of paramount importance in enhancing production and promoting sustainability. Nevertheless, there are certain obstacles concerning the accessibility of technology, particularly for small-scale farmers. Digital agriculture technologies have the potential to mitigate food waste throughout all stages of the supply chain (Lim, 2022).

Data gaps are a critical concern since they hinder the ability to make well-informed judgments in the field of agriculture. Nevertheless, there exists a dearth of dependable and punctual data pertaining to many facets of sustainable agriculture. Policy gaps refer to the absence of inadequacy of regulatory measures that can impact the widespread and equitable adoption of digital technologies in the agricultural sector. It is crucial for policies to consider the potential impact of these technologies on the required skill sets within the industry (Ehlers *et al.,* 2021). The resolution of these gaps necessitates a collaborative endeavor from several stakeholders, encompassing scholars, policymakers, farmers, non-governmental organizations (NGOs), and industry representatives.

9.3. Social and Cultural Factors

Sustainable agriculture is subject to the effect of diverse social and cultural elements. The significance of farm family dynamics and socio-cultural values is of considerable importance. The process of farm transitions is shaped by various elements, including socio-cultural values, land tenure, succession, community dynamics, and economic conditions. The cultural and historical history of a producer has a significant impact on the overarching motivations and values that shape the structure of a farm and guide the decision-making process during transitions. Likewise, the farm business is directly affected by social concerns, including the financial implications of healthcare expenses and childcare costs on farm household economics.

Additionally, the factors of race, ethnicity, and gender also exert influence. Social and cultural variables are shaped by the race and ethnicity of farmers. The growing ethnic diversity among farmers, including individuals of Hispanic, Asian, Native American, and African American backgrounds, underscores the importance of comprehending the impact of cultural factors on farm structure and transition. Moreover, the pursuit of social fulfillment by engaging in farming and ranching activities constantly emerges as a key factor that motivates individuals to persist in ranching endeavors, even in the face of financial challenges and the encroachment of urban expansion.

Decisively, the role of political influences is equally significant. The functioning of agricultural systems is significantly shaped by the social and political contexts in which they are embedded. In order to achieve sustainability in agricultural systems, a comprehensive understanding of the ways in which social and political variables exert impact upon them is of utmost importance. The consideration of social and cultural aspects is of paramount importance in facilitating the shift towards more sustainable agriculture methods.

10. CONCLUSION

This chapter acknowledges the interconnectedness between several objectives, such as promoting sustainable agriculture, empowering small-scale farmers, advancing gender equality, eradicating rural poverty, promoting healthy lifestyles, addressing climate change, and other topics encompassed by the 17 Sustainable Development Goals. The matter at hand holds significant significance as it serves to address not only the issues of hunger, poverty, and malnutrition, but also plays a crucial role in shaping the national security and political and economic autonomy of nations. The primary objective of the Sustainable Development Goals (SDGs) is to eradicate hunger and guarantee universal access to safe, nutritious, and adequate food throughout the year, with a special focus on impoverished individuals and those in vulnerable circumstances, such as infants. The practice not only facilitates the preservation of water resources but also entails a reduction in energy consumption, mitigation of environmental consequences, and a positive contribution to the long-term sustainability of the agricultural industry. The integration of sustainable lifestyles is a fundamental component of this method, as it aims to achieve a harmonious equilibrium between human need, agricultural efficiency, and the preservation of the environment. The promotion of sustainable agriculture as a means to improve livelihoods is in line with the primary goal of SDG 1, which aims to eradicate poverty and enhance the welfare of individuals residing in rural regions. The utilization of this approach provides small-scale farmers, who are frequently the fundamental support system of rural societies, with the means to engage in market activities, enhance their agricultural output, and attain stable and sustainable sources of income.

10.1. Future Prospects for Sustainable Agriculture and the SDGs

In an era characterized by urgent global difficulties, including but not limited to food security, climate change, and environmental degradation, sustainable agriculture has emerged as a pivotal factor in effectively tackling these concerns. The Sustainable Development Goals (SDGs), which were endorsed by the United Nations in 2015, serve as a comprehensive framework for worldwide development, placing significant emphasis on the concept of sustainability. This chapter examines the potential future outcomes for sustainable agriculture in relation to the Sustainable Development Goals (SDGs). It explores the changing environment of sustainable agriculture and its alignment with and contribution to the attainment of these goals.

The agricultural sector is currently undergoing a significant shift as a result of the implementation of innovative methodologies. Agroecology and regenerative agriculture play a significant role in the achievement of Sustainable Development Goal 15, which focuses on promoting life on land, as well as Sustainable Development Goal 13, which aims to address climate action. These approaches prioritize the restoration of biodiversity and the sequestration of carbon, so making valuable contributions to the aforementioned goals. Precision agriculture and technological advancements, such as data analytics and artificial intelligence, have the potential to optimize resource utilization and are in line with the objectives of Sustainable Development Goal 2, which aims to eradicate hunger, and Sustainable Development Goal 12, which promotes responsible consumption and production. In addition to the agricultural sector, the implementation of farm-to-table initiatives and the attainment of sustainability certifications serve to promote conscientious consumption and sustainable land management practices, thereby bolstering the objectives outlined in Sustainable Development Goal 12 and Sustainable Development Goal 15.

The influence of government policy and international cooperation is crucial in determining the course of sustainable agriculture. The alignment of regulatory frameworks, subsidies, and support mechanisms with Sustainable Development Goal 1 (No Poverty) and Sustainable Development Goal 8 (Decent Work and Economic Growth) can be observed. At the global level, organizations such as the Food and Agriculture Organization (FAO) play a crucial role in advancing sustainable agriculture on a worldwide scale, making substantial contributions towards the achievement of the Sustainable Development Goals (SDGs). Despite the persistent obstacles of resource access and opposition to change, the future of sustainable agriculture holds significant potential. The concept encapsulates the capacity for a global food system that is more sustainable, resilient, and equitable, highlighting its crucial significance in tackling worldwide concerns.

REFERENCES

Aslam, A. A., Aslam, M. S., & Aslam, A. A. (2022). An Overview on Green synthesis of Nanoparticles and their Advanced Applications in Sustainable Agriculture. *International Journal of Applied Chemical and Biological Sciences*, *3*(2), 70–99.

Baiyeri, P. K., Foleng, H. N., Machebe, N. S., & Nwobodo, C. E. (2019). Crop-livestock interaction for sustainable agriculture. *Innovations in Sustainable Agriculture*, 557-582.

Barker, R., & Chapman, D. (2020). The economics of sustainable agricultural systems in developing countries. In *Sustainable Agricultural Systems* (pp. 478–494). CRC Press. doi:10.1201/9781003070474-32

Brooks, J., Deconinck, K., & Giner, C. (2019). *Three key challenges facing agriculture and how to start solving them*. Agriculture and Fisheries.

Bucci, G., Bentivoglio, D., & Finco, A. (2018). Precision agriculture as a driver for sustainable farming systems: State of art in literature and research. *Calitatea, 19*(S1), 114–121.

Choudhary, M., Ghasal, P. C., Kumar, S., Yadav, R. P., Singh, S., Meena, V. S., & Bisht, J. K. (2016). Conservation agriculture and climate change: an overview. *Conservation Agriculture: An Approach to Combat Climate Change in Indian Himalaya*, 1-37.

Corsi, S., & Muminjanov, H. (2019). *Conservation Agriculture: Training guide for extension agents and farmers in Eastern Europe and Central Asia*. Food and Agriculture Organization.

Czyzewski, B., & Kryszak, L. (2022). Definitions and Origins of the Concept of Sustainable Agriculture. In *Sustainable Agriculture Policies for Human Well-Being: Integrated Efficiency Approach* (pp. 1–21). Springer International Publishing. doi:10.1007/978-3-031-09796-6_1

Diakosavvas, D., & Frezal, C. (2019). *Bioeconomy and the sustainability of the agriculture and food system: Opportunities and policy challenges*. Academic Press.

Haq, S. M., Pieroni, A., Bussmann, R. W., Abd-ElGawad, A. M., & El-Ansary, H. O. (2023). Integrating traditional ecological knowledge into habitat restoration: Implications for meeting forest restoration challenges. *Journal of Ethnobiology and Ethnomedicine, 19*(1), 33. doi:10.1186/s13002-023-00606-3 PMID:37559120

Hendriks, S., Soussana, J. F., Cole, M., Kambugu, A., & Zilberman, D. (2023). Ensuring Access to Safe and Nutritious Food for All Through the Transformation of Food Systems. *Science and Innovations for Food Systems Transformation, 31*.

Kufeoglu, S. (2022). SDG-15: Life on Land. In *Emerging Technologies: Value Creation for Sustainable Development* (pp. 469–486). Springer International Publishing. doi:10.1007/978-3-031-07127-0_17

Lemaire, G., Franzluebbers, A., de Faccio Carvalho, P. C., & Dedieu, B. (2014). Integrated crop–livestock systems: Strategies to achieve synergy between agricultural production and environmental quality. *Agriculture, Ecosystems & Environment*, *190*, 4–8. doi:10.1016/j.agee.2013.08.009

Lemaire, G., Gastal, F., Franzluebbers, A., & Chabbi, A. (2015). Grassland–cropping rotations: An avenue for agricultural diversification to reconcile high production with environmental quality. *Environmental Management*, *56*(5), 1065–1077. doi:10.1007/s00267-015-0561-6 PMID:26070897

Lim, X. (2022). *How technology can help address challenges in agriculture. World Economic Forum.*

Lipinski, B., Hanson, C., Waite, R., Searchinger, T., & Lomax, J. (2013). *Reducing food loss and waste*. Academic Press.

Omilola, B., & Robele, S. (2017). *The Central position of agriculture within the 2030 Agenda for sustainable development* (Vol. 1683). Intl Food Policy Res Inst.

Paroda, R. S., & Joshi, P. K. (2019). Sustainable development goals: role of agriculture. *2030 Agenda and India: Moving from Quantity to Quality: Exploring Convergence and Transcendence*, 17-40.

Rehman, A., Farooq, M., Lee, D. J., & Siddique, K. H. (2022). Sustainable agricultural practices for food security and ecosystem services. *Environmental Science and Pollution Research International*, *29*(56), 84076–84095. doi:10.1007/s11356-022-23635-z PMID:36258111

Shafiee-Jood, M., & Cai, X. (2016). Reducing food loss and waste to enhance food security and environmental sustainability. *Environmental Science & Technology*, *50*(16), 8432–8443. doi:10.1021/acs.est.6b01993 PMID:27428555

Umrani, R. (2010). *Agroforestry systems and practices*. Academic Press.

Velten, S., Leventon, J., Jager, N., & Newig, J. (2015). What is sustainable agriculture? A systematic review. *Sustainability (Basel)*, *7*(6), 7833–7865. doi:10.3390/su7067833

Chapter 2
Climate-Smart Agriculture for Resilience and Profitability

Mian Muhammad Ahmed
University of Agriculture, Faisalabad, Pakistan

Umer Sharif
Muhammad Nawaz Shareef University of Agriculture, Multan, Pakistan

Aamir Raza
https://orcid.org/0009-0001-1867-2660
University of Agriculture, Faisalabad, Pakistan

Muhammad Safdar
https://orcid.org/0009-0006-1779-6967
University of Agriculture, Faisalabad, Pakistan

Waqar Ali
Sindh Madresstual Islam University, Karachi, Pakistan

Muhammad Asim
University of Agriculture, Faisalabad, Pakistan

Hafsa Muzammal
University of Agriculture, Faisalabad, Pakistan

Jaffar Sattar
https://orcid.org/0009-0006-6107-9156
Khwaja Fareed University of Engineering and Information Technology, Rahim Yar Khan, Pakistan

Sheraz Maqbool
University of Agriculture, Faisalabad, Pakistan

Malaika Zaheer
Ondokuz Mayıs Üniversitesi, Turkey

ABSTRACT

The agricultural sector faces considerable challenges as a result of climate change requiring the implementation of new and sustainable methods. This chapter delves into climate-smart agriculture (CSA), an all-encompassing strategy for boosting agricultural resilience and profitability that incorporates environmental, economic, and social considerations. Various agricultural settings are taken into account with topics including soil, water, crop, livestock, forest, and fisheries management.

DOI: 10.4018/979-8-3693-2011-2.ch002

Precision farming, reforestation, and animal husbandry are all examples of CSA methods. Increased productivity, stable food supplies, lessened vulnerability, and lower expenses are just some of the ways it boosts economic benefits. The results of successful CSA programs can be learned from case studies. Challenges and future prospects are discussed in the latter section of the chapter, along with calls for further study, technical advancements, and policy changes.

1. INTRODUCTION

1.1. What Is Climate-Smart Agriculture?

Climate-smart agriculture (CSA) represents a paradigm shift in agricultural practices, recognizing that traditional methods are increasingly vulnerable to the impacts of climate change. It embodies an integrated approach that aims to transform agriculture into a more sustainable, resilient, and adaptable system in the face of evolving climate patterns. CSA is a holistic and forward-thinking approach to farming that aims to address the significant challenges posed by climate change in the agricultural sector (Angadi, 2019). In this chapter, we will delve deeper into the concept of CSA, its core principles, and the key components that define it.

1.1.1 Three Pillars of Climate-Smart Agriculture

CSA is founded upon three pivotal three pillars as shown in figure 1. Firstly, it places a strong emphasis on Sustainable Productivity, striving to elevate agricultural output in a manner that is both ecologically sustainable and economically viable. This approach involves the careful selection of crop varieties tailored to local conditions, the implementation of efficient irrigation methods, and the integration of precision agriculture technologies to maximize yield potential while minimizing detrimental environmental impacts. Secondly, CSA prioritizes the crucial need for Adaptation to Climate Change within the agricultural sector. This entails the implementation of strategies like crop and livestock diversification, adjusting planting schedules, and utilizing climate-resilient varieties. Additionally, it advocates for the development of early warning systems and the establishment of climate-resilient infrastructure to fortify against the impacts of extreme weather events. Lastly, CSA is dedicated to Mitigating Greenhouse Gas Emissions, recognizing the significant contribution of agriculture to these emissions. Through practices such as conservation tillage, agroforestry, and enhanced livestock management, CSA aims to reduce emissions (van Wijk *et al.,* 2020). Furthermore, by promoting carbon sequestration in soils and advocating for sustainable land use, CSA plays a pivotal role in mitigating the effects of climate change.

1.1.2 Key Principles of Climate-Smart Agriculture

CSA is a holistic approach to sustainable agriculture that emphasizes the importance of considering various aspects of agriculture, including soil health, water management, crop cultivation, livestock care, and ecosystem vitality (Scherr *et al.,* 2012). It emphasizes the need for local context and collaboration with local communities to develop tailored solutions. CSA also advocates for multi-stakeholder engagement, involving farmers, researchers, policymakers, and communities to foster a culture of knowledge-sharing and capacity-building. This collective effort is crucial in navigating the complexities of climate change in agriculture. CSA also emphasizes the importance of adaptability and continuous learning in a world shaped by dynamic climate patterns. It encourages ongoing research, innovation, and adaptation to equip farmers with the necessary knowledge and resources to make informed decisions in the face of environmental challenges. CSA serves as a dynamic learning process, guiding a resilient and sustainable agricultural future.

1.2. Importance of Climate-Smart Agriculture

CSA is of paramount importance due to the unprecedented threat posed by climate change to global food security and agricultural livelihoods. As temperatures rise and extreme weather events become more frequent, traditional farming practices are being strained. Figure 2 illustrates the multifaceted impacts of climate change on agriculture, showcasing the intricate web of challenges including altered precipitation patterns, shifting growing seasons, and increased vulnerability to extreme weather events. It provides a visual representation of the complex relationship between climate change and agriculture's resilience and sustainability. CSA represents a critical paradigm shift in response to these challenges. It plays a multifaceted role, starting with ensuring food security (Lipper *et al.,* 2014).

With the global population projected to reach 9.7 billion by 2050, CSA practices enhance the resilience of agricultural systems, enabling them to withstand climate-related stresses and ensure a steady and sufficient food supply. Moreover, CSA is instrumental in preserving livelihoods and sustaining rural communities, as agriculture not only drives the global economy but also forms the backbone of many rural areas. Smallholder farmers, a significant portion of the agricultural workforce, are particularly vulnerable to climate impacts, and CSA empowers them by providing the knowledge and tools needed to adapt to changing conditions, safeguarding their livelihoods and promoting sustainable rural development. Furthermore, CSA addresses environmental sustainability by mitigating degradation caused by traditional farming practices.

Techniques such as conservation tillage, agroforestry, and organic farming are prioritized, preserving ecosystem services, safeguarding biodiversity, and mitigating agriculture's negative impacts on the environment. CSA also enhances resilience and adaptation by incorporating practices that improve soil health, optimize water use, and promote crop diversity. This equips farmers to weather extreme weather events and prolonged periods of stress and emphasizes adaptive strategies to respond to evolving climatic conditions. Additionally, CSA contributes significantly to climate change mitigation by reducing greenhouse gas emissions. Practices like agroforestry, cover cropping, and precision agriculture not only sequester carbon but also minimize emissions from agricultural activities. This dual benefit of adaptation and mitigation places CSA as a pivotal component of global efforts to combat climate change. As climate challenges escalate, the adoption of CSA practices becomes imperative for a resilient and sustainable agricultural future.

1.3. The Benefits of Climate-Smart Agriculture

- CSA offers numerous benefits beyond agriculture, including economic, environmental, and social aspects. It aims to address the challenges of climate change and ensure sustainable food production systems. CSA practices, such as precision irrigation and improved seed varieties, significantly boost agricultural productivity, ensuring food security and enhancing the economic viability of farming communities (McCarthy *et al.,* 2012).
- Soil conservation techniques like conservation tillage and cover cropping foster nutrient retention, microbial activity, and soil structure, resulting in healthier soil that is more resilient to environmental stresses. CSA also champions sustainable land use and conservation measures, countering environmental degradation associated with conventional practices.
- Water efficiency, climate resilience, and poverty reduction are crucial aspects of CSA. Practices like drip irrigation and rainwater harvesting optimize water use, ensuring crops receive the right amount of moisture for optimal growth.
- CSA also equips farmers with strategies to adapt to changing climate conditions, such as diversifying crops and using climate-resilient seed varieties. This contributes to poverty reduction and improved livelihoods in rural communities.
- Moreover, CSA plays a crucial role in mitigating climate change impacts. Agroforestry sequesters carbon in trees and precision agriculture reduces emissions from fertilizer use. These practices foster community engagement,

knowledge sharing, and empowerment of marginalized groups, including women and indigenous communities.

- This inclusive and sustainable approach not only addresses climate change challenges but also strengthens the resilience of food production systems, ensuring a more sustainable and prosperous future for all.

1.4. How to Implement Climate-Smart Agriculture Practices

CSA is a complex process that requires a combination of knowledge, resources, and community engagement. It involves adopting techniques and technologies that enhance the resilience of agricultural systems to changing climatic conditions.

- To implement CSA, governments and institutions must create an environment conducive to climate-smart agriculture through policies, regulations, and incentives. Financial support, such as subsidies, incentives, and tax breaks, offsets initial investment costs and sets the foundation for responsible agricultural practices. Figure 3 illustrates the diverse pathways to Climate Smart Agriculture interventions, emphasizing agro-technological and institutional approaches. It also underscores the underutilized potential of the business development pathway, encompassing entrepreneurship and value chain enhancements for enhancing Climate Smart Agriculture practices.
- Investments in research and development drive innovation in CSA technologies and practices, ensuring agriculture remains adaptable in the face of changing climatic conditions. There are various pathways to CSA interventions, emphasizing agro-technological and institutional approaches. The underutilized potential of the business development pathway, encompassing entrepreneurship and value chain enhancements, is also highlighted. Access to appropriate technologies is crucial for successful CSA implementation. This includes promoting climate-resilient crop varieties, precision agriculture tools, and access to timely weather forecasts and early warning systems. Diverse technological approaches associated with CSA are essential for addressing climate-related challenges in agriculture. Figure 4 showcases the varied technological approaches linked to CSA, highlighting the diverse strategies employed to address climate-related challenges in agriculture
- Community engagement and knowledge sharing are vital for the widespread adoption of CSA practices. Building spaces for collaboration and learning among farmers and stakeholders is essential for the widespread adoption of CSA practices. Farmers' cooperatives and groups foster a culture of mutual support and shared knowledge. Knowledge-sharing platforms, both

online and offline, provide a space for farmers, experts, and stakeholders to exchange information, experiences, and best practices. Peer-to-peer learning networks foster a culture of continuous improvement and adaptation within the agricultural community, creating a strong foundation for sustainable and resilient agricultural systems in the face of a changing climate.

2. CLIMATE-SMART AGRICULTURE PRACTICES

CSA is an approach to farming that aims to address the challenges of climate change while ensuring sustainable agricultural productivity. CSA practices help farmers adapt to the changing climate, reduce greenhouse gas emissions, and enhance food security. Here are some key climate-smart agriculture practices.

2.1. Soil Management

Soil management lies at the core of climate-smart agriculture, recognizing soil as a vibrant, vital ecosystem pivotal to agricultural productivity. Conservation tillage, a key practice, diverges from traditional methods by minimizing soil disruption, thereby preserving crucial organic matter and microbial life. These fosters improved soil structure, moisture retention, and nutrient accessibility, while simultaneously curbing erosion for safeguarding water quality. Cover cropping proves equally essential, shielding the soil during fallow periods, suppressing weeds, and enhancing nutrient cycling. Moreover, it bolsters soil with organic matter, enriching fertility, and microbial diversity, offering a protective shield against extreme weather events. Incorporating organic matter, like crop residues or compost, is fundamental, acting as a natural fertilizer that bestows essential nutrients to plants and boosts water-holding capacity, while also supporting beneficial microorganisms for nutrient cycling and disease control.

This practice plays a pivotal role in climate change mitigation through carbon sequestration. Balanced nutrient management is crucial for optimal crop growth and environmental conservation, emphasizing judicious fertilizer use based on soil type, crop variety, and local climate conditions. Figure 5 depicts an illustrative representation of sustainable agricultural soil management practices. Precision agriculture technologies, such as soil testing, remote sensing, and variable rate application, revolutionize soil management by tailoring inputs to specific conditions and crop needs, maximizing resource efficiency and minimizing environmental impact. Effective soil management is the linchpin of climate-smart agriculture, empowering farmers to bolster soil health, fortify resilience against climate impacts, and contribute to sustainable food production, while precision agriculture technologies further amplify these efforts towards a more climate-resilient agricultural future.

2.2. Water Management

Water management plays a pivotal role in climate-smart agriculture, particularly given the growing influence of climate change on the availability and distribution of water resources. Sustainable practices are vital to bolster the resilience of agricultural systems in light of shifting precipitation patterns, droughts, and floods. Among the array of strategies, drip irrigation stands out as a highly efficient method, delivering water directly to crop root zones. This approach not only conserves water by minimizing evaporation and wastage but also allows precise control over water delivery, optimizing both usage and crop yield. Especially valuable in arid and semi-arid regions grappling with water scarcity, drip irrigation proves indispensable. Rainwater harvesting emerges as another crucial technique, involving the collection and storage of rainwater for agricultural purposes. This practice not only provides a supplementary water source during dry spells but also reduces dependence on groundwater and surface water supplies.

Ranging from simple container-based collection to advanced storage tank systems, rainwater harvesting effectively buffers against water scarcity, enhancing overall crop production. Proper drainage systems also play a vital role; they manage excess water during heavy rainfall, preventing detrimental waterlogging that can lead to root damage and crop loss. Climate-smart agriculture advocates for well-designed drainage systems that efficiently remove excess water, ensuring crops receive the right balance of water and oxygen (Keyhanpour *et al.,* 2021). Soil moisture monitoring technologies are invaluable tools, enabling farmers to accurately assess soil moisture content. Real-time data empowers informed decisions about irrigation, preventing overuse and subsequently conserving water, reducing energy costs, and averting soil degradation like salinization and erosion.

Figure 6 illustrates the mechanism of Integrated Water Resource Management (IWRM), which is predicated on the integration of water demand and supply. Embracing agroecological practices, which tailor farming to local environmental conditions, is paramount. This encompasses considering the water needs of crops and adjusting planting schedules to align with optimal moisture levels. By selecting crop varieties suited to local conditions and fostering biodiversity within agricultural ecosystems, agroecological practices enhance water efficiency and crop resilience. Equally critical is sustainable land use planning, which involves thoughtful decisions about where to establish crop fields and water storage facilities to mitigate flooding and erosion risks, all while maximizing water use efficiency. Ultimately, effective water management not only augments agricultural productivity but also reduces the environmental footprint of farming. Climate-smart agricultural practices in water management are designed to ensure the sustainable use of water resources while bolstering agricultural resilience in the face of climate change. Implementing these strategies is instrumental in addressing the challenges posed by water scarcity and variability, which are increasingly prevalent in our changing climate.

2.3. Crop Management

Crop management is a crucial aspect of climate-smart agriculture, emphasizing the cultivation of crops in a manner that maximizes productivity while minimizing environmental impacts. This encompasses a range of techniques and practices aimed at ensuring sustainable and resilient crop production in the face of evolving climatic conditions. One pivotal consideration is crop selection, where farmers must factor in local climate conditions, soil type, and market demand. Utilizing climate-resilient crop varieties adapted to specific agro-ecological zones becomes essential for ensuring successful harvests in unpredictable weather patterns. Additionally, diversifying crops in a given area proves vital in mitigating risks associated with monoculture farming, which can increase vulnerability to pests, diseases, and climate-related challenges. This approach not only spreads risk but also enhances soil health and overall farm resilience.

Agroecological practices play a pivotal role in this context, applying ecological principles to agriculture. Emphasizing biodiversity, natural processes, and sustainable resource management, these practices promote long-term soil fertility, effective pest management, and water conservation. Conservation agriculture, which involves minimal soil disturbance through techniques like no-till farming, cover cropping, and crop rotation, aids in preserving soil structure and reducing erosion. This, in turn, bolsters soil health and water retention, aligning well with the principles of climate-smart agriculture. Organic farming further complements these efforts by eschewing synthetic inputs in favor of natural alternatives, with a focus on cultivating healthy soil ecosystems. This not only enhances nutrient retention but also aids in pest control, reducing environmental impacts and promoting sustainable land management.

Precision agriculture harnesses technology and data-driven approaches to optimize resource utilization, enhance crop yields, and minimize environmental impact. Techniques like GPS-guided planting, variable rate fertilization, and remote sensing empower farmers to make informed decisions based on real-time data. Satellite and drone-based imagery, coupled with Geographic Information Systems (GIS), offer valuable insights into crop health, soil moisture levels, and pest infestations (Gupta *et al.,* 2020). This information enables targeted interventions, such as precision irrigation or pest management, resulting in resource savings and improved yields.

Moreover, advances in crop breeding have led to the development of climate-resilient varieties that are better suited to changing environmental conditions. These varieties may exhibit traits like drought tolerance, disease resistance, and improved heat tolerance. Through selective breeding programs, researchers work towards developing crops that thrive amidst climate challenges. Emerging technologies like CRISPR-Cas9 hold promise for precise genetic modification of crops, allowing scientists to introduce desired traits for climate resilience, such as enhanced drought tolerance, nutrient efficiency, and disease resistance. By integrating these crop management practices, farmers can significantly enhance the resilience, productivity, and sustainability of their agricultural systems, ultimately contributing to the goals of climate-smart agriculture.

2.4. Livestock Management

Climate-smart livestock management is crucial for sustainable agriculture, given the significant role livestock play in providing essential protein and dairy products globally. To address environmental concerns, practices such as rotational grazing are promoted, allowing for land recovery and reducing overgrazing, soil erosion, and greenhouse gas emissions. Sustainable feed sources like grasses and legumes are encouraged to minimize the carbon footprint of livestock production (Giro and Kumar, 2022). Moreover, prioritizing animal health through genetic selection, disease management, and optimal nutrition enhances productivity and resilience to climate stressors, reducing resource inputs. Proper manure management techniques like composting and anaerobic digestion convert it into valuable resources, mitigating greenhouse gas emissions. Additionally, methane capture from livestock facilities prevents this potent greenhouse gas from entering the atmosphere. Integrating livestock into mixed farming systems enhances resilience and efficiency, as crop residues become valuable feed, reducing waste. Ethical animal welfare practices not only benefit the animals but also align with climate-smart agriculture principles, ultimately leading to more productive and less emission-intensive livestock. Emphasizing climate-resilient traits in breeding, such as heat tolerance and disease resistance, reduces vulnerability to climate change impacts, ensuring the long-term sustainability of livestock systems. By incorporating these strategies, a harmonious balance can be achieved between livestock production and environmental conservation within the framework of climate-smart agriculture, leading to reduced emissions, enhanced resource efficiency, and greater profitability in livestock-based agricultural systems.

2.5. Forestry Management

Forests hold a pivotal role in climate-smart agriculture, acting as vital carbon sinks, fostering biodiversity, and providing an array of essential ecosystem services. Effective forestry management is imperative in striking a balance between agricultural production and conservation endeavors. This encompasses various practices geared towards sustainable forestry within the framework of climate-smart agriculture. Afforestation, entailing the creation of forests on previously non-forested lands, and reforestation, the restoration of previously deforested areas, are integral practices for carbon sequestration, biodiversity enhancement, and soil erosion prevention. The prudent selection of native tree species and consideration of local environmental conditions are crucial factors for the success of afforestation and reforestation projects. Agroforestry systems, which harmoniously integrate trees or shrubs with agricultural crops or livestock, not only offer supplementary income sources but also contribute to carbon sequestration, enhance soil fertility, and provide habitat for beneficial organisms. Alley cropping, silvopasture, and windbreaks are common agroforestry systems tailored to specific ecological and socio-economic conditions. In regions where logging is a significant economic activity, sustainable logging practices are indispensable for preserving forest health and biodiversity. This involves meticulous planning and execution of logging operations to minimize ecosystem damage. Selective cutting, reduced-impact logging, and post-harvest reforestation efforts are key components of sustainable timber production within the context of climate-smart agriculture. Preserving natural forests and shielding them from illegal logging and land conversion is paramount in climate-smart agriculture. Protected areas, community-managed forests, and stringent enforcement of conservation laws are crucial strategies. This not only safeguards biodiversity but also maintains their role as carbon sinks and climate regulators. Carbon offset programs offer a significant avenue for forestry's involvement, involving the sequestration of carbon in forests to offset emissions in other sectors. Forest owners and communities can participate in these programs, providing an additional source of revenue for sustainable forestry practices. Ongoing research is fundamental in advancing sustainable forestry practices within the realm of climate-smart agriculture. Innovations in remote sensing, forest monitoring technologies, and genetic research for tree breeding are critical areas of focus, improving the efficiency and effectiveness of forestry management strategies. Forestry management is a cornerstone of climate-smart agriculture, contributing to both climate mitigation and adaptation efforts. Through the integration of sustainable forest practices with agricultural activities, we can forge resilient and environmentally sustainable agricultural systems for the future.

2.6. Fisheries Management

Fisheries management plays a crucial role in climate-smart agriculture, particularly for coastal communities reliant on aquatic resources. This comprehensive approach encompasses various strategies to ensure the sustainability and resilience of fisheries amidst a changing climate. Central to this effort are sustainable harvesting practices, which entail setting catch limits, regulating fishing seasons, and implementing gear restrictions to prevent overfishing and safeguard the long-term health of fish populations. Additionally, habitat conservation and restoration efforts are integral, involving the protection of critical spawning grounds, coastal wetlands, and the restoration of degraded aquatic environments. These measures contribute to the provision of essential breeding and feeding grounds for fish, supporting sustainable population levels.

Promoting aquaculture, with a focus on minimizing environmental impacts and employing sustainable techniques, complements wild fisheries and eases pressure on natural fish stocks. Integrating aquaculture with traditional fisheries allows communities to diversify income sources and enhance food security. Community-based fisheries management is pivotal, empowering local stakeholders and relevant authorities to collectively make decisions about resource use, conservation, and enforcement. This participatory approach fosters a sense of ownership and accountability, leading to more effective and equitable fisheries management. Adapting to climate change is imperative for the long-term viability of fisheries, necessitating the adoption of climate-resilient fishing practices such as adjusting fishing seasons and employing more flexible gear types. Real-time monitoring systems track changing conditions, enhancing communities' resilience to environmental fluctuations. Furthermore, robust monitoring and enforcement mechanisms, utilizing technology like GPS tracking and satellite imagery, alongside penalties for illegal fishing, are essential for compliance with fisheries regulations. Sound fisheries management hinges on accurate and up-to-date data, underlining the significance of data collection, analysis, and sharing to inform decision-making processes. By incorporating these climate-smart fisheries management practices into agricultural systems, we can ensure the sustainability of aquatic resources while bolstering the resilience of coastal communities in the face of climate change. This holistic approach, considering ecological, social, and economic dimensions, sets the stage for thriving fisheries that provide essential benefits for both humans and the environment.

3. CLIMATE-SMART AGRICULTURE FOR RESILIENCE

CSA for resilience is a multifaceted approach aimed at enhancing the ability of agricultural systems to withstand and recover from the impacts of climate change (Steenwerth *et al.,* 2014). CSA can help to achieve resilience by increasing yields, reducing risks, conserving resources, and protecting the environment as shown in figure 7. CSA technologies include climate-resilient crop varieties, precision agriculture tools, timely weather forecasts and early warning systems, conservation agriculture, agroforestry, and integrated pest management. By integrating adaptive measures and building on the principles of sustainability, this approach seeks to ensure the long-term viability of agricultural practices in an unpredictable climate.

3.1. Building Resilience to Climate Change

CSA practices can help farmers build resilience to climate change by improving soil health, water management, and crop diversity. For example, cover cropping can help to improve soil moisture retention and reduce erosion, both of which are important for coping with drought. Agroforestry can provide shade and wind protection for crops, making them more resilient to extreme weather events. And planting a variety of crops can help to reduce the risk of crop failure due to pests, diseases, or climate change impacts. CSA practices can also help to build resilience at the community level. For example, community-based irrigation systems can help to ensure that all farmers have access to water, even during drought years. And farmer field schools can provide farmers with the knowledge and skills they need to adapt to climate change.

Building resilience to climate change is crucial for farmers and communities. Diversification of crops and livestock is a fundamental strategy for enhancing agricultural resilience. This involves cultivating a variety of crops and integrating different types of livestock to mitigate the impacts of adverse weather conditions on a single crop or animal. This not only fosters ecosystem stability but also reduces vulnerability to pests and diseases. A robust disaster preparedness plan is essential for safeguarding agricultural assets. This includes early warning systems, strategic stockpiling of resources, and community response networks. These measures minimize losses in the event of a disaster and facilitate swift recovery.

Agroecological approaches, which integrate ecological principles into agricultural practices, harness natural processes like nutrient cycling and pest control to enhance the resilience of farming systems. Techniques like intercropping, agroforestry, and the use of beneficial insects are integral components of agroecological farming. By working in harmony with nature, farmers can build a robust system that is less reliant on external inputs and better equipped to withstand environmental stresses. Agroecology promotes sustainability and resilience in the face of changing conditions.

3.2. Reducing Greenhouse Gas Emissions

Agriculture is a major source of greenhouse gas emissions, but CSA practices can help to reduce these emissions by improving nutrient management, reducing deforestation, and increasing the use of renewable energy sources. For example, composting manure and using crop residues can help to reduce the need for synthetic fertilizers, which produce nitrous oxide, a powerful greenhouse gas. Planting trees and other vegetation can help to absorb carbon dioxide from the atmosphere. And using solar or wind power to run irrigation pumps can help to reduce reliance on fossil fuels. Climate change is primarily driven by greenhouse gas emissions from agriculture, which are primarily released through livestock digestion and manure management. Reducing these emissions is crucial for climate-smart agriculture. Precision farming and soil health are key components of this approach, using advanced technologies like GPS-guided machinery, drones, and remote sensing to minimize excess usage and reduce emissions. Cover cropping and crop rotation are also essential for soil health management, as cover crops like legumes can fix atmospheric nitrogen, reducing reliance on synthetic fertilizers and decreasing nitrous oxide emissions.

Agroforestry systems are a powerful strategy for carbon sequestration, integrating trees and shrubs with traditional crops and livestock. Trees act as nature's carbon captors, sequestering carbon dioxide from the atmosphere and providing additional income streams through the production of valuable resources like timber, fruits, and nuts. This dual-purpose approach contributes to climate change mitigation and strengthens the economic resilience of agricultural operations. Effective waste management is a critical aspect of sustainable agriculture, with anaerobic digestion transforming organic matter into biogas and improved manure management practices like composting and controlled application reducing methane emissions from livestock operations. These practices emphasize the importance of waste management in mitigating emissions from agricultural activities.

3.3. Adapting to Climate Change Impacts

CSA practices can also help farmers to adapt to the specific impacts of climate change that they are facing. For example, in areas that are experiencing more frequent and severe droughts, farmers can plant drought-tolerant crops or use drip irrigation to reduce their water needs. In areas that are experiencing more flooding, farmers can plant crops on raised beds or build drainage canals to protect their fields. And in areas that are experiencing more extreme temperatures, farmers can plant crops that are adapted to heat or cold stress. CSA is crucial in addressing the inevitable impacts of climate change. Various strategies and practices are being explored to enhance agricultural adaptation, helping farmers cope with the effects of climate change. Climate-resilient agricultural practices, such as the establishment of robust early warning systems using modern technologies, are essential for farmers to take proactive measures to safeguard their crops and livestock. Investment in climate-resilient infrastructure, such as flood-resistant irrigation systems, wind-resistant storage facilities, and climate-smart housing for livestock, is also crucial.

Adaptive crop management and water resources are also essential in coping with climate change. This includes selecting crop varieties that are better suited to changing climate conditions, planting drought-resistant or heat-tolerant crops in regions facing increased aridity and high temperatures and adjusting planting and harvesting dates to align with shifting climate patterns. Efficient water management is also vital, with advanced irrigation techniques such as drip or precision irrigation optimizing water use and enhancing crop resilience. Rainwater harvesting systems provide supplementary water sources during dry periods, while efficient drainage systems prevent waterlogging in areas prone to excessive rainfall. By prioritizing water management, crops receive the necessary moisture even as precipitation patterns become less predictable.

4. CLIMATE-SMART AGRICULTURE FOR PROFITABILITY

Climate-smart agriculture not only benefits the environment and resilience to climate change but also offers substantial economic advantages for farmers and stakeholders in the agricultural sector (Chandra *et al.,* 2018). This section examines various aspects of making agriculture more profitable while adopting climate-smart practices.

4.1. Increasing Productivity

CSA practices can help farmers increase productivity by improving soil health, water management, and crop resilience to climate change. For example, a study in Malawi found that farmers who adopted CSA practices such as cover cropping, agroforestry, and conservation tillage increased their maize yields by an average of 17%. Another study in India found that farmers who adopted CSA practices such as drip irrigation and precision agriculture increased their water use efficiency by up to 50%. This led to significant reductions in production costs and increases in profitability. In the pursuit of climate-smart agriculture, enhancing productivity is a central goal. This section delves into a range of strategies and practices aimed at maximizing agricultural output while minimizing environmental impacts. Sustainable agricultural practices encompass a range of strategies aimed at optimizing resource use while minimizing environmental degradation as shown in figure 8. One key approach is sustainable intensification, which involves the careful application of inputs like fertilizers and water to achieve higher yields without compromising the ecosystem. This method places a strong emphasis on precision agriculture techniques, including the use of GPS-guided machinery, remote sensing technologies, and data-driven decision-making (Gawande *et al.,* 2023). By judiciously allocating resources, farmers can attain higher levels of productivity per unit of land. This not only enhances food production but also contributes to long-term environmental sustainability.

Agroecological farming is a holistic approach that emphasizes the interconnectedness of crops, livestock, and the environment. It uses natural processes and biodiversity to boost productivity through techniques like companion planting, crop rotation, and agroforestry. This approach not only increases agricultural output but also promotes a sustainable farming system. Empowering farmers with knowledge and resources to implement modern agricultural techniques is crucial for increased productivity and resilience. Training programs on climate-smart practices, such as integrated pest management and efficient irrigation methods, help farmers make informed decisions. Access to finance and quality inputs is also essential for smallholder farmers to adopt climate-smart practices. This ensures food security, economic prosperity, and sustainable livelihoods in the face of a changing climate.

4.2. Reducing Costs

CSA practices can also help farmers reduce costs by reducing their reliance on external inputs such as fertilizers and pesticides. For example, a study in Kenya found that farmers who adopted CSA practices such as integrated pest management and crop rotation reduced their pesticide costs by up to 70%.Another study in Bangladesh found that farmers who adopted CSA practices such as composting and manure management reduced their fertilizer costs by up to 30%. This led to significant increases in net incomes. Incorporating climate-smart agricultural practices not only contributes to environmental sustainability but also holds the potential to significantly reduce production costs for farmers (Sardar *et al.,* 2020). This section delves into various strategies that can be employed to optimize resource utilization and minimize expenses, ultimately bolstering the profitability of agricultural operations. The benefits and promises of CSA are illustrated in Figure 9. There has been an observed rise in crop productivity, accompanied by a simultaneous decrease in water usage and associated expenses.

Sustainable agriculture focuses on reducing costs and promoting climate-smart practices (Sarker *et al.,* 2019). Precision agriculture uses advanced technologies like GPS-guided machinery and remote sensing to manage inputs like water, fertilizers, and pesticides, resulting in higher yields and lower costs. IPM, on the other hand, combines biological control methods, cultural practices, and targeted pesticide use to manage pest populations effectively. Organic farming techniques, such as crop rotation, composting, and biological pest control, focus on natural processes and avoid synthetic inputs like chemical fertilizers and pesticides.

Conservation tillage and agroecological design are crucial in reducing costs and promoting sustainable agriculture. Conservation tillage techniques minimize soil disturbance, improving soil health and water retention. Agroecological design focuses on creating self-sustaining, biodiverse agricultural systems that rely on natural processes and ecological principles. These practices lead to significant cost savings and enhance system resilience, ensuring long-term viability and profitability for farmers. Efficient resource management is a cornerstone of cost reduction in climate-smart agriculture. This includes optimizing irrigation schedules, using locally sourced organic matter for soil improvement, and employing crop diversification to maximize resource use efficiency. This proactive approach leads to economic sustainability and aligns with environmentally conscious agriculture principles, contributing to a more resilient and profitable farming operation in the face of changing climate conditions.

4.3. Improving Market Access

Climate-smart agriculture is a crucial aspect of sustainable farming practices that contributes to profitability and sustainability. To enhance market access, farmers must adopt various strategies, tools, and approaches. These include value chain integration, market information systems, post-harvest handling and storage techniques, farmer cooperatives, and market diversification (McCarthy *et al.*, 2019). Value chain integration links farmers into the broader agricultural value chains, streamlining distribution and ensuring products reach consumers at competitive prices. Market information systems provide real-time data on market trends, allowing farmers to adapt their strategies in response to price fluctuations. Post-harvest handling and storage techniques reduce losses after harvest, enhancing product quality and extending shelf life. Farmer cooperatives and collective marketing empower small-scale farmers to pool resources, negotiate better prices, and gain access to markets beyond their reach.

Market diversification is essential to mitigate risks associated with relying solely on one crop or product. It spreads risk and provides a more stable income stream, allowing farmers to navigate market fluctuations and ensure consistent revenue sources. Adhering to established quality standards and obtaining certifications like organic or fair trade can open doors to premium markets, increasing the value of farmers' products and aligning with consumer preferences for sustainably sourced and high-quality goods. The integration of improved market access with climate-smart agriculture yields numerous benefits for farming communities and the broader agricultural sector. This includes increased income, financial stability, and diversification against market fluctuations. However, challenges such as transportation constraints, gaps in market information, and trade barriers pose significant hurdles. Policymakers and practitioners must collaborate to create an enabling environment that supports improved market access for all farmers, regardless of their scale or geographical location. This collaborative effort is essential for realizing the full potential of climate-smart agriculture in ensuring sustainable and resilient farming practices.

5. CASE STUDIES OF SUCCESSFUL CLIMATE-SMART AGRICULTURE PROJECTS

In this section, we will examine real-world examples of climate-smart agriculture projects from various regions around the world. These case studies showcase the practical implementation of CSA practices and highlight the positive impacts on agricultural systems, livelihoods, and the environment. Additionally, we will draw valuable lessons from these successful projects to inform future efforts in promoting climate-smart agriculture.

5.1. Climate-Smart Agriculture in the Dry Corridor of Central America

The Dry Corridor of Central America is highly susceptible to the impacts of climate change, characterized by high temperatures, prolonged periods of dryness, and severe weather events. To combat these challenges, the Food and Agriculture Organization of the United Nations (FAO) and the International Center for Tropical Agriculture (CIAT) collaborated on a climate-smart agriculture project. The project aimed to strengthen farmers' capacity to withstand and recover from climate change challenges while increasing their access to a reliable food supply (FAO and CIAT, 2018). One of the primary strategies used was the implementation of agroforestry systems, which integrated trees with traditional crops within a shared land area. Trees not only enhance soil fertility and water retention but also serve as a revenue source for farmers through the use of timber and non-timber forest products. Conservation agriculture, which minimizes soil disturbance and ensures continuous crop cover, was another crucial element. This method enhanced soil health, mitigated erosion, and reduced greenhouse gas emissions, contributing to the development of sustainable and resilient agricultural systems.

The project also focused on improving water management practices to address water scarcity in the Dry Corridor. Farmers were trained in rainwater gathering and efficient irrigation systems, which helped them manage drought and other water-related difficulties. The implementation of climate-smart agriculture has resulted in significant improvements in food security for farmers in the region. The increase in agricultural yields can be attributed to the adoption of agroforestry systems, conservation agriculture, and improved water management techniques. A comprehensive study by CIAT showed a 30% increase in maize yields among farmers who adopted these climate-smart techniques, demonstrating the project's effectiveness in promoting sustainable and climate-resilient agriculture methods.

In conclusion, the climate-smart agriculture project in the Dry Corridor of Central America serves as a significant example of effective adaptation to climate change in a region highly susceptible to its impacts. By integrating agroforestry systems, conservation agriculture, and improved water management techniques, farmers have increased their ability to withstand climate-related problems and experienced significant advancements in their food security.

5.2. Sloping Agricultural Land Technology (SALT) in the Philippines

The Philippines, due to its unique geology and proximity to the Pacific Ring of Fire, is highly susceptible to climate change's effects. Natural calamities like landslides and flooding pose significant challenges to agricultural operations, necessitating the implementation of innovative and flexible agricultural approaches. One such strategy is the Sloping Agricultural Land Technology (SALT), which aims to address soil erosion and water retention on sloping terrains (CIAT, 2018). The SALT technique involves the establishment of terraces on agricultural land with a sloping topography, which serve as both a protective barrier against swift water currents and a means for fostering plant species growth. The core of SALT's success lies in its dual functionality: protecting the land's integrity and optimizing its agricultural capacity.

The cultivation of a combination of crops and trees is a fundamental aspect of the SALT approach. Trees, found in many ecosystems, have a significant impact on soil fertility by securing the soil and enhancing the land's ability to retain water. This is particularly important in areas susceptible to drought or erratic precipitation patterns. The integration of smart terracing techniques with the growth of various crops and trees has the potential to significantly enhance the productivity of sloping terrains, transforming them into highly productive agricultural landscapes. The application of the SALT approach in many communities within the Philippines has resulted in notable and concrete favorable results. One notable accomplishment is the significant decrease in soil erosion, as the terraces serve as natural obstacles that efficiently impede the velocity of water, mitigating the rapid erosion of topsoil. Research conducted by the International Center for Tropical Agriculture

(CIAT) revealed an 80% decrease in soil erosion in specific regions where SALT was implemented.

Another significant result is the improvement in water retention, as the implementation of terraces and judicious tree planting has significantly enhanced the land's ability to retain moisture. This component is crucial in areas where water scarcity is a prevalent problem, contributing to the overall resilience of the ecosystem. In conclusion, the SALT effort in the Philippines has emerged as a noteworthy case study, showcasing not only local achievements but also providing a source of optimism and motivation for the wider adoption of climate-smart agriculture practices worldwide.

6. CHALLENGES AND FUTURE PROSPECTS

Climate-smart agriculture (CSA) faces several challenges, including limited financial resources, knowledge gaps, climate variability, socioeconomic factors, and market constraints. Smallholder farmers often struggle with financial constraints, such as limited access to credit and microfinance options, and lack of supportive policies related to land tenure, agricultural subsidies, and regulatory frameworks. Governments must implement policy reforms to create an enabling environment for CSA, providing financial support and incentives. Technical know-how and awareness are also crucial for farmers, as many lack the necessary training and knowledge to effectively implement climate-smart practices. Extension services and farmer training programs play a crucial role in knowledge dissemination, helping farmers acquire the skills needed for sustainable practices. Raising awareness among farmers and policymakers about the advantages of climate-smart agriculture is essential for its widespread adoption and success.

Climate variability, socioeconomic factors, and market constraints further hinder the adoption of CSA practices. Climate change introduces unpredictability in weather patterns, making it challenging for farmers to plan and adapt their practices accordingly. Extreme weather events like droughts, floods, and storms can have devastating effects on agricultural production. Inequitable access to resources, gender disparities, and socioeconomic differences further hinder the adoption of climate-smart practices. To address these challenges effectively, a multi-faceted approach involving various stakeholders is necessary. Governments, NGOs, research institutions, and local communities must collaborate to implement policy reforms, provide capacity-building programs, invest in research and innovation, and offer financial support. By recognizing and tackling these obstacles, we can pave the way for a more sustainable and resilient agricultural future in the face of a changing climate.

Technological advancements in precision agriculture have revolutionized the way we monitor and manage agricultural systems, providing valuable data on crop health, soil conditions, and environmental variables. Advanced algorithms analyze this data to offer accurate yield forecasts, aiding in harvest planning and market strategies. Climate-resilient crop varieties and livestock are being developed to address the challenges posed by changing climates. Techniques like trait-based breeding, genomic selection, and precision livestock farming are enabling precise modification of specific traits, accelerating the development of resilient crop varieties. Digital agriculture platforms are also transforming modern farming practices by integrating data analytics, artificial intelligence, and connectivity solutions to optimize farm management. These platforms offer tools for planning, monitoring, and analyzing operations, from planting decisions to irrigation management and pest control strategies. IoT technologies and smart sensors provide real-time data on critical factors like soil moisture, temperature, and crop conditions, which are processed by advanced algorithms to generate actionable recommendations, improving resource efficiency and overall productivity.

A comprehensive roadmap for sustainable agriculture in a changing climate includes research and development for climate-smart agriculture, building climate-resilient infrastructure, and policy, finance, and community empowerment. A collaborative approach is necessary between scientists, agronomists, and agricultural practitioners to conduct rigorous research on climate-resilient crop varieties, innovative farming techniques, and sustainable land management practices. Investments in climate-resilient infrastructure are crucial for safeguarding agricultural productivity, including robust irrigation systems, erosion control measures, and sustainable energy solutions tailored to farming operations. Infrastructure must be designed with climate adaptation in mind to ensure its capacity to withstand extreme weather events.

Policy, finance, and community empowerment are essential for successful climate-smart agriculture. Governments play a pivotal role in enacting policies that incentivize and facilitate the adoption of climate-resilient practices, such as providing subsidies for climate-smart technologies, implementing climate-smart zoning regulations, and integrating climate considerations into land-use planning. Access to finance is also critical for farmers to invest in climate-smart technologies and practices, with financial institutions, government programs, and international development agencies developing innovative financing models tailored to smallholder farmers' needs. Engaging local communities in the decision-making process fosters a sense of ownership and ensures that climate-smart interventions align with their needs and priorities. This triad of policy, finance, and community engagement forms the bedrock of a resilient and sustainable agricultural sector, poised to thrive in the face of a changing climate.

7. CONCLUSION

Climate-smart agriculture (CSA) is a crucial approach to address the challenges posed by climate change on agriculture. It involves the use of remote sensing and Geographic Information Systems (GIS) to provide farmers with accurate, up-to-date information for informed decision-making. This data-driven approach enhances resource efficiency, allowing for precise resource allocation based on spatial data, reducing waste and improving economic returns. Additionally, these technologies serve as early warning systems, enabling the timely detection of environmental stressors like drought or pest outbreaks, facilitating proactive responses. CSA's impact extends to resilience-building, emphasizing diversification, innovative crop and livestock management, and disaster preparedness to equip farmers to cope with unforeseen climatic events. Technological advancements are set to shape the future of CSA, with the integration of cutting-edge tools such as artificial intelligence, blockchain for supply chain transparency, and climate modeling. These innovations promise to optimize resource utilization and enhance decision-making capabilities.

Sustainable resource management forms a cornerstone of CSA, with practices like precision farming and organic agriculture gaining widespread adoption. National and international policies will be instrumental in driving the CSA movement forward. Governments must incentivize and support climate-smart practices, opening up new market opportunities for climate-resilient products. Access to climate finance and investment is expected to increase, providing crucial support for the implementation of CSA projects. Both public and private sectors are likely to allocate more resources to initiatives promoting sustainable and climate-resilient agriculture. Education and awareness are poised to be key drivers in the widespread adoption of CSA. As knowledge-sharing and awareness campaigns proliferate, farmers, communities, and policymakers will become more informed about the manifold benefits of CSA. This growing understanding and support will further solidify CSA's position as a transformative force in agriculture, ensuring a more sustainable and climate-resilient future for the sector.

Addressing the multifaceted challenges posed by climate change on agriculture necessitates a proactive approach in the form of climate-smart agriculture (CSA) strategies. Policymakers and practitioners must be steadfast in their commitment to integrating these principles into national agricultural and environmental policies, ensuring coherence with overarching climate and sustainability objectives. The provision of financial and technical support is paramount, particularly for smallholder farmers who may encounter barriers in implementing CSA practices. Allocating sufficient resources and establishing training programs to facilitate seamless adoption are essential.

A critical facet of CSA lies in bolstering market linkages for climate-smart agricultural products. This involves advocating for value-added processing and streamlining access to markets, with a specific focus on uplifting smallholder farmers. Enhancing the accessibility and dissemination of climate information to farmers is essential for informed decisions regarding crucial practices such as planting and harvesting times, as well as water management. Implementing social protection programs tailored to vulnerable communities affected by climate change is vital, including climate-resilient livelihood diversification initiatives.

An emphasis on ecosystem-based approaches to agriculture forms a cornerstone of CSA, encompassing practices like agroecology and agroforestry. Strengthening extension services is pivotal in disseminating knowledge and best practices in climate-smart agriculture. Establishing robust Monitoring, Reporting, and Verification (MRV) systems is vital for tracking the implementation and impact of CSA projects. Fostering platforms for knowledge exchange and policy learning among diverse stakeholders is instrumental, as well as addressing gender-specific challenges and empowering women in agriculture. Long-term planning and vision are essential for CSA, fostering a vision for resilient, productive, and sustainable agricultural systems.

REFERENCES

Ahmed, N., Masood, S., Ahmad, S., Bashir, S., Hussain, S., Hassan, W., & Ali, M. A. (2019). *Soil management for better crop production and sustainable agriculture. Agronomic Crops* (Vol. 2). Management Practices.

Angadi, R. (2019). Introduction to Smart City and Agricultural Revolution: Big Data and Internet of Things (IoT). In Smart Cities and Smart Spaces: Concepts, Methodologies, Tools, and Applications (pp. 135-176). IGI Global.

Bessou, C., Ferchaud, F., Gabrielle, B., & Mary, B. (2011). Biofuels, greenhouse gases and climate change. *Sustainable Agriculture, 2*, 365–468.

Chandra, A., McNamara, K. E., & Dargusch, P. (2018). Climate-smart agriculture: Perspectives and framings. *Climate Policy, 18*(4), 526–541. doi:10.1080/1469306 2.2017.1316968

CIAT. (2018). *Sloping Agricultural Land Technology (SALT) in the Philippines: A Case Study*. Bogor, Indonesia: CIAT. Available online at https://lib.icimod.org/record/21492/files/Sloping%20Agricultural%20Land%20Technology%20(SALT).pdf

FAO & CIAT. (2018). *Climate-Smart Agriculture in the Dry Corridor of Central America: Case Studies.* Available online at https://www.fao.org/3/CA2386EN/ca2386en.pdf

Gawande, V., Saikanth, D. R. K., Sumithra, B. S., Aravind, S. A., Swamy, G. N., Chowdhury, M., & Singh, B. V. (2023). Potential of Precision Farming Technologies for Eco-Friendly Agriculture. *International Journal of Plant and Soil Science, 35*(19), 101–112. doi:10.9734/ijpss/2023/v35i193528

Giro, A., & Kumar, N. (2022). Climate Smart Livestock System [Review]. *Journal of Agricultural Research Pesticides and Biofertilizers, 3*(1).

Gupta, D., Gujre, N., Singha, S., & Mitra, S. (2022). Role of existing and emerging technologies in advancing climate-smart agriculture through modeling: A review. *Ecological Informatics, 71*, 101805. doi:10.1016/j.ecoinf.2022.101805

Gupta, M., Abdelsalam, M., Khorsandroo, S., & Mittal, S. (2020). Security and privacy in smart farming: Challenges and opportunities. *IEEE Access : Practical Innovations, Open Solutions, 8*, 34564–34584. doi:10.1109/ACCESS.2020.2975142

Kazmi, D. H., Afzaal, M., Ali, S., & Nasim, W. (2022). Agro-meteorological Aspect of Climate Change. *Building Climate Resilience in Agriculture: Theory, Practice and Future Perspective*, 17-24.

Keyhanpour, M. J., Jahromi, S. H. M., & Ebrahimi, H. (2021). System dynamics model of sustainable water resources management using the Nexus Water-Food-Energy approach. *Ain Shams Engineering Journal, 12*(2), 1267–1281. doi:10.1016/j.asej.2020.07.029

Lipper, L., Thornton, P., Campbell, B. M., Baedeker, T., Braimoh, A., Bwalya, M., Caron, P., Cattaneo, A., Garrity, D., Henry, K., Hottle, R., Jackson, L., Jarvis, A., Kossam, F., Mann, W., McCarthy, N., Meybeck, A., Neufeldt, H., Remington, T., ... Torquebiau, E. F. (2014). Climate-smart agriculture for food security. *Nature Climate Change, 4*(12), 1068–1072. doi:10.1038/nclimate2437

McCarthy, N., Lipper, L., & Zilberman, D. (2018). Economics of climate smart agriculture: An overview. *Climate smart agriculture: Building resilience to climate change*, 31-47.

Osman, R. S. (2019). *Enhanced biogas production by anaerobic co-digestion from food waste and agriculture byproducts* [Master's thesis]. PAUWES.

Sardar, A., Kiani, A. K., & Kuslu, Y. (2021). Does adoption of climate-smart agriculture (CSA) practices improve farmers' crop income? Assessing the determinants and its impacts in Punjab province, Pakistan. *Environment, Development and Sustainability*, *23*(7), 10119–10140. doi:10.1007/s10668-020-01049-6

Sarker, M. N. I., Wu, M., Alam, G. M., & Islam, M. S. (2019). Role of climate smart agriculture in promoting sustainable agriculture: A systematic literature review. *International Journal of Agricultural Resources, Governance and Ecology*, *15*(4), 323–337. doi:10.1504/IJARGE.2019.104199

Scherr, S. J., Shames, S., & Friedman, R. (2012). From climate-smart agriculture to climate-smart landscapes. *Agriculture & Food Security*, *1*(1), 1–15. doi:10.1186/2048-7010-1-12

Steenwerth, K. L., Hodson, A. K., Bloom, A. J., Carter, M. R., Cattaneo, A., Chartres, C. J., Hatfield, J. L., Henry, K., Hopmans, J. W., Horwath, W. R., Jenkins, B. M., Kebreab, E., Leemans, R., Lipper, L., Lubell, M. N., Msangi, S., Prabhu, R., Reynolds, M. P., Sandoval Solis, S., ... Jackson, L. E. (2014). Climate-smart agriculture global research agenda: Scientific basis for action. *Agriculture & Food Security*, *3*(1), 1–39. doi:10.1186/2048-7010-3-11

van Wijk, M. T., Merbold, L., Hammond, J., & Butterbach-Bahl, K. (2020). Improving assessments of the three pillars of climate smart agriculture: Current achievements and ideas for the future. *Frontiers in Sustainable Food Systems*, *4*, 558483. doi:10.3389/fsufs.2020.558483

Chapter 3
Agriculture Supply Chains

James Kanyepe
https://orcid.org/0000-0002-6871-8831
University of Botswana, Botswana

Tinashe Musasa
https://orcid.org/0000-0002-3180-0399
University of Botswana, Botswana

Katlego Mahupa Ketlhaetse
https://orcid.org/0009-0007-1013-292X
University of Botswana, Botswana

Brave Zizhou
https://orcid.org/0000-0003-3775-8213
Loughborough University, UK

ABSTRACT

This chapter explores the emerging issues in the agriculture supply chain. The chapters also aims to address the following objectives: (1) to examine sustainable agriculture practices that can reduce the environmental footprint, (2) to examine the environmental, logistical, and market-related challenges faced by agricultural supply chains, (3) to determine the role of digitalization and data analytics, including IoT applications and precision agriculture, and blockchain technology within agriculture supply chains, (4) to explore the future trends and opportunities in agriculture supply chain. The chapter employed a critical review of academic journals and other articles, with content analysis as the analytical approach. This chapter serves as a comprehensive resource for stakeholders, facilitating informed decision-making and promoting sustainability and resilience.

DOI: 10.4018/979-8-3693-2011-2.ch003

INTRODUCTION

Approximately one-third of the world's population faces malnutrition, hunger, micronutrient deficiencies, overweight, and obesity (Negra et al., 2020). They also asserted that global health issues and chronic non-infectious diseases are caused by poor diet. In response to this, agricultural supply chains (ASCs) have become critical in countries that produce agricultural products (Guritno, 2018; Boyabatlı et al., 2018; 2022). This chapter is a timely response to the Sustainable Development Agenda of the United Nations as outlined in the Sustainable Development Goals (SDGs). For example, SDG 2 focuses on ending hunger, achieving food security, improving nutrition, and promoting sustainable agriculture. Considering this goal, this chapter will educate farmers and other policymakers on how to address challenges in agricultural value chains to ensure food security and promote sustainability.

The SDG 13 aims to address climate change and its effects. This chapter examines how agricultural value chains are influenced by climate change and various policy imperatives that can be suggested to mitigate its effects. Furthermore, SDG 15 seeks to protect and promote the sustainability of terrestrial ecosystems. Therefore, this chapter advocates for the adoption of sustainable agriculture practices (such as organic agriculture) to protect terrestrial ecosystems. Furthermore, this chapter identifies areas for growth and innovation in ASCs.

1. BACKGROUND TO THE STUDY

The history of agriculture (which dates back to the Neolithic period) represents a dynamic story of development change that spans a millennium. The Third Agricultural Revolution (1950-1960s) was acknowledged as a significant turning point in the history of agriculture (Jing et al. 2020). This period was characterized by research and development (RD) and technological advances aimed at increasing agricultural production. Genetic engineering, biotechnology, and information technology have been shown to significantly influence crop yield, quality, and diversity. Additionally, the use of mechanization, irrigation systems, precision agriculture (involving the use of drones, sensor technology and global positioning systems), and genetically modified organisms (GMOs) has increased crop yields (Goodwin & Piggott, 2020; Kedisso et al., 2022; Amari, 2023; Grits, Rostovtsev & Dichensky, 2023; Elashmawy & Uysal, 2023). Despite the impact of these technologies and practices on crop yield, the extent of their adoption remains low in developing countries.

To intensify efforts vital for low carbon development, a global treaty was endorsed during the 2015 COP21 Climate Conference (Rhodes, 2017). The Paris Agreement stresses a reduction in carbon emissions using evolving technologies and capturing carbon emissions from the atmosphere (Cifuentes-Faura, 2022). Furthermore, the 2019 COP 26 climate conference and the 2021 European (EU) climate law amplify the importance of acting on the provisions of the agreement. The law has established a significant reduction target for greenhouse gas emissions by 2030, which considers the removal of carbon from agricultural activities and the attainment of carbon neutrality by 2030 (Cifuentes-Faura, 2022). Several studies have emphasized that environmental concerns significantly influence decisions and practices used in the agricultural industry. For example, regenerative agricultural activities result in the retention of organic carbon in the soil and a reduction in atmospheric carbon dioxide (CO_2) concentrations (O'Donoghue et al., 2022).

In most countries, farmers benefit from what is referred to as carbon agriculture, or carbon farming, by trading carbon credits (White, 2022). Sustainable agriculture aims to limit the use of natural resources and energy, thus limiting possible adverse effects on the environment (Wrzaszcz & Prandecki, 2020). Agroecological techniques that support the use of environmental concepts in agricultural systems are becoming increasingly important. This includes methods such as permaculture, agroforestry, and organic agriculture, which seek to improve ecological balance and reduce their negative effects on the environment. Environmental issues have increased the importance of supply chain traceability and transparency. Multinational corporations have widely recognized the importance of the sustainability agenda as more than a trend but a responsibility (Veit et al., 2018).

Consequent to the globalization of supply chains, focal firms have expanded their operations globally and outsourced their noncore activities, mainly to developing economies, to take advantage of cost benefits (Ortas et al., 2018). As a result, opportunistic and unethical behaviours have been uncovered in upstream supply chains, particularly in manufacturing and factory plants, due to the non-stringent environmental and labour laws in these developing economies (Ashby, 2016). At the same time, the expansion strategies of global companies and supply chains have consistently been driven by the need to maintain competitiveness by achieving a cost-productivity advantage in their upstream supply chains (Veit et al., 2018).

Additionally, scandals or socioenvironmental concerns uncovered in agricultural value chains of developing countries have resulted in product boycotts, public scrutiny and criticism, and most products are sold in developed countries (Gualandris et al., 2014). Although the production process outsourced to manufacturing or factory plants often belongs to vendors, consumers often link these scandals to local firms and criticize firms for adopting unsustainable sourcing practices (Molet et al., 2013). Consequently, environmentally conscious customers place pressure on local firms by demanding increased transparency regarding the sustainability of their sourcing practices and insisting that these firms should have greater control over their outsourced activities (Bastian & Zentes, 2013). With the business case of sustainability agendas in supply chains that have been developed and are gaining increased attention, the demand for the integration of sustainability in global settings has also grown (Quarshie et al., 2016). Studies have also shown that although environmentally friendly and sustainable agricultural practices are becoming more popular, the degree of adoption remains questionable in most agricultural supply chains. This chapter seeks to inform policymakers on how to address these challenges and emerging issues in agricultural supply chains.

2. OBJECTIVES OF THE CHAPTER

The objectives of this chapter are to:

1. Examine sustainable agriculture practices that can reduce the environmental footprint.
2. Explore the environmental, logistic and market-related challenges faced by agricultural supply chains.
3. Determine the role of digitalization and data analytics, including IoT applications and precision agriculture, and blockchain technology within agricultural supply chains.
4. Establish future trends and opportunities in agriculture supply chains.

3. LITERATURE REVIEW

3.1 Overview of the Agriculture Supply Chain

Although lacking a universally accepted definition, the agricultural supply chain (ASC) broadly encapsulates the spectrum of products and services necessary for transporting agricultural products from the farm to the end consumer. Throughout scholarly research, various terms have emerged for the agricultural supply chain, including food supply chain, agricultural value chain, post-harvest supply chain, fruit supply chain, agribusiness supply chain, perishable produce supply chain, fresh produce supply chain and horticulture supply chain (Some & Prusty, 2020; Jaleta, 2021; Nagurney, 2022; Heldman, 2022). ASC, characterized by fundamental operations such as production, cyclical and continuous processing, and distribution, exhibits an increasingly intricate and dynamic nature (Ahumada & Villalobos, 2009). Central to the agricultural value chain paradigm is the interconnectedness of various entities involved in a linked sequence of activities aimed at producing and delivering products to customers.

Agricultural value chains tend to recognize the addition of value to a product at every transformation stage, as do all participants engaged in producing a natural product until the end of use (Joshua et al., 2021). Additionally, each activity node within each transformational stage is characterized by vertical, horizontal, backward, and forward linkages and action networks (Yanti et al., 2021). From an economic point of view, agricultural value chains are described as a smaller economic system, whereby the actors of the upstream and downstream supply chain are connected by financial, social, organizational, and technical relationships (Junankar, 2016). Developing economies acknowledge the crucial role agriculture plays in social and economic growth. Consequently, this has required the development of value chains with respect to crucial resources, such as infrastructure, knowledge, finance, technology, and necessary policies (Donovan et al., 2015). The integration of the value chain concept is perceived as an operational tool for sustainable development, with economic benefits derived from agriculture-dependent countries (Kumar & Rajeev, 2016).

3.2 Sustainable Agricultural Practices

The sustainability agenda has gained increased prominence in agrifood supply chains due to environmental sustainability concerns relating to the climate change crisis, the loss of biodiversity, and deforestation (Foley, 2011). There are also social sustainability issues that plague agrifood supply chains, especially since the sector is known to employ a significant share of vulnerable populations worldwide (Lambin & Thorlakson, 2018). These social issues comprise of sweatshop labour practices, i.e. the use of child labour, and unfair and poor working conditions. Additionally, there are other social issues related to human trafficking and modern slavery, of which the agrifood sector has been scrutinized (Swinnen, 2016). Considering the socioenvironmental challenges plaguing supply chains, the agrifood sector prioritizes the development of sustainable practices, as well as integration standards and certification (Clapp,2017). Sustainable agricultural practices are essential to ensure long-term protection of the planet, preservation of plant and animal habitats, promoting environmental stewardship, and meeting the food needs of a growing global population. Considering the increasing concern about climate change and its potential harm to global socio-environmental sustainability, significant efforts are necessary to address climate change related issues (Kundzewicz, 2017). The agricultural value chain has the potential to reduce these challenges through the adoption of sustainable agricultural practices that reduce excessive levels of greenhouse gas (GHG) emissions. For instance, the capacity of organic farming to reduce GHG emissions is evolving into a key area of discussion with arguments that this type of farming can enhance organic sequestration by eliminating synthetic fertilizers, thus reducing agricultural GHG emissions (Holka et al., 2022).

3.2.1 Crop Rotation

It involves growing crops on the same piece of land in a predictable sequence to maintain soil fertility and prevent farmers from making losses (Chamberlain et al., 2020). They added that rotating crops helps reduce nutrient depletion and soil deterioration over a given period. When a single crop is planted in a given field year after year, the crop's roots draw nutrients from the same depth of soil each time, depleting the rhizosphere, and lowering crop yield (Savchuk et al., 2020). Furthermore, monocultures can alter the nutritional balance at different soil depths by using an excessive amount of some crop nutrients and leaving an excessive amount of other nutrients unused (Haruna & Nkongolo, 2020). As a result, agricultural production consistently declines and eventually becomes unprofitable. One of the most significant factors that influence crop rotation is climate (which includes light, temperature, humidity, precipitation, and wind). In high-temperature regions, farmers should rotate their crops to heat-tolerant crops (Amiri et al., 2020).

3.2.2 Agroforestry

Agroforestry is defined as a combination of land use systems that integrate trees and shrubs on farmland and rural landscapes, with or without livestock, to improve the sustainability, diversity, productivity, and profitability of the ecosystem (Das et al., 2020). Many studies on agroforestry have indicated that agroforestry plays a crucial role in sequestering carbon dioxide and storing carbon biomass above and below ground (Nair et al., 2010; Mukherjee et al., 2015; Nguyen et al., 2020; Wani et al., 2020). It provides great potential to improve the economic and environmental benefits of the local community from substantial agricultural distribution to provide food, fuel, feed, lumber and fiber for the growing socioeconomic status (Castañeda-Ccori et al., 2020). It is asserted that one of the most effective ways to combat climate change is to increase the amount of forest and tree cover by implementing an agroforestry system and strengthening the resilience of the agricultural system through farmers' adoption of lone crop production (Dhyani, 2014).

3.2.3 Conservation Tillage

Conservation tillage is defined as a residue management method that involves minimizing soil disturbance while planting a crop in mulched or herbicide-killed soil (Wang et al., 2020). In the same way, conservation tillage can also be described as a tillage and planting strategy that aims to cover three-tenths of the soil surface with mulch or crop residue from the previous year during planting. The benefits of soil surface mulch, combined with minimized soil disturbance, include the capacity to cultivate steeper slopes, preserve soil moisture, increase the timeliness of field operations, and use less fuel and labour (Baghel et al., 2020). According to Lamptey et al. (2020), conservation tillage may improve crop quality because it reduces the chance that young plants are damaged by blown sand (sandblasting) and reduces the amount of dirt that is splashed onto plant parts during rainfall. Conservation tillage has also been found to inhibit weed growth by shadowing the surface of the soil when cover crop residue degrades, and chemicals that are allelopathic to weed growth are released (Pareja-Sánchez et al., 2019; Zhang et al., 2019).

3.2.4 Cover Cropping

Sustainable intensification of crop production can be greatly aided by covering crops through several mechanisms, such as improving soil structure and quality, improving the establishment of cash crops, retaining more water and nutrients, lowering the risk of surface runoff, soil erosion, and diffuse pollution by creating soil cover and controlling weeds and soilborne pests (Jones et al., 2020). Evidence shows that cover cropping provides benefits such as decreased erosion and weed management, improved soil water infiltration, decreased nitrate leaching, and loosened subsoil and appears to be achievable with cover crops (Adetunji et al., 2020; Griffith et al., 2020). Although cover crops can have several secondary benefits, the decision to seed them is typically motivated by the desire to achieve one or two main goals. Numerous variables, such as establishment methods, when to plant and destroy the crop, and other crops in the rotation, influence the choice of cover crop (Demir & Işik, 2020).

3.2.5 Organic Farming

Organic farming is a production system that excludes synthetic fertilizers, pesticides, growth regulators, and feed additives (El-Shafie, 2020). According to Murshed and Uddin (2020), organic farming is an agricultural approach that relies primarily on crop rotation, crop residues, animal manure, legumes, green manure, organic waste from the farm, mineral rocks, and bio-fertilizers to maintain soil productivity to the greatest extent possible. Moreover, it circumvents the usage of synthetic fertilizers and pesticides. Using natural means to control weeds and pests instead of harmful fertilizers and pesticides, organic farming preserves and replaces soil fertility. According to Yuvaraj et al. (2020), organic farming provides a sustainable substitute for conventional methods of recycling organic materials, minimizing nutrient losses, and minimizing trash accumulation. Crop management techniques used in organic agriculture can reduce CO_2 emissions. Nielsen (2019) contends that organic farming not only safeguards the environment but also improves soil quality, contributing to overall resilience and longevity of the farm. Furthermore, compared to synthetic systems, organic farming is believed to lead to a higher concentration of organic carbon in the soil (Shukla et al., 2019). They added that there may be observable decreases in greenhouse gas emissions when organic farming is combined with other environmentally beneficial methods.

3.2.6 Integrated Pest Management (IPM)

To ensure yields and quality in agriculture, one of the essential components of crop protection is pest management. Pests in agricultural production cause reductions in yield and product quality in both the field and after harvest (Galea, 2010). They noted that, in contrast to the eradication strategies employed in traditional practices, integrated pest management (IPM) adopts a combination of methods to manage insect populations, considering both economic efficiency and environmental implications. IPM reduces the dependency on chemical pesticides by managing pests in an environmentally conscious way by combining biological, cultural, physical, and chemical strategies. IPM has been used successfully in crop protection in numerous models around the world (Peshin et al., 2009; Way & van Emden, 2000). Modern agriculture now incorporates IPM more effectively due to recent technological advances in pest management and increased public knowledge of food safety and healthy living environments. Anon (1996) claims that the growth of IPM and the improved use of traditional techniques, which offer a greater range of management options and preventive measures, have been aided by new technologies. As an alternative to straightforward tables or charts, sophisticated forecasting computer models have been employed as a decision support system (Anon, 1996). Furthermore, advancements in biotechnology have led to the creation of semi-chemicals, natural enemies, and cultivars resistant to pests.

3.2.7 Precision Farming

Precision farming emphasizes the collection, organization and analysis of individual temporal and spatial data (Mani et al., 2021). Subsequently, it integrates these data with other available information to facilitate management decisions based on estimated variability, aiming to increase productivity, quality, profitability, and sustainability in agricultural production (Ahmad & Dar, 2020). Although this concept is built on continuous interactions between farmers (users) and various IT tools (machines), it nevertheless makes it clear how important human resources are. Precision agriculture can be described as a management approach that leverages extensive data collection and analysis to guide targeted activities aimed at improving the sustainability, productivity, and efficiency of agricultural operations (Karunathilake et al., 2023). It involves the use of technologies such as GPS-guided tractors and sensors, which help optimize the use of resources such as water, fertilizers, and pesticides, improving efficiency and reducing environmental impact (Gawande et al., 2023).

There is increasing evidence that precision agriculture exhibits fundamental differences between developed and developing countries. Larger farms in developed countries employ a wider array of accessible elements, while adoption in developing countries faces constraints due to capacity issues, budgetary limitations, and challenges in accessing technology (Osrof et al., 2023). Furthermore, socioeconomic constraints (such as age, availability of funding, cost and complexity of technology, openness to innovation, knowledge exchange and farm size) are also important factors (Adams et al., 2021). Despite the potential cost implications of precision farming technologies, it is evident that the pursuit of higher profits remains the predominant motivator for the overall adoption of precision farming (Kendall et al., 2022).

3.2.8 Crop Diversity

Crop diversity is commonly associated with the addition of more crops to an established cropping system (Hufnagel et al., 2020). They added that it can be linked to the substitution of high-value commodities (typically fruits and vegetables for the export market) for low-value commodities. Other forms of diversification include producing cash crops (such as cashews and coffee) or integrating crops and livestock, which is typically referred to as mixed farming (Hufnagel et al., 2020). Alternatively, the integration of crops and trees is commonly termed agroforestry (see, e.g., Altieri et al., 2015). A growing number of scholars believe that crop diversification contributes to agroecosystem sustainability (Moonen & Barberi, 2008). Increasing crop diversity would support risk aversion techniques, reduce farmers' dependency on outside chemical inputs, and control pest populations. One of the most economical methods to reduce income uncertainty for farmers is through crop diversification, particularly for small-holder farmers who are struggling financially.

3.2.10 Agribusiness Recycling

Foster (2020) defines agribusiness recycling as the systematic collection, processing and reuse of agricultural by-products and wastes to reduce environmental impact and promote sustainability. In the same way, Adegbeye et al. (2020) explained that agribusiness recycling plays a crucial role in the conservation of resources and the elimination of waste. They further emphasized that it increases organic matter in the soil, strengthens soil structure, and lessens the need for outside fertilizers. The use of organic materials such as crop waste can help provide biomass energy, thus contributing to a more sustainable energy balance (Sharma et al., 2019). Recycling mulch films, tubing, and packaging helps to create a more sustainable waste management system and reduces pollution (Barros et al., 2020). Furthermore, conversion of agricultural waste into energy through gasification or incineration promotes the production of renewable energy while also aiding in waste management.

3.3 Challenges in Agricultural Supply Chains

3.3.1 Environmental Challenges

Climate change and global warming have severe effects on agricultural activities and affect crop and livestock productivity (Hatfield et al., 2020). The diverse impacts of climate change on agriculture include changes in temperature, rainfall, weather, and harsh weather, all of which contribute to these effects. Continuously increasing temperatures, leading to heat stress, have a significant impact on the growth, productivity, and yields of both animals and crops. Due to the severe consequences of climate change, numerous agricultural regions are expected to face water shortages, exacerbating the frequency and intensity of droughts, particularly in many developing countries (Nhemachena et al., C2020). On the contrary, certain areas may experience increased and heavy rainfall, which would cause flooding and soil erosion, while other areas might see higher crop yields because of extended growing seasons and higher carbon dioxide levels. However, the adverse effects of climate change, such as heat stress, erratic rainfall, and extreme weather events, often outweigh its benefits and pose significant challenges to the agricultural sector.

Frequent and intensified occurrences of storms, cyclones, and other extreme weather events can result in crop failure and financial losses for farmers. In addition, rising temperatures and shifting climate patterns may lead to the proliferation of pests and the introduction of invasive species, posing adverse effects on both livestock and crops (Sivakumar, 2020). The rise of temperatures can cause heat stress in cattle, which can affect their well-being, the ability to reproduce, and the ability to provide milk or meat (Wankar et al., 2021).

3.3.2 Logistical Challenges

According to Borzacchiello, Torrieri, and Nijkamp (2009), transportation is an essential and dynamic component of any contemporary space economy, providing significant financial advantages and promoting economic growth. This view is supported by Ghiani, Laporte, and Musmanno (2013) who noted that transportation plays a crucial role in agriculture value chains. They added that inadequate transportation management strategies can lead to low farmer production. Similarly, Parikh et al. (2017) pointed out that one of the primary factors contributing to farming expenses is the transportation of agricultural products. Furthermore, they added that an increase in transportation costs is exacerbated by the state of the road network. Likewise, according to the (Sekaran et al., 2021), poor infrastructure and support services increase transportation costs, which have a detrimental effect on farmers' productivity. Moyo (2014) added that the provision of sound infrastructure and public roads improves farmers' performance.

Farmers encounter substantial transactional costs in gathering information (Shepherd, 1997), and this information, which is not freely available, carries concomitant consequences for contracts and transactions (Goetz, 1992). Makhura et al. (2001) highlight that smallholder farmers often lack reliable information on markets and potential exchange partners. According to the Mapiye et al., 2023), the widespread distribution of small farmers makes it difficult to establish telecommunication lines and extension services, resulting in inadequate information dissemination on agriculture. Furthermore, Mwakaje (2010) and Bosana and Gebresennbet (2013) point out that there are significant barriers to access to the market for small farmers due to limited market information. Moreover, Ariong et al. (2023) revealed that small farmers do not have secure post-harvest storage facilities, negatively affecting harvest quality. The use of improved storage structures leads to post-harvest loss reduction, product quality conservation, and increased storage duration (Olorunfemi & Kayode, 2021).

3.3.3 Market and Price Volatility

The key variables that can affect the market and price volatility in agriculture are market power and market access to agricultural supply chains (Hadachek et al., 2024). In addition, changes in trade regulations and geopolitical issues can also affect farmers' capacity to export and trade agricultural produce to international markets (Abrhám et al., 2021). Moreover, the concentration of market power among a few large and powerful agribusinesses, especially chain retailers, can cause imbalances in negotiations, resulting in the exploitation of farmers who sell their produce at lower and unsustainable prices (Borsellino et al., 2020). Large agribusinesses and chain retailers generally possess considerable market power to dominate and control several phases of the supply chain, including production, distribution, and retail, which significantly affects farmers' ability to negotiate or secure fair prices for their produce. This dominance of large agribusinesses and chain retailers in the supply chain significantly impacts pricing dynamics and competition, which affects small and independent farmers by securing fair compensation for their products (Nematollahi, Tajbakhsh & Sedghy, 2021).

Market inefficiencies may also be caused by a lack of openness to demand, pricing, and market trends (de Janvry & Sadoulet, 2020). Global factors such as changes in weather patterns, geopolitical development, commodity price fluctuations, and shifts in demand around the world significantly influence commodity prices, thereby affecting farmers' incomes and the profitability of agricultural enterprises (Khatri et al., 2023). Price volatility in commodity markets often stems from financial speculation, leading to commodity price swings that may not accurately reflect the fundamental supply and demand of agricultural commodities (Zhang & Broadstock, 2020). In addition, government initiatives, such as support and subsidy schemes, can affect market dynamics. Prices and market conditions for agricultural products may be affected by abrupt changes in these regulations (Ali & Khan, 2020).

According to Godfray et al. (2016), smallholder farmers' lack of financial resources prevents them from having the technical knowledge and skills needed to boost production. Furthermore, the pricing structure is adversely affected by the high transaction costs associated with coordinating farmers (Chikazunga & Paradza, 2017). According to Baiyegunhi and Fraser (2014), small-holder farmers cannot provide the collateral required by commercial banks.

3.4 Technology and Innovation In Agricultural Supply Chains

3.4.1 Internet of Things (IoT) in Agriculture

There is no standard definition for IoT in the literature, however, IoT is about the interconnectedness of things or applications that may or may not be connected to the Internet (Piramuthu, 2022). Similarly, IoT is a network connection of physical items (sensors and actuators/enablers), software items, and the connectivity of the network with the potential to be integrated with the environment for sustainable control and monitoring (Polymeni et al., 2022). In essence, from an ASC point of view, 'things' now become plants, livestock, agricultural inputs, fields, transport, etc. All linked with sensors to gauge humidity, temperature, or location (Wolfert & Isakhanyan, 2022).

Some of these "things" have sensors that gauge the surrounding conditions in supply chains. A prime example is radio frequency identification (RFID) tags in supply chains. To achieve the SDGs, the implementation of IoT is recommended at the production level throughout the supply chain (Piramuthu, 2022). This is so, since environmental sustainability not only addresses the reduction of footprint after production but also in production (Bellemare et al., 2017; Sala, 2017). From this perspective, application of IoT in ASCs is wide ranging. The use of IoT during production would aim to improve both quality and quantity output, efficient use of resources, and ultimately environmental sustainability (Nayal et al., 2021). Instances of implementation of IoT in production according to the literature range from the use of soil and plant sensors to monitor and act on sustainable weather conditions, efficient use of resources (water, energy, and other inputs), pest reduction, livestock feeding, regulation of CO_2 emissions (Alonso et al., 2020; Tao et al., 2021; Verdouw et al., 2016).

Furthermore, soon after production, the use of IoT in agriculture should aim to maintain edible foods for the end user (thereby reducing the degradation of items to be consumed), maximizing traceability and limiting footprint (Wolfert & Isakhanyan, 2022; Maraveas et al., 2022). The literature advocates for RFID and sensor network information in managing the shelf life of foods and information for the consumer (Sangeetha & Vijayalakshmi, 2020). The implementation of IoT in supply chains is evident in RFID technology. The usefulness of RFID information is seen in improved inventory accuracy, traceability, targeted promotion, and reduced touch points in the supply chain (Piramuthu, 2022; Tan & Sidhu, 2022).

The application of RFID technology is wide ranging. For example, Pirsa, Sani, and Mirtalebi (2022) focused on the use of intelligent packaging for foods in transit enabled by RFID tag sensors to maintain food temperature. Such technologies facilitate effective communication down the supply chain of food items, reaching a reasonable shelf life for faster consumption. In addition, intelligent packaging enhances traceability. These benefits contribute to the sustainability goals in ASCs by reducing waste, increasing food security, and combating climate change. Alternatively, RFID information can be used to efficiently transport items and pallets in storage or transport facilities to reduce emissions in these environments.

Another perspective observes the use of RFID technology (through IoT) in accounting for several expiration dates even of the same batch of items (Nagarajan, Deverajan & Chatterjee, 2022), identifying extremely unhealthy additives (Tan & Sidhu, 2022). A single expiration date for a batch of items is no longer effective in supply chains, as several issues contribute to the perishability of an item (eg, distance from the ripening fluid of the cooling device from the next immediate item) (Grunow & Piramuthu, 2013; Wang et al., 2016). Such recording is impossible without the use of such automated technology. In general, RFID information enables better and more efficient identification and recall of affected items to guarantee the achievement of the SDGs in a supply chain.

IoT devices are used to monitor the health and well-being of livestock. For example, wearable sensors can track parameters such as temperature, movement, and feeding patterns, helping farmers detect early signs of disease and improve overall animal welfare (Kumar, 2021). In addition, artificial intelligence (AI) algorithms process large datasets from IoT devices to monitor crop conditions and predict potential problems. This allows farmers to take proactive measures, such as adjusting irrigation or applying pesticides only when necessary, contributing to resource efficiency and sustainable practices.

3.4.2 Precision Agriculture and Smart Agriculture

Part of the operationalization of IoT at the production level includes precision agriculture. Precision agriculture is a welcome development in response to the demanding labour and monitoring required in today's vast land farming. Automated aids are therefore used to pinpoint the exact areas that require attention in agriculture. In essence, precision agriculture is the contrast to traditional blanket farming, where crops are handled as homogeneous units, rather than independent and precise units of farming that require attention (Piramuthu, 2022). Consequently, with precision farming, environmental sustainability is achieved in reduced blanket treatment (Auernhammer, 2021; Gaikwad, 2021).

According to Piramuthu (2022), precision farming is the combined and simultaneous use of global information systems (GIS) and global positioning system (GPS). The former would deliver information on soil moisture, erosion, pest presence, and the latter would deliver the exact location in the field. Despite the expensive implementation of these automated systems to facilitate precision agriculture, it is believed that the benefits of precision agriculture (ie reduced treatment costs, footprint, time, and effort required) are invaluable. Moreover, the use of IoT in precision agriculture is far-reaching to also include the rearing of animals. Modern marketing demands a consumer orientation such that consumers demand information on the diet of livestock, the medications used and their potential side effects, the dates of slaughter and the general health of livestock (Mangla et al., 2019).

3.4.3 Artificial Intelligence (AI)

Another digital 4.0 approach in sustainable food chains focuses on the use of artificial intelligence (AI). AI encompasses machine, software, or robotic learning to make smart and informed decisions similar to humans (Ben Ayed & Hanana, 2021). The ultimate of AI would therefore be to automate ASC processes for sustainability. This would lead to faster, food security, quality, forecasting, and mitigation of potential disasters. Konfo et al. (2023) propose that AI has the potential to automate ASC processes such as sorting, grading, packaging, forecasting, and identifying possible food risks. At the production level of the ASC, automation may occur in planting, weeding, harvesting, and cutting (Botta et al., 2022). However, this innovative approach in sustainable ASC has sceptics about possible job losses and possible shifts of market power to fewer platforms (Bahn et al., 2021).

3.4.4 Big Data

Differentiated by the 5 Vs of volume, velocity, variety, veracity, and value, big data stores enormous quantities of data that traditional technology cannot manage (Konfo et al., 2023). From this perspective, data is stored in unstructured vast quantities that can be promptly shared and exploited, made available in several forms, and is accurate and valuable for decision-making. For sustainable supply chains, big data facilitates customer knowledge and personalization. Similarly, literature demonstrates the use of big data to better service customers, develop new knowledge and improve revenues (Ben Ayed et al., 2022; Lezoche et al., 2020).

3.4.6 Case Studies of Technology Applications in Agricultural Supply Chains

Studies report the successful use of digital technologies in ASC (Konfo et al., 2023; Bacco et al., 2019). Although the list is endless, some of the key highlights include efforts by IBM, Famosa, Blue River Technology, Taranis, and Carrefour. IBM's blockchain technologies in ASCs are revered for transparency and traceability from farms to the shopping floor (Singh et al., 2021). Furthermore, the IBM blockchain initiative is commended for promoting food safety and reducing inventory waste by promptly identifying contaminated sources (Rejeb et al., 2020).

Famosa of Italian origin provides wireless sensor networks for effective irrigation (ie precision) and has been credited with successfully watering sensitive fruits such as the kiwi fruit (Hassine, 2022). Fennimore and Cutule (2019) identify the Blue River Technology purchased by the John Deere company for precise weed removal on crops by applying herbicides while minimising chemical use and maximising yields. As part of AI, Taranis drones have been effectively used in scanning farms and warning farmers of potential dangers to ultimately increase yields (Bacco et al., 2019). Lastly, at the end of the value chain, Carrefour, a French-based multinational retailer, successfully uses blockchain technology to generate accurate customer data (Bacco et al., 2019).

3.5 Future Trends and Opportunities in Agriculture Supplies Chains

Approximately a third of the world's population is in need of food, experiencing hunger, micronutrient deficiencies, overweight, or obesity (Negra et al., 2020). Poor diets are the primary cause of global health problems and chronic noninfectious diseases (Negra et al., 2020). Food systems, including agriculture, transport, logistics and disposal, collectively pose the greatest threat to biodiversity loss (Dudley &

Alexander, 2017) and contribute to almost one-third of global greenhouse gas emissions (Niles et al., 2017; Negra et al., 2020). Food security is significant for human well-being, both inherently and as a critical means of fostering thriving livelihoods (Barrett, 2021).

Biotechnology and genetically modified organisms (GMO) offer promising solutions to improve crop resilience, increase yields, mitigate the environmental impact of agriculture and combat climate change (Tariq, 2021). GMOs are modified or engineered to express specific (beneficial) traits such as resistance to pests, disease, and herbicides, drought tolerance, and nutritionally enhanced traits to increase crop yields and combat malnutrition (Gbashi et al., 2021). Despite the contentious debates surrounding their adoption, GMOs have the potential to revolutionize ASCs by fostering sustainable agriculture through reduced pesticide use, improved drought tolerance, and improved nutritional profiles in crops (Arslan, Sutan & Yanik, 2019; Gbashi et al., 2021; Chandra, 2023). For example, the introduction of biotech cotton has significantly boosted economies in various countries.

Biotech cotton, also known as Bt cotton (Bacillus thuringiensis), is a pest resistant genetically engineered crop that has gained significant traction in the agricultural sector. Since its introduction into the market around 1996, biotech cotton has been quickly adopted by many countries including the United States, China, and India (Tokel, Genc & Ozyigit, 2022). The introduction of biotech cotton has brought a sustainable solution to agriculture, decreasing chemical spraying, reducing the costs of farming, preserving workers' health, and substantially increasing the yield of cotton production (Tokel, Genc & Ozyigit, 2022). In China, it generated more than \$15 billion in economic benefits between 1996 and 2012, with \$2.2 billion in 2021 alone (Bakhsh et al., 2021). India saw an increase in farm income of \$5.1 billion from 2002 to 2008 and \$1.8 billion in 2008 alone (Brookes & Barfoot, 2009) and Pakistan has also reported \$1.7 billion in economic gains from biotech cotton (Bakhsh et al., 2021).

Several studies have also analyzed the effects of the adoption of biotech plants on crop yield and chemical use (Juan et al., 2023; Deshmukh, 2021; Roohi et al., 2018; Bakhsh et al., 2021; Berman et al., 2013; Arslan, Sutan & Yanik, 2019; Chandra, 2023; Qaim, 2020). Qaim (2020) has found that the adoption of biotech plants can increase crop yield by 22% and reduce chemical use by 37%, highlighting the impact of biotech farming on sustainability. The literature has several case studies where the adoption of biotech plants has significantly increased crop yield, food nutrition, and food security. For example, the genetic modification of maize (corn) variety with improved drought resilience, allowing farmers in regions with erratic rainfall patterns to achieve higher crop yields even during periods of drought (Bakhsh et al., 2021), the introduction of rainbow papaya in Hawaii, which was genetically engineered to resist the ringspot virus, saving the industry and securing a vital source of nutrition for the local population (Virus, 2014), the innovative engineering of Golden rice enriched with Vitamin A to fight vitamin A deficiency, a serious health issue affecting millions of people in more than 118 countries, especially in developing countries (Dubock, 2019; Berman et al., 2013). These case studies, among many others in the literature (see Bakhsh et al., 2021; Wamatsembe, Asea & Haefele, 2017) show that genetically modified crops have the potential to play a significant role in addressing food cost and affordability as well as improving food security, reducing malnutrition, and addressing climate change.

On the other hand, the concept of circular economy intersects with the initiative of sustainable agriculture, aiming to reduce waste generation in ASCs. Previous studies have indicated that the world already produces enough food to feed everyone (Gbashi et al., 2021); however, 820 million people in the world suffer from hunger (FAO, 2019). The treatment of the global food waste epidemic that according to Mouat (2022) accounts for 2.625 trillion USD is a crucial goal in combating global food insecurity and combating climate change. By minimizing food waste throughout the ASCs, resources are utilized more efficiently, contributing to the reduction of hunger and starvation globally (Dhanya et al., 2023). Redirecting surplus food to those in need (Sundgren, 2022), implementing innovative technologies for food preservation, and fostering collaboration among stakeholders can help bridge the gap between food surplus and food insecurity (Nadeau & Koebele, 2023).

The ASC presents substantial investment opportunities and avenues for sustainable development, combating climate change, and alleviating hunger and poverty (Mouat, 2022). Innovative practices such as repurposing organic waste as fertilizers, the application of precision agricultural methodologies to optimize resource allocation, and the adoption of closed-loop systems to recycle materials collectively contribute to fostering a more sustainable and efficient ASC (Mouat, 202). The convergence of increasing pressure to address sustainability issues in the agricultural sector (Colombo et al., 2023) and the urgent need to address global food insecurity challenges have intensified the interest of scholars and other stakeholders in sustainable agriculture practices (Ma, Mau & Sharbel, 2018). This growing focus on sustainability not only drives innovation, but also serves as a robust business model within ASCs.

Embracing circular economy principles in ASCs becomes an imperative initiative to minimize waste production and efficient utilization of resources (Dhanya et al., 2023). Collaboration among stakeholders such as farmers, processors, distributors, consumers, nongovernmental organizations, and policy makers is vital in implementing sustainable practices across ASCs (Moreno-Miranda & Dries, 2022; Kornher & Kalkuhl, 2019). In their study, Moreno-Miranda and Dries (2022) highlighted the need for committed collaborative efforts to unlock the full potential of investment opportunities in ASCs. Chandra (2023) emphasizes that the use of biotechnological advances, the acceptance of circularity and the adaptation to evolving consumer demands present avenues for sustainable growth and innovation in agricultural supply chains.

Interestingly, approximately one-third of croplands are used to grow feed for livestock production rather than human consumption (van Huis & Gasco, 2023), while livestock farming has been cited as one of the major contributors to greenhouse gas emissions, contributing 14.5% according to Twine (2021). This highlights the opportunities for alternative meat production methods, such as cultured (lab-grown) and plant-based meats (Negowetti, 2020). These are promising sustainable biotechnology solutions, with plant-based meat already gaining traction in the market. As the agriculture sector embraces these advances, future trends are expected to address the inefficiencies of animal feed production through innovations in alternative and lab-grown meat production, thus optimizing land use for human food production.

While previous evolutions in the agricultural sector have predominantly been influenced by advances in mechanical technologies and biotechnology genetic enhancements, contemporary agricultural innovation is increasingly driven by digital tools and solutions such as automation (Dartey et al., 2021), IoT (Memon et al., 2016), digital twins (Purcell & Neubauer, 2023), robotic systems (Gonzalez-de-Santos et al., 2017), drones and artificial intelligence (Spanaki et al., 2022), precision farming (Lee et al., 2023), among others. These digital tools hold great promise in revolutionizing the ASC by enhancing efficiency, addressing food insecurity, increasing harvest yield, and combating climate change. Additionally, the integration of digital technologies enables real-time monitoring and management of agricultural processes, allowing for timely interventions and improved decision making.

4. METHODOLOGY

This study adopted a systematic analysis of synthesis. Academic databases including Google Scholar, Scopus, Web of Science, and IEEE Xplore were used in the study. Academic publications, reports, and related materials were obtained using keywords. Furthermore, the English language was used to search for study material, and all articles written in languages other than English were excluded.

5. CONCLUSION

This study established that sustainable agricultural practices such as crop rotation, agroforestry, conservation, tillage, cover cropping, integrated pest management, water conservation, crop diversity, and agribusiness recycling can contribute to reducing the environmental impact of agriculture while ensuring long-term productivity and resilience in the face of environmental challenges. Additionally, there is substantial evidence that ASCs are negatively influenced by environmental, logistical, and market-related challenges. To achieve the SDGs, it is necessary to accelerate the implementation of IoT in different ways throughout the supply chain. Furthermore, using IoT during the production process improves both the quality and quantity of output, the efficient use of resources, and, ultimately, the sustainability of the environment. Furthermore, there is evidence in the literature that biotechnology can offer promising solutions to improve crop resilience, increase yield, and mitigate the environmental impacts of agriculture. Additionally, by integrating sustainable practices (such as regenerative farming techniques and reduced chemical inputs), agricultural businesses can redefine their success metrics to include ecological impacts and profitability.

6. DISCUSSION AND IMPLICATIONS

This chapter explores emerging issues in agricultural supply chains. Discussion of the findings was guided by the research objectives. There is strong agreement in the literature that sustainable agricultural practices play a crucial role in ensuring the long-term health of the planet, promoting environmental stewardship, and meeting the food needs of the growing global population. Nadeem et al. (2019) highlighted that crop rotation helps reduce nutrient depletion and soil deterioration by gradually changing the type of plants planted in a particular area over a given period. Similarly, Piramuthu (2022) explained that one of the most efficient ways to combat climate change is to increase forest and tree cover. Similarly, Kornher and Kalkuhl (2019) noted that cover cropping lowers the risk of surface runoff, soil erosion, and diffuse pollution by creating soil cover and controlling weeds and soilborne pests. Consequently, Holka et al. (2022) explained that organic farming improves organic sequestration by eliminating synthetic fertilizers, thus reducing agricultural GHG emissions.

Hatfield et al. (2020) have observed that climate change affects crop and livestock productivity. Skendi et al. (2021) emphasised that temperature changes, altered precipitation patterns, harsh weather, and changes in the dynamics of pests and diseases all contribute to these effects. This is echoed by Elahi et al. (2022), who emphasized that storms, cyclones, and other extreme weather events that occur more frequently and with greater intensity can cause crop failure and financial losses for farmers. The study also found that agricultural supply chains are influenced by logistics, market, and environmental challenges. Ghiani et al. (2013) noted that transportation is essential to agricultural value chains. Therefore, poor transportation management strategies can result in low farmer productivity. Similarly, Parikh et al. (2017) emphasized that a greater portion of the expenses incurred by farmers is related to the transportation of agricultural products.

Additionally, Magesa et al. (2020) observed that limited access to market information leads to a significant barrier to market access for farmers, particularly smallholder farmers. This view is supported by Ariong et al. (2023), who revealed that smallholder farmers do not have secure post-harvest storage facilities, which negatively affects the quality of the harvest. Furthermore, this study found that contemporary technologies such as IoT and blockchain play a critical role in shaping the trajectory of agricultural value chains. This view is supported by Zhang et al. (2021), who noted that IoT devices can be used to monitor the health and well-being of livestock. For example, wearable sensors can track parameters such as temperature, movement, and feeding patterns, helping farmers detect early signs of disease and improve overall animal welfare. Gbashi et al. (2021) noted that biotechnology and GMOs have emerged as dynamic forces that shape agricultural landscapes to combat climate change and hunger.

7. LIMITATIONS AND FUTURE RESEARCH DIRECTIONS

This chapter examines emerging issues in agricultural supply chains. As shown in the previous sections, this chapter provides a spectrum of findings that are expected to be interesting and informative for farmers, policymakers, and other players in the agricultural value chain. This study can be seen as an effective vehicle for future research on agricultural supply chains, as it informs policymakers in designing and implementing the most effective intervention schemes to reduce environmental impact. Although this study offers significant information, several limitations and opportunities for future research still exist (for example, research methodology, context, and analysis). These limitations provide a platform for researchers to build on and extend the understanding developed in this study. Therefore, more research should be done to strengthen the findings of this study and extend the practical understanding and academic knowledge of the agricultural supply chain.

REFERENCES

Abrhám, J., Vošta, M., Čajka, P., & Rubáček, F. (2021). The specifics of selected agricultural commodities in international trade. *Agricultural and Resource Economics: International Scientific E-Journal, 7*, 5-19.

Adams, A., Jumpah, E. T., & Caesar, L. D. (2021). The nexuses between technology adoption and socioeconomic changes among farmers in Ghana. *Technological Forecasting and Social Change*, *173*, 121133. doi:10.1016/j.techfore.2021.121133

Ahmad, S. F., & Dar, A. H. (2020). Precision farming for resource use efficiency. *Resources Use Efficiency in Agriculture*, 109-135.

Alexanderson, M. S., Luke, H., & Lloyd, D. J. (2023). Regenerative farming as climate action. *Journal of Environmental Management*, *347*, 119063. doi:10.1016/j.jenvman.2023.119063 PMID:37804633

Auernhammer, H. (2021). Precision farming — The environmental challenge. *Computers and Electronics in Agriculture*, *30*(1-3), 31–43. doi:10.1016/S0168-1699(00)00153-8

Bacco, M., Barsocchi, P., Ferro, E., Gotta, A., & Ruggeri, M. (2019). The digitisation of agriculture: A survey of research activities on smart farming. *Array (New York, N.Y.)*, *3*, 100009. doi:10.1016/j.array.2019.100009

Bahn, R. A., Yehya, A. A. K., & Zurayk, R. (2021a). Digitalization for sustainable agri-food systems: Potential, status, and risks for the MENA region. *Sustainability (Basel)*, *13*(6), 3223. doi:10.3390/su13063223

Bakhsh, A., Sırel, I. A., Kaya, R. B., Ataman, I. H., Tillaboeva, S., Dönmez, B. A., & Duru, E. (2021). Contribution of Genetically Modified Crops in Agricultural Production: Success Stories. In *Policy Issues in Genetically Modified Crops* (pp. 111–142). Academic Press. doi:10.1016/B978-0-12-820780-2.00006-6

Barrett, C. B. (2021). Overcoming global food security challenges through science and solidarity. *American Journal of Agricultural Economics*, *103*(2), 422–447. doi:10.1111/ajae.12160

Bastian, J., & Zentes, J. (2013). Supply chain transparency as a key prerequisite for sustainable agri-food supply chain management. *International Review of Retail, Distribution and Consumer Research*, *23*(5), 553–570. doi:10.1080/09593969.2013.834836

Bellemare, M. F., Çakir, M., Peterson, H. H., Novak, L., & Rudi, J. (2017). On the measurement of food waste. *American Journal of Agricultural Economics,* Ben Ayed, R., & Hanana, M. (2021). Artificial intelligence to improve the food and agriculture sector. *Journal of Food Quality*, *2021*, 1–7.

Bharti, A. (2020). Supply Chain Management and Strategy Implementation for Perishable Goods *Supply Chain and Logistics Management*, 284–302. doi:10.4018/978-1-7998-0945-6.ch014

Bolfe, É. L., Jorge, L. A. D. C., Sanches, I. D. A., Luchiari Júnior, A., da Costa, C. C., Victoria, D. D. C., Inamasu, R. Y., Grego, C. R., Ferreira, V. R., & Ramirez, A. R. (2020). Precision and digital agriculture: Adoption of technologies and perception of Brazilian farmers. *Agriculture, 10*(12), 653. doi:10.3390/agriculture10120653

Borsellino, V., Schimmenti, E., & El Bilali, H. (2020). Agri-food markets towards sustainable patterns. *Sustainability (Basel), 12*(6), 2193. doi:10.3390/su12062193

Boyabatlı, O., Kazaz, B., & Tang, C. S. (2022). Agricultural Supply Chain Management Research. *Springer Series in Supply Chain Management.* doi:10.1007/978-3-030-81423-6

Castañeda-Ccori, J., Bilhaut, A., Mazé, A., & Fernández-Manjarrés, J. (2020). Unveiling Cacao Agroforestry Sustainability through the Socio-Ecological Systems Diagnostic Framework: The Case of Four Amazonian Rural Communities in Ecuador. *Sustainability, 12*(15), 5934. doi:10.3390/su12155934

Chamberlain, L. A., Bolton, M. L., Cox, M. S., Suen, G., Conley, S. P., & Ané, J. (2020). Crop rotation, but not cover crops, influenced soil bacterial community composition in a corn-soybean system in southern Wisconsin. *Applied Soil Ecology, 154*, 103603. doi:10.1016/j.apsoil.2020.103603

Chandra, N. (2023). Significant advancements in agriculture science. *Dr. Ajay B. Jadhao*, 72.

Chathuranika, I., Khaniya, B., Neupane, K., Rustamjonovich, K. M., & Rathnayake, U. (2022). Implementation of water-saving agro-technologies and irrigation methods in agriculture of Uzbekistan on a large scale as an urgent issue. *Sustainable Water Resources Management, 8*(5), 155. doi:10.1007/s40899-022-00746-6

Cho, C., Lim, H., Kim, B., Jung, H., & Park, S. (2022). Current status of research and market in alternative protein. *Food and Life, 2022*(1), 9–18. doi:10.5851/fl.2022.e4

Cifuentes-Faura, J. (2022). European Union policies and their role in combating climate change over the years. *Air Quality, Atmosphere & Health, 15*(8), 1333–1340. doi:10.1007/s11869-022-01156-5 PMID:35035604

Colombo, S. M., Roy, K., Mraz, J., Wan, A. H., Davies, S. J., Tibbetts, S. M., Øverland, M., Francis, D. S., Rocker, M. M., Gasco, L., Spencer, E., Metian, M., Trushenski, J. T., & Turchini, G. M. (2023). Towards achieving circularity and sustainability in feeds for farmed blue foods. *Reviews in Aquaculture, 15*(3), 1115–1141. doi:10.1111/raq.12766

Dartey, Y. O. O., Antwi, E. B. O., Bassit, M. M., & Oduro-Koranteng, E. A. (2021). Small-scale Automated Drip Circulation System. *JEEE-U (Journal of Electrical and Electronic Engineering-UMSIDA), 5*(2), 186-196, doi:10.21070/jeeeu.v5i2.1576

Das, A. K., Rahman, A., Keya, S. S., Saha, S. R., & Rahman, M. (2020). Malta-based agroforestry system: an emerging option for improving productivity, profitability and land use efficiency. *Environmental Sustainability, 3*(4), 521-532. doi:10.1007/s42398-020-00139-5

de Janvry, A., & Sadoulet, E. (2020). Using agriculture for development: Supply- and demand-side approaches. *World Development, 133*, 105003. doi:10.1016/j.worlddev.2020.105003

Elashmawy, R., & Uysal, I. (2023). Precision Agriculture Using Soil Sensor Driven Machine Learning for Smart Strawberry Production. *Sensors, 23*(4), 2247. doi:10.3390/s23042247

Fennimore, S. A., & Cutulle, M. (2019). Robotic weeders can improve weed control options for speciality crops. *Pest Management Science, 75*(7), 1767–1774. doi:10.1002/ps.5337 PMID:30653830

Gaikwad, S. V., Vibhute, A. D., Kale, K. V., & Mehrotra, S. C. (2021). An innovative IoT-based system for precision farming. *Computers and Electronics in Agriculture, 187*, 106291.

Gawande, V., Saikanth, D. R. K., Sumithra, B. S., Aravind, S. A., Swamy, G. N., Chowdhury, M., & Singh, B. V. (2023). Potential of precision farming technologies for eco-friendly agriculture. *International Journal of Plant and Soil Science, 35*(19), 101–112. doi:10.9734/ijpss/2023/v35i193528

Gonzalez-de-Santos, P., Ribeiro, A., Fernandez-Quintanilla, C., Lopez-Granados, F., Brandstoetter, M., Tomic, S., Pedrazzi, S., Peruzzi, A., Pajares, G., Kaplanis, G., Perez-Ruiz, M., Valero, C., del Cerro, J., Vieri, M., Rabatel, G., & Debilde, B. (2017). Fleets of robots for environmentally-safe pest control in agriculture. *Precision Agriculture, 18*(4), 574–614. doi:10.1007/s11119-016-9476-3

Haruna, S. I., & Nkongolo, N. V. (2020). Influence of Cover Crop, Tillage, and Crop Rotation Management on Soil Nutrients. *Agriculture, 10*(6), 225. doi:10.3390/agriculture10060225

Hassine, T. B. (2022). A Language & an Approach for the Development of IoT Solutions. *American Journal of Electrical and Computer Engineering*, *6*(1), 1–14. doi:10.11648/j.ajece.20220601.11

Hatfield, J. L., Antle, J., Garrett, K. A., Izaurralde, R. C., Mader, T., Marshall, E., Nearing, M., Philip Robertson, G., & Ziska, L. (2020). Indicators of climate change in agricultural systems. *Climatic Change*, *163*(4), 1719–1732. doi:10.1007/s10584-018-2222-2

Heldman, D.R. (2022). Sustainability of the food supply chain: energy, water, and waste. *Food Engineering Innovations Across the Food Supply Chain*, 47-60. doi:10.1016/B978-0-12-821292-9.00021-2

Juan, L. (2023, August). Ramos., Patricia, Bernal., Davinia, Salvachúa. (2023). Microbial biotechnology in the effort to end hunger. *Microbial Biotechnology*, *16*(8), 1581–1583. Advance online publication. doi:10.1111/1751-7915.14295

Junankar, P. (2016). *Development Economics: The Role of Agriculture in Development*. Palgrave Macmillan.

Karunathilake, E. M. B. M., Le, A. T., Heo, S., Chung, Y. S., & Mansoor, S. (2023). The path to smart farming: Innovations and opportunities in precision agriculture. *Agriculture*, *13*(8), 1593. doi:10.3390/agriculture13081593

Kremen, C. (2020). Ecological intensification and diversification approaches to maintain biodiversity, ecosystem services and food production in a changing world. *Emerging Topics in Life Sciences*, *4*(2), 229–240. doi:10.1042/ETLS20190205 PMID:32886114

Kshetri, N. (2022). Supply chain management. *The Rise of Blockchains*. Edward Elgar Publishing. doi:10.4337/9781802208177.00013

Kumar, D., & Rajeev, P. (2016). Value Chain: A Conceptual Framework. *International Journal of Engineering and Management Sciences*, *7*(1), 74–77.

Kumar, R. (2021). IoT and Deep Learning for Livestock Management. In Examining the Impact of Deep Learning and IoT on Multi-Industry Applications (pp. 80-96). IGI Global. doi:10.4018/978-1-7998-7511-6.ch006

Kundzewicz, Z. (2017). Large-scale climate change (observations, interpretation, projections). In Climate Change and Its Impact on Selected Sectors in Poland. Ridero IT Publishing.

Lambin, E., & Torlakson, T. (2018). Sustainability standards: Interactions between private actors, civil society, and governments. *Annual Review of Environment and Resources*, *43*(1), 369–393. doi:10.1146/annurev-environ-102017-025931

Lee, C. L., Strong, R., Briers, G., Murphrey, T., Rajan, N., & Rampold, S. (2023). A Correlational Study of Two US State Extension Professionals' Behavioral Intentions to Improve Sustainable Food Chains through Precision Farming Practices. *Foods*, *12*(11), 2208. doi:10.3390/foods12112208 PMID:37297453

Lezoche, M., Hernandez, J. E., Díaz, M. D. M. E. A., Panetto, H., & Kacprzyk, J. (2020). Agri-food 4.0: A survey of the supply chains and technologies for the future agriculture. *Computers in Industry*, *117*, 103187. doi:10.1016/j.compind.2020.103187

Ma, X., Mau, M., & Sharbel, T. F. (2018). Genome editing for global food security. *Trends in Biotechnology*, *36*(2), 123–127. doi:10.1016/j.tibtech.2017.08.004 PMID:28893405

Magesa, M. M., Michael, K., & Ko, J. (2020). Access and use of agricultural market information by smallholder farmers: Measuring informational capabilities. *The Electronic Journal on Information Systems in Developing Countries*, *86*(6), e12134. doi:10.1002/isd2.12134

Maraveas, C., Piromalis, D., Arvanitis, K. G., Bartzanas, T., & Loukatos, D. (2022). Applications of IoT for optimized greenhouse environment and resources management. *Computers and Electronics in Agriculture*, *198*, 106993. doi:10.1016/j.compag.2022.106993

McNunn, G., Karlen, D. L., Salas, W., Rice, C. W., Mueller, S., Muth, D. Jr, & Seale, J. W. (2020). Climate smart agriculture opportunities for mitigating soil greenhouse gas emissions across the US Corn-Belt. *Journal of Cleaner Production*, *268*, 122240. doi:10.1016/j.jclepro.2020.122240

Meixell, M., & Gargeya, V. (2005). Global supply chain design: A literature review and critique. *Transportation Research Part E, Logistics and Transportation Review*, *41*(6), 531–550. doi:10.1016/j.tre.2005.06.003

Memon, M. H., Kumar, W., Memon, A. R., Chowdhry, B. S., Aamir, M., & Kumar, P. (2016). Internet of Things (IoT) Enabled Smart Animal Farm. *Proceedings of the 3rd International Conference on Computing for Sustainable Global Development*, 2067-2072.

Mishra, B. K., Kumar, P., Saraswat, C., Chakraborty, S., & Gautam, A. (2021). Water security in a changing environment: Concept, challenges and solutions. *Water (Basel), 13*(4), 490. doi:10.3390/w13040490

Moreno-Miranda, C., & Dries, L. (2022). Integrating coordination mechanisms in the sustainability assessment of agri-food chains: From a structured literature review to a comprehensive framework. *Ecological Economics, 192*, 107265. doi:10.1016/j.ecolecon.2021.107265

Mouat, A. R. (2022). Sustainability in food-waste reduction biotechnology: A critical review. *Current Opinion in Biotechnology, 77*, 102781. doi:10.1016/j.copbio.2022.102781 PMID:36029714

Nayal, K., Raut, R., Lopes de Sousa Jabbour, A. B., Narkhede, B. E., & Gedam, V. V. (2021). Integrated technologies toward sustainable agriculture supply chains: Missing links. *Journal of Enterprise Information Management.* Advance online publication. doi:10.1108/JEIM-09-2020-0381

Negowetti, N. E. (2020). A Planetary Health Approach to the Labeling of Plant-Based Meat. *Food and Drug Law Journal, 75*, 142.

Negra, C., Remans, R., Attwood, S., Jones, S., Werneck, F., & Smith, A. (2020). Sustainable agri-food investments require multi-sector co-development of decision tools. *Ecological Indicators, 110*, 105851. doi:10.1016/j.ecolind.2019.105851

Nematollahi, M., Tajbakhsh, A., & Sedghy, B. M. (2021). The reflection of competition and coordination on organic agribusiness supply chains. *Transportation Research Part E, Logistics and Transportation Review, 154*, 102462. doi:10.1016/j.tre.2021.102462

Nguyen, M. P., Vaast, P., Pagella, T., & Sinclair, F. (2020). Local Knowledge about Ecosystem Services Provided by Trees in Coffee Agroforestry Practices in Northwest Vietnam. *Land, 9*(12). doi:10.3390/land9120486

Osrof, H. Y., Tan, C. L., Angappa, G., Yeo, S. F., & Tan, K. H. (2023). Adoption of smart farming technologies in field operations: A systematic review and future research agenda. *Technology in Society, 75*, 102400. doi:10.1016/j.techsoc.2023.102400

Pirsa, S., Sani, I. K., & Mirtalebi, S. S. (2022). Nano-bio composite based colour sensors: Investigation of structure, function, and applications in intelligent food packaging. *Food Packaging and Shelf Life, 31*, 100789. doi:10.1016/j.fpsl.2021.100789

Polymeni, S., Athanasakis, E., Spanos, G., Votis, K., & Tzovaras, D. (2022). IoT-based prediction models in the environmental context: A Systematic Literature Review. *Internet of Things : Engineering Cyber Physical Human Systems, 20*, 100612. doi:10.1016/j.iot.2022.100612

Purcell, W., & Neubauer, T. (2023). Digital Twins in Agriculture: A State-of-the-art review. *Smart Agricultural Technology, 3*, 100094. doi:10.1016/j.atech.2022.100094

Quarshie, A., Salmi, A., & Leuschner, R. (2016). Sustainability and corporate social responsibility in supply chains: The state of research in supply chain management and business ethics journals. *Journal of Purchasing and Supply Management, 22*(2), 82–97. doi:10.1016/j.pursup.2015.11.001

Rejeb, A., Keogh, J. G., Zailani, S., Treiblmaier, H., & Rejeb, K. (2020). Blockchain technology in the food industry: A review of potentials, challenges and future research directions. *Logistics, 4*(4), 27. doi:10.3390/logistics4040027

Rhodes, C. (2016). The 2015 Paris Climate Change Conference: Cop21. *Science Program*, 97–104.

Sala, S., Anton, A., McLaren, S. J., Notarnicola, B., Saouter, E., & Sonesson, U. (2017). In quest of reducing the environmental impacts of food production and consumption. *Journal of Cleaner Production, 140*(2), 387–398. doi:10.1016/j.jclepro.2016.09.054

Scott, C., Lundgren, H., & Thompson, P. (2018) Guide to Return in Supply Chain Management *Guide to Supply Chain Management*, 107–127. doi:10.1007/978-3-319-77185-4_6

Sekaran, U., Lai, L., Ussiri, D. A., Kumar, S., & Clay, S. (2021). Role of integrated crop-livestock systems in improving agriculture production and addressing food security–A review. *Journal of Agriculture and Food Research, 5*, 100190. doi:10.1016/j.jafr.2021.100190

Shukla, P., Skea, J., Calvo Buendia, E., Masson-Delmotte, V., Pörtner, H.-O., Roberts, D. C., Zhai, P., Slade, R., Connors, S., & Van Diemen, R. (2019). Climate Change and Land: An IPCC Special Report on Climate Change, Desertification, Land Degradation, Sustainable Land Management, Food Security, and Greenhouse Gas Fluxes in Terrestrial Ecosystems. Intergovernmental Panel on Climate Change (IPCC).

Sundgren, C. (2022). Circular supply chain relationships for food redistribution. *Journal of Cleaner Production, 336*, 130393. doi:10.1016/j.jclepro.2022.130393

Tan, W. C., & Sidhu, M. S. (2022). Review of RFID and IoT integration in supply chain management. *Operations Research Perspectives*, *9*, 100229. doi:10.1016/j.orp.2022.100229

Tao, W., Zhao, L., Wang, G., & Liang, R. (2021). Review of the Internet of Things communication technologies in smart agriculture and challenges. *Computers and Electronics in Agriculture*, *189*, 106352. doi:10.1016/j.compag.2021.106352

Virus, R. (2014). Agribiotechnology: costs and benefits of genetically modified papaya. Encyclopedia of agriculture and food systems, 35.

Wamatsembe, I. M., Asea, G., & Haefele, S. M. (2017). A survey: Potential impact of genetically modified maize tolerant to drought or resistant to stem borers in Uganda. *Agronomy (Basel)*, *7*(1), 24. doi:10.3390/agronomy7010024

Wang, M., Zhong, R. Y., Dai, Q., & Huang, G. Q. (2016). A MPN-based scheduling model for IoT- enabled hybrid flow shop manufacturing. *Advanced Engineering Informatics*, *30*(4), 728–736. doi:10.1016/j.aei.2016.09.006

Wang, Z., Li, Y., Li, T., Zhao, D., & Liao, Y. (2020). Tillage practices with different soil disturbance shape the rhizosphere bacterial community throughout crop growth. *Soil and Tillage Research, 197*. doi:10.1016/j.still.2019.104501

Wolfert, S., & Isakhanyan, G. (2022). Sustainable agriculture by the Internet of Things–A practitioner's approach to monitor sustainability progress. *Computers and Electronics in Agriculture*, *200*, 107226. doi:10.1016/j.compag.2022.107226

Wrzaszcz, W., & Prandecki, K. (2020). Agriculture and the European Green Deal. *Problems of Agricultural Economics*, *365*, 156–179.

Yanti, N., Carl, S., & Ammar, A. (2021). A systematic review of the contribution of system dynamics to value chain analysis in agricultural development. *Agricultural Systems*, 189.

Zhang, D., & Broadstock, D. C. (2020). Global financial crisis and rising connectedness in the international commodity markets. *International Review of Financial Analysis*, *68*, 101239. doi:10.1016/j.irfa.2018.08.003

Zhang, D., Sial, M. S., Ahmad, N., Filipe, A. J., Thu, P. A., Zia-Ud-Din, M., & Caleiro, A. B. (2020). Water scarcity and sustainability in an emerging economy: A management perspective for future. *Sustainability (Basel)*, *13*(1), 144. doi:10.3390/su13010144

Zhang, M., Wang, X., Feng, H., Huang, Q., Xiao, X., & Zhang, X. (2021). Wearable Internet of Things enabled precision livestock farming in smart farms: A review of technical solutions for precise perception, biocompatibility, and sustainability monitoring. *Journal of Cleaner Production, 312*, 127712. doi:10.1016/j.jclepro.2021.127712

Zhang, S., Cui, S., McLaughlin, N. B., Liu, P., Hu, N., Liang, W., Wu, D., & Liang, A. (2019). Tillage effects outweigh seasonal effects on soil nematode community structure. *Soil and Tillage Research, 192*, 233-239. doi:10.1016/j.still.2019.05.017

Chapter 4
Fair Trade and Ethical Sourcing in Agriculture:
Paving the Way for Sustainable Supply Chains

Mohamed Salah El Din
https://orcid.org/0000-0002-9360-8589
Majan University College, Oman

Masengu Reason
Middle East College, Oman

ABSTRACT

This chapter examines the contemporary practices and implications of fair trade and ethical sourcing within the agricultural supply chain, highlighting its pivotal contribution in achieving sustainability. Fairtrade scheme is considered an alternative to traditional trade systems that supports farmers in developing nations by promoting sustainable development and anti-poverty efforts. The usefulness of fair-trade programs has been questioned. Ethical sourcing is a comprehensive concept that focuses on incorporating and fostering environmental and social standards in all supply chain activities. However, the application of fair trade and ethical sourcing can be challenging. By conducting an extensive literature review, the chapter seeks to identify the principles and benefits of fair trade and ethical sourcing in developing a sustainable agricultural supply chain, investigate the challenges associated with fair trade and ethical sourcing, and finally, to present case studies and recommend best practices for implementing fair trade and ethical sourcing.

DOI: 10.4018/979-8-3693-2011-2.ch004

INTRODUCTION AND BACKGROUND

The modern agricultural supply chain, a vital component of international food security, is a sophisticated and interconnected form of supply chain that involves all the stages associated with production, storage and delivery of agricultural goods starting from farming till they reach consumers (Denis et al., 2020). Fairtrade in sustainable agriculture investigates an array of issues which includes ethical trade, standardization of value chain, fair pricing, market access and relevant technical support of all players along the supply chain (Alluvione et al., 2011; Talukder et al., 2019). Notably, over 1.2 million farmers across the globe have subscribed to the fairtrade framework and they have been incorporating fair trade practices into their operations (Mark et al., 2018). Resultantly, consumers have shown positive perception towards fairtrade labeled products (Alluvione et al., 2011).

The fair-trade movement operates as a voluntary organization which uses the acronym 'FINE'. Originally, it came out as a combination of four institutions, namely Fair Trade labeling (F), International Fair Trade Organization(I) Network of European World Shops (NEWS!) "N" and European Fair Trade Association (EFTA) (E) (Food and Agriculture Organization of the United Nations, 2008; International Trade Centre (ITC), 2011). The movement was extended its scope to encompass a variety of agricultural product, from cotton to textiles, handicrafts, coffee, cocoa, fruits, vegetables, and flowers (Bonisoli et al., 2019; Hilson et al., 2018). These products are noticed by specific and authentic labeling to ensure traceability and transparency throughout the supply chain (Luna et al., 2021; Partzsch & Kemper, 2019). In Bacon, (2005) and Bartley (2007), fair trade covers wide range of aspects such as child labor, wages, human trafficking, exploitation of vulnerable persons, consumer attitudes. For Jaffee, (2012 and Shorette (2014), fair trade also includes aspects of ecological issues (fertilizer, pesticides, organic farming, trade inequality and ethical concerns).

Despite the progression in fair trade initiatives across the globe, there has been some debates regarding the fair-trade initiatives among small-scale farmers. This debate emanates from the fact that small-scale farmers in developing countries remain marginalized. The small-scale farmers in Africa, Asia and some parts of Latin America receive low prices for their exported commodities and import at exorbitant prices. Despite much research being done and organizations being formed to advocate for fair trade, critics argue that it is leading to marginalization and extreme poverty. This has caused some scholars such as Hilson et al. (2018) to propose that there is need to establish a fair trade that addresses the concerns of small-scale farmers. Bartley, (2007) and Bonisoli et al., (2019) have noted in the current fair trade system large scale farmers are getting preferential treatment at the expense of small-scale farmers who constitute majority of growers across the globe.

The same sentiments were echoed by (Bigirimana & Masengu, 2015) who note that small-scale farmers in Zimbabwe have been failing to sustain their farms due to inequalities in international coffee markets. Other critiques question the "fairness of fair trade" given the fact that owing to the length of the value chain, there is no direct link between the producers and the consumers. Hence little money reaches developing countries and the farmers (Andorfer & Liebe, 2012a; Bissinger, 2019; Durevall, 2020). Bassett, (2010), did not find any difference in price offered to fairtrade and non-fairtrade farmers. Relatedly, Fairtrade encouraged Nicaraguan farmers to switch to organic coffee, but this resulted into higher costs and lower yields. This resulted in lower net income for the farmers (Bissinger, 2019). Another weakness of the Fairtrade movement is that fair trade cooperatives use inefficient marketing systems (Andorfer & Liebe, 2012a).

Unethical practices within the fairtrade movement such as labeling of coffee as Fairtrade by retailers and by packers in the developing countries is noted by (Baughman & Barclay, 2011). They highlighted the payment of low prices to farmers and failure to provide credit facilities to vulnerable farmers. In importing countries, the inner workings of the fairtrade movement are veiled with secrecy and lack of information may mislead consumers and volunteers (International Labour Organisation, 2021; Van Baar, n.d.).

There are many complaints of poor enforcement problems and a general failure to monitor standards. For instance, farm workers in Peru are paid less than the minimum wage while substandard labor and living conditions for seasonally hired labor in coffee production have been observed(Bissinger, 2019). Moreover, there are no reinforcement mechanisms for supermarkets and retailers. From a broader macroeconomic point of view, the fairtrade movement fails to achieve trade justice because if support to fair trade cooperatives leads to increased production, there is a risk of market flooding and consequently a decrease in price. It is relevant to note that there is need to have trade policy interventions (Durevall, 2020), that focuses on disadvantaged farmers(Bacon, 2005) to benefit from the fairtrade initiatives which are being implemented across the globe. The major question remains unanswered 'Is fair trade fair'.

This chapter seeks to give readers a thorough grasp of fair trade in the context of the agricultural supply chain, emphasizing its capacity to advance sustainability while also recognizing its many facets and difficulties.

Research Objectives

1. To identify the principles and benefits of fair trade and ethical sourcing in developing a sustainable agricultural supply chain.
2. To investigate the challenges Associated with Fair Trade and Ethical Sourcing

3. To Present case studies and recommend best practices for Implementing Fair Trade and Ethical Sourcing.
4. To recommend strategies that can be implemented to improve ethical sourcing in sustainable agricultural supply chain

RESEARCH METHODOLOGY

To accomplish the research objectives of this research, a comprehensive literature review was carried out, encompassing relevant research papers that had recently been published and supplemented by some secondary data and information obtainable from reputable websites and reports. The collection of pertinent information entailed conducting a thorough search of published research. According to Snyder (2019), literature review methodology is usually utilized to assess the current level of knowledge on a certain topic and can be utilized to determine research gaps, generate new research possibilities, and provide guidance for the creation of theoretical frameworks and methods. It makes it possible to organize a lot of data into smaller portions for analysis. It also makes it easier to integrate and cross-examine data to pinpoint the main themes. Several keywords were utilized in this investigation to mark out the literature: Fair trade, ethical sourcing, sustainable agricultural supply chain, fair trade case study, and fair-trade sustainability. An overview of the relevant analysis of literature and its significance for sustainable supply chain management is provided in the following section.

Literature Review

The incorporation of fair-trade principles and ethical sourcing initiatives into the agricultural supply chain can play an integral role in fostering sustainability. Numerous scholars have devoted their attention to this transdisciplinary intersection, offering valuable perspectives on its manifold aspects.

The Principles of Fair Trade and Ethical Sourcing

The traditional global agriculture supply chain has proved to be unsustainable and negatively impacting small-scale farmers and their families. This globalized sourcing system increases social injustice, undermines rural communities, and may disrupt food value chains (Sharma et al., 2018). Fair trade is considered one of the prominent examples of unconventional trade mechanisms. Its origins date back to the middle of the 20th century, when it started as a movement to support sales of poor craftsmen who are deprived of big markets' access. The initiative developed in the following decades, especially the last thirty years, due to economic and social reasons, to cultivate an alternative trade model, that focus on promoting equality, sustainability, and anti-poverty practices, particularly in developing countries (Ribeiro-Duthie et al., 2021a; Sharma et al., 2018).

The international effort for fair trade is organized and facilitated by two prominent organizations; The World Fair Trade Organization (WFTO) which oversees the involvement of business organizations in fair trade initiatives, and Fair-Trade Labelling Organization (FLO), the party responsible for maintaining the labeling and certifying system (Kemper & Partzsch, 2019; Nicholls & Opal, 2006). Organizations that embrace fair trade practices, often known as "fair trade organizations", including wholesalers and retailers buy certified goods either directly from farmer groups or via local importers at a premium price. Additionally, food producers can benefit from a shorter supply chain which cuts down the costs by removing intermediaries and ensures fair prices. The role of fair-trade organizations is also extended to training and networking. Figure 1 illustrates a typical fair trade supply chain (Zhang et al., 2020).

The fundamental concept of fair trade is supported by an extensive set of principles that have been stated by the World Fair Trade Organization (WFTO). These principles act as a basic framework for manufacturers, traders, and other stakeholders to ensure fair and open trade practices (WFTO, 2023).

At the heart of these practices is the emphasis on opportunities for marginalized small-scale producers, as well as their labor communities and local communities, with the aim of securing their livelihoods and improving their exposure to international marketplaces. This is accompanied by a focus on transparent and open trade processes

Figure 1. A typical fair trade supply chain

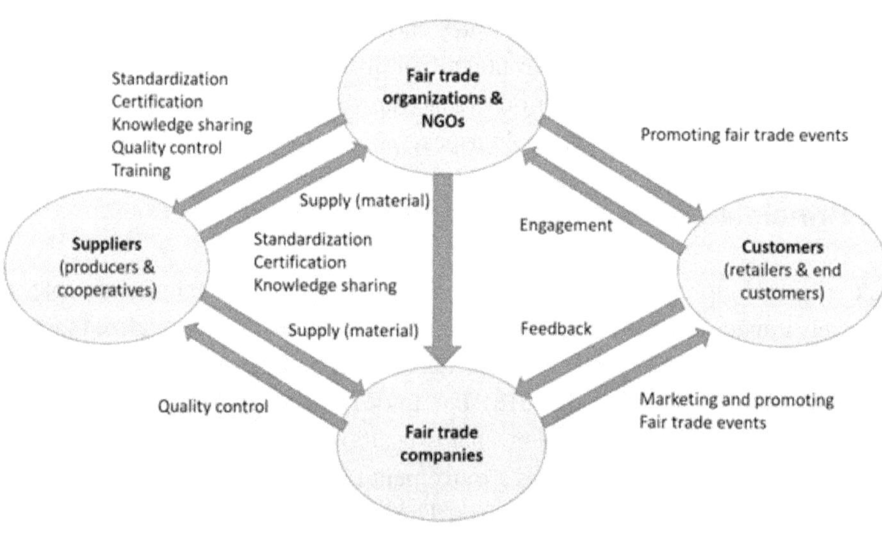

throughout the entire supply chain. Organizations committed to fair trade initiatives should be dedicated to upholding fair-trade practices, which include offering living and reasonable wages, respecting agreements, and fostering cooperative relationships while protecting cultural identities and traditional skills (WFTO, 2023).

The ethical foundation of fair trade is strongly evident. It clearly rejects all forms of child and forced labor, and embraces the principles of non-discrimination, whether based on gender, religion or ethnicity. In addition, it advocates maintaining employees' health and safety by ensuring fair working conditions (WFTO, 2023).Fair trade takes an interest in developing the competencies and skillsets of marginalized farmers and their labor by encouraging manufacturers and retailers to engage in skill building activities with them. Moreover, there is an obligation on organizations to create awareness about the importance of fair trade and its possible impact on communities (WFTO, 2023)Lastly, fair trade principles are not only concerned about economic and social issues, but they extend to deeply include the environmental elements by promoting sustainability and respect for the environment (WFTO, 2023).

Fairtrade organizations are pivotal in supporting and encouraging fair trade practices globally. They focus on improving work conditions and providing fair compensation for the entire agricultural supply chain in developing nations. Also, they foster awareness of consumers and boost sales of fair-trade goods (Koos, 2021; Parvathi et al., 2017). Fair trade allows small-scale farmers take greater portions of the markets for high value products exports (Balineau, 2012; Herrell et al., 2017). Farmers are increasingly participating in the fair-trade certification programs due to the rise in the premiums received, as in the case of the Honduran coffee farmers (Herrell et al., 2017).

Agricultural products that are "fair-trade" certified and labeled are usually sold with higher prices as compared to other products (Abbasi & Din, 2019; Lawrence et al., 2023). The global certification system is organized by the non-for-profit, membership-based Fair-Trade Labeling Organization (FLO). The organization also set standards for fair trade and systemize the global strategy (Fairtrade, 2023).The certification system proved to have powerful positive impact on the lives of farmers and their families (White & Samuel, 2016). Although the fair-trade movement is intricate and multidimensional, it is obvious that the Sustainable Development Goals (SDGs) and fair trade are strongly related (Zysk, 2020).

Ethical sourcing is procuring materials and products from ethical suppliers, who fairly treat their workers and care about sustainability (Mboga, 2017). According to Kim & Chae (2022), ethical sourcing initiatives encompass all the efforts of the entire supply chain partners, whether formal or informal, to align their operations with CSR targets, such as fair trade, environmental sustainability, diversity, human rights, health and safety, as well as animal welfare.

Ethical sourcing is more extensive than fair trade as assessing and choosing suppliers according to how well they follow moral principles like social responsibility, environmental sustainability, and labor rights (Barrientos & Dolan, 2012; Goff, 2018; Kim et al., 2018). Compared to fair trade practices, ethical sourcing has multiple certification systems and standards. Moreover, while fair trade is limited to specific agricultural products, ethical sourcing extends to all products (Goff, 2018; Kim et al., 2018; Ladhari & Tchetgna, 2017). Both practices can be influences by consumers behavior (Goff, 2018). In other words, fair trade can be considered as one form of ethical sourcing, which may include other forms such as organic certification, sustainable sourcing, and ethical labor practices (Ladhari & Tchetgna, 2017). Lastly, as it focuses on the entire supply chain, ethical sourcing is not limited to supplier – buyer relationships which is the focus of fair trade (Neilsno & Pritchard, 2010).

Fair Trade Consumption Theories

Fair trade consumption is multifaceted and shaped by multiple theoretical frameworks. Plenty of studies have discussed these various theories. From an economic perspective, customers' choice to buy fair trade goods is influenced by additional benefits they get over the cost of traditional options (Bürgin & Wilken, 2021).

The Norm Activation Theory (NAM) relates fair trade consumption to moral obligation and willingness to be involved in social initiatives (Schenk, 2019). It posits that a person's environmentally conscious actions are guided by their awareness of responsibility (AR) and knowledge of consequences (AC) (Liu et al., 2017). In the context of fair trade, it could imply that customers are more likely to purchase fair trade products if they became aware of unfair trade practices, leading to more ethical buying norms (Le & Nguyen, 2022).

Value - Belief Norm (VBN) theory is similar to NAM in the sense that both attribute customer awareness as a main drive to purchase fair trade goods. In addition, VBN considers individual's moral compass and perceived social pressure as determining factors (Schenk, 2019). This aligns with Gohary et al. (2023), who argues that ethical principles held by consumers, including sustainability and social equality, motivate fair trade purchases. VBN theory considers the interplay between personal values, beliefs and purchasing habits that involve fair trade products.

Value-Identity-Personal (VIP) norm model brings attention to the environmental self-identity in boosting fair trade consumption (Schenk, 2019). The VIP model proposes the existence of correlation between the degree people believe they are environmentally responsible (van der Werff & Steg, 2016) and fair-trade consumption. The model was found effective in predicting fair trade purchases, addressing the relationship between values, norms and fair trade (Schenk, 2019).

Although Resource Advantage (R-A) is a general theory of competition, it could be applied to understand fair trade consumption. It showcases the significance of identifying unique resources to gain competitive advantage in niche markets, ultimately improving their bottom line (Doherty, 2011). In the context of fair trade, ethical responsibility, innovation, supply networks and fair-trade certification could be leveraged as their focal resources, which supports their quest for sustainable competitive advantage (Doherty, 2011; Thao Tran et al., 2012).

Brunner (2014) employed neutralization theory to justify the gap between pro-fair-trade beliefs and actual purchasing behavior. According to this theory, customers use neutralization strategies to give reasons for their lack of action. These strategies could be internal, such as questioning the effectiveness of fair trade) or external, like being concerned with the actual impact on farmers and their families. The neutralization theory emphasizes the psychological complexities that face converting ethical values of fair trade into actual consumption norms (Brunner, 2014; Gruber & Schlegelmilch, 2014). This corresponds with Shih-Tse Wang & Chen (2019) who highlighted customers' perception of organizations' fairness as a leading factor in their decision to buy fair trade products.

While the beforementioned theories mostly consider norm as a driving factor to purchase fair trade products, there are other papers that related fair trade consumption to economic reason (Dragusanu et al., 2014), and sociocultural factors including lifestyle (Schenk et al., 2021), social dominance (Gohary et al., 2023), cross-country differences (Andorfer & Liebe, 2012b) or using other approaches such as post-colonial theory (Blanchet, 2012; Dam, 2015) and feminist approach (Sen, 2018).

Benefits of Fair Trade and Ethical Sourcing

There has been a rapid adoption of ethical sourcing and fair-trade principles over the past few years (2015-2023). According to Ndiaye et al., (2009) ethical sourcing and fairtrade certification in the fisheries industry has enable to have market access to international players who may want to protect their brands reputation. Fair trade certification allows small-scale players to benefit from supply chain promotion of sustainable practice and increased investment in market development. Pertinent to the concept of ethical sourcing is the fact that unified standard and practices that it links the product to the state of resources and its related components that guarantees compliance (Ndiaye et al., 2009).

The is a deliberate integration of food production and retailing into the global ethical sourcing system. The major challenge arising is that can the global lobbying and civil society pressure lead to global compliance to fair trade. Research has indicated that fair trade will result in better conditions for small-scale producers and security of workers who constitute the backbone of global food supply chains.

The globalization of food production and the dominance of large food producers and retailers have resulted in a reduction in consumer bargaining power. This has raised a lot of concerns about unethical behaviors through media houses and civil society groups calling for the improvement in agriculture product supply chain visibility improvement in the way food is produced. In response to the wave of ethical condemnations and outcry big retailers have tried to rebrand and reposition themselves as socially responsible supply chain actors. This has led to the incorporation of fair-trade labelled products that have standardized codes of labor of employees across supply chain actors. The repositioning strategy is also coming from the fact that consumers are now also concerned with the social and environmental impact of the products they consume. For Dai et al., (2020) citing Nelsen (2014) "Consumers around the world are saying loud and clear that a brand's social purpose is among the factors that influence purchase decisions,". All these different aspects have resulted in increased attention to ethical sourcing and fair-trade practices that has resulted in improved supply chain visibility and traceability.

Crane et al., (2010, 2014), stipulate that supply chain actors that subscribe to the principles of ethical sourcing and fair trade have the potential to remain competitive in the market through good reputational management. This is since socially responsible consumers have shown unwavering commitment to products that are produced ethically (Agarwal et al., 2015; Bansal & Song, 2017) .Bansal & Song, (2017) state that results in increased opportunities for market access and demand for sustainable products that can reduce supply chain crisis.

Due to increased activism and consumer concerns, governments have put in place regulations to guide the sourcing of products. Agarwal et al., (2015) reiterates that supply chain actors who align their practice to ethical sourcing mitigate the risk of disruptions and bad publicity. For, Gold et al., (2010) ethical sourcing and fair trade increase the employee's morale and may result in increased productivity. It provides an environment that is reap for innovation in product development that is key in having a competitive edge (Bansal & Song, 2017).

Challenges of Fair Trade and Ethical Sourcing

The fair-trade movement and ethical sourcing have risen in global significance as mechanisms to rectify injustices in global supply chains. However, several challenges are obstructing their complete realization. In Pazderka, (2010), the integration of ethical sourcing in United Kingdom enables the inclusion of small-holder farmers into the value chain and assist in the alleviation of poverty. Pazderka concur with Guthure (2010) study on ethical sourcing in Kenya flower industry. The study revealed a paradoxical situation in Kenya floriculture, in which global business appear to add no value to the lives of small-scale farmers. Relatedly Hilson et al.,

(2018), state that in Europe, ethical issues have been raised regarding working conditions of employees in different stages of the food supply chain. This has made the enactment of regulations oblige companies to comply with several standards and codes that guarantee ethical sourcing. For Roy et al., (2012), to guarantee ethical sourcing retailers need to have third party certifications quality assurance system. Ethical consumerism has been increasing and consumers know what retailers need to authenticate the products production process. Ezedike, (2013), note that international corporation is Nigeria agriculture sector have been falling short of ethical standards based on expected principles, fairness, rights, trust telling and honest of the supply chain activities. The study recommends that, in the face of lapses of ethics among the multinational corporations there is need to have a comprehensive framework to guide and protect the supply chain actors. Githui, (2012) raised concerns that supply chain actors were devoid of ethical inclination and recommends that the supply chain needs to re-invent itself through ethical consideration that could reengineer the whole system and (Kumar Sharma, 2020), cites supply chain dilemma of food products due to the fact that corporates are always optimizing on profits.

The main concern is the debate over the financial viability of fair trade. (Balineau, 2012) questioned the effectiveness on fair trade initiatives in guaranteeing higher yields for small-scale farmers, when compared to conventional trade schemes, especially with the high labor costs and the additional certification fees. Furthermore, whereas technical support strengthens the producers' ability to negotiate and bargain, it is doubtful if this empowerment extends outside the fair-trade umbrella.

Even if financial viability is evident, the distribution of the premium yields remains questionable. González-Prieto (2018) raised a concern on the ability to track the share of marginalized farmers received from the received premiums. In addition, Oya et al. (2017) noted that some buyers do not pay the premiums to farmers, particularly with the inability of FTO to ensure the applications of its standards. The ultimate outcome of these challenges could be limited impact on the profitability of small-scale farmers (González-Prieto, 2018; Naegele, 2020).

The mainstreaming and commercializing of fair-trade concept present a notable challenge. According to Doherty et al. (2013) and Ribeiro-Duthie et al. (2021), mainstreaming can dilute the standards of fair trade, despite its effectiveness in supporting poor farmers who suffer from inequality and unfairness. This is shown by Jaffee & Howard (2015), who found numerous rivaling fair-trade products in US market, where all of them have different fair-trade labels. There is always a risk that retailers use fair trade branding to greenwash their reputation, which could confuse customers (Ballet & Carimentrand, 2010; Doherty et al., 2013). Moreover, mainstreaming could push farmers to rely heavily on big purchases, due to pre-financing, which could affect them negatively (Doherty et al., 2013). At the same time, the limited market size also represents an obstacle. If fair trade products are not leading the market, the objective of sustainable economic growth for the marginalized communities will be jeopardized (Utting, 2008).

Another complication for the efficacy of fair-trade practices is consumer behavior. Benson & Connell (2014) noted that a sizable market segment, baby bloomers, found it difficult to identify and buy fair trade items. This is a result of inconsistent labeling, limited market accessibility, unavailability for pre-purchase trials (which could be a decisive factor in purchasing), and lack of variety in nearby stores. Their findings are also aligned with their conviction that fair-trade goods were out of reach financially or in terms of lifestyle. In addition, there are doubts on the ability of fair trade to actively assist small-scale farmers to access premium markets (Balineau, 2012).

In the context of fair trade, gender equity is both an opportunity and a difficulty. Although fair trade programs are praised for promoting gender parity (Lyon et al., 2010; Sen, 2018), several obstacles continue to exist. It is a complex task to monitor and ensure gender equality in mainstream marketplaces (Das, 2011). Furthermore, research to date emphasizes how important it is to carefully integrate gender, labor, and well-being paradigms into the fair-trade frameworks (Terstappen et al., 2012). The intricate and diverse fair-trade procedures frequently present obstacles for female labor, hindering their comprehension and manifestation of their rights (Lyon et al., 2010; Makita, 2011).

Fair-trade products compete with other fair-trade products, as well as other conventional products (Geysmans & Hustinx, 2016). The latter type is always under pressure of lower prices for normal products which could be more appealing to customers (Bürgin & Wilken, 2021). This competition is affected by consumer preferences and purchasing habits (Benson & Connell, 2014).

As agricultural supply chains become more sophisticated and extended, fair trade adoption becomes more challenging (Chen et al., 2022). With the increased globalization of production and consumption, it can be complicated to ensure adherence to fair trade rules, especially in nations with weak governance systems. For example, retailers may lack the necessary resources to track the sustainability of the fair-trade products (Ribeiro-Duthie et al., 2021b; Simangunsong et al., 2016).

CASE STUDIES

Starbucks

Developing a positive influence on communities requires businesses to commit to becoming more sustainable. Starbucks has embraced fair trade, sustainability and ethical sourcing through procuring processed products such as tea, coffee, and cocoa (Mboga, 2017). Starbucks recognizes that the future success of its business is closely linked to the future of farmers and their families, since it purchases around three percent of the coffee produced worldwide, which comes from over 400,000 farmers across 30 nations (Starbucks, 2020). Starbucks made its choice to work directly with farmers by investing in the community and incorporating innovation and technology (Mboga, 2017). In 2022, Starbucks cut ties with the certification system of the fair trade in UK and Europe and shifted to its unique C.A.F.E. Practices standard (Independent, 2022). CAFÉ refers to "Coffee and Farmers Equalities", set of standards and practices that imitate the fair-trade standards (Rizqiyanto, 2017). (Dragusanu et al., 2022) found that there is a positive effect of fair-trade certification scheme on coffee farmers in Costa Rica.

On the other hand, there is a negative side to the story of Starbucks. (Dragusanu et al., 2022) emphasized the importance of continuous support and enhancement of fair-trade initiatives to optimize its performance and solve current challenges faced by coffee farmers. Rizqiyanto (2017) doubts that farmers get the promised yields under CAFÉ standards. Starbucks continually asserts that Fair Trade does not ensure that the products it sources meet its requirements. In other words, suppliers' nonconformance with CAFÉ standards does not mean that Starbucks will not buy their products.

Pepsi

PepsiCo's sustainable agriculture framework integrates trade fairness, ethical sourcing, and integrity in the entire agricultural supply chain of the organization to meet the standards and the expectations of stakeholders, especially consumers. To overcome the difficulties of global enforcement of this framework, PepsiCo developed the Supplier Code of Conduct (SCoC) and other related policies. The SCoc has four key cornerstones; maintaining the organization's reputation, safeguarding the environment and social concerns, developing and sustaining supplier capabilities, and enhancing sustainability performance (Mboga, 2017; Pepsico, 2023).

The three main focuses of PepsiCo's sustainable sourcing initiative are people, planet and agriculture. Regarding people, PepsiCo pledges to uphold human rights at every stage of its supply chain. This involves being certain that suppliers refrain from employing child or forced labor as well as ensuring they offer their employees a healthy and secure work environment. Safeguarding the environment is targeted by collaborating with suppliers on reduction of emissions and carbon footprint. At the same time, the company encourage environmentally friendly farming, by cooperating with suppliers to have more healthy soils and less usage of fertilizer and pesticides in farming (Pepsico, 2023). PepsiCo conducts rigorous auditing process for suppliers by using sustainability performance scorecard for transparent tracking and review (Kashmanian, 2017).

Coca-Cola

Coca-Cola considers that sustainable agriculture provides answers to interconnected problems like women's empowerment, climate adaptation, accessibility to water, human rights, and reducing greenhouse gas emissions. Coca-Cola is one of the premier organizations in coffee purchases globally, financing farmer loans to Fairtrade cooperatives exceeding $14 million (Coca-Cola, 2023). It understands the severity of unethical sourcing on reputation and sustainability of its agricultural supply chain. The company oversees all workers in partner organizations to avoid such issues. In case of any discrepancy in terms of ethical practices, Coca-Cola suppliers are given time to implement corrective measures. Failure to conform and correct their performance could lead to contract termination (Mboga, 2017).

Unilever

In 2007, Unilever collaborated with the Rainforest Alliance, a nonprofit organization, to launch an ambitious plan of sourcing 100% of for its world best-selling tea brand, Lipton, in a sustainable way. As a result, east African tea farmers, especially in Kenya, joined the fair-trade certification program to ensure their access to Unilever purchases. The collaboration with Rainforest Alliance extended to implementing training and development programs to elevate the skills of the farmers to comply with the set standards of Unilever. This had a huge impact on the Kenyan farmers. 2018 research stated that more than 85% of Kenyan tea producers held Rainforest Alliance certification (Archer & Elliott, 2021).

DAGRIS

Under FLO guidance, DAGRIS and Max Havelaar France initiated fair-trade cotton in Mali in 2003. They introduced an independent certification agency, necessitating the adherence to sustainability standards in the accepted products. The system was based on buyers paying fair prices to producers and a premium fee to their cooperatives to support their economic growth. The sustainability standards involved embracing inclusivity, democracy, non- discrimination, and environmentally responsible farming, in addition to allocation of premium amounts for the purpose of future growth. The results of the program were positive as it managed to improve the sustenance of the Malian farmers, foster the strength of their organizations, and enhance the quality of the cotton itself (Balineau, 2012).

DISCUSSION

Integrating Principles With Practice in Fair Trade and Ethical Sourcing

The literature review showcased the way of fair trade and ethical sourcing principles place a strong emphasis on the value of transparent, fair-trading procedures, and human rights respect in the value chain for agricultural products. In a globalized sourcing framework, such concepts are essential for promoting sustainability and resolving the difficulties small-scale farmers and their families confront. Due to its unsustainable nature, the conventional global agriculture supply chain has exacerbated social inequity and threatened rural communities. Fair trade and ethical sourcing frameworks signify a shift from the traditional capitalist trading system towards an ethical and responsible version of capitalism that prioritizes social and environmental responsibility.

Benefits and Challenges: A Dual Perspective

Substantial advantages have resulted from the recent quick embrace of fair trade and ethical sourcing principles, notably higher investments in growth of markets, greater access to markets for small-scale businesses, and support for sustainable practices. In particular, ethical sourcing goes beyond the concept of fair trade and includes efforts to match operations with corporate social responsibility (CSR) goals like diversity, human rights, environmental sustainability, health and safety, and animal welfare. Nevertheless, there are a number of obstacles facing the fair-trade movement and ethical sourcing, such as the integration of responsible sourcing across national borders, contradictory circumstances in which international trade fails to provide value to small-scale agricultural producers, and problems with worker conditions at various points in the food supply chain. Those challenges demonstrate how difficult it may be to enforce compliance with fair trade laws, particularly in countries with inadequate governance frameworks.

Policy and Regulatory Frameworks

The review of the literature highlights the significance of policy and regulatory frameworks that are favorable in advancing ethical sourcing and fair trade. Through trade agreements, regulations, and policies, governments and international organizations are essential in fostering contexts that support these behaviors. In order to level up competition and encourage more enterprises to consider ethical sourcing, policies that sanction predatory actions and reward ethical business practices are crucial.

Case Studies

The real-life case studies, which feature Starbucks, Pepsico, Unilever, Coca Cola and Dagris, offer useful insights into how companies can positively impact societies by adhering to fair trade and ethical sourcing initiatives. These cases provided an illustration of how the concepts of fair trade can be applied in practical situations, providing insightful guidance for prospective businesses considering adopting comparable policies.

Consumer Education and Engagement

The literature also emphasizes the importance of raising consumer involvement and education. Even while items with ethical sourcing are becoming more and more popular, many customers are still ignorant of what fair trade and ethical sourcing really mean. Increasing consumer awareness of the effects of their purchases and ways to support fair trade practices is essential to increasing the market for goods supplied ethically.

Towards a Sustainable Future

The importance of a comprehensive strategy for fair trade and ethical sourcing is highlighted in this conversation. Although the principles in question offer a framework, applying them in practice presents advantages as well as difficulties that need to be meticulously considered. The case studies illustrate the necessity of ongoing development and modification while also providing insight into the way these principles might be used successfully. Future studies should concentrate on creating models that direct companies to incorporate these ideas into their key business processes while continuing to be flexible in response to changing social and market environments.

CONCLUSION

This chapter explores fair trade and ethical sourcing throughout the agricultural supply chain, highlighting the crucial connection between social justice, sustainability, and economic viability. Despite the respectful and noble intention of fair trade and ethical sourcing frameworks, they display a complicated landscape filled with possibilities and obstacles. The intricate connections between social justice, environmental conservation, fair trade, and ethical sourcing in agriculture are highlighted in this research. Driven by customer demand for goods offered in an ethical way, these techniques offer benefits like sustainable agriculture and market access for small-scale farmers. But there are considerable obstacles to international trade, such as logistical complexity and the requirement for sophisticated methods. The possibilities and challenges of putting these approaches into practice are demonstrated by case studies like Starbucks. In the end, building a sustainable, equitable agricultural supply chain necessitates constant innovation, cooperation from all stakeholders, and information exchange to realize the goals of fair trade and ethical sourcing.

REFERENCES

Abbasi, J. Y., & Din, M. (2020, June). Fairtrade and Workers' Welfare: A Case Study of the Soccer Industry in Pakistan. *Journal of Asian and African Studies*, *55*(4), 536–551. Advance online publication. doi:10.1177/0021909619885968

Agarwal, J., Osiyevskyy, O., & Feldman, P. M. (2015). Corporate Reputation Measurement: Alternative Factor Structures, Nomological Validity, and Organizational Outcomes. *Journal of Business Ethics*, *130*(2), 485–506. doi:10.1007/s10551-014-2232-6

Alluvione, F., Moretti, B., Sacco, D., & Grignani, C. (2011). EUE (energy use efficiency) of cropping systems for a sustainable agriculture. *Energy*, *36*(7), 4468–4481. doi:10.1016/j.energy.2011.03.075

Andorfer, V. A., & Liebe, U. (2012a). Research on Fair Trade Consumption-A Review. *Journal of Business Ethics*, *106*(4), 415–435. doi:10.1007/s10551-011-1008-5

Andorfer, V. A., & Liebe, U. (2012b). Research on Fair Trade Consumption-A Review. In Journal of Business Ethics (Vol. 106, Issue 4, pp. 415–435). doi:10.1007/s10551-011-1008-5

Archer, M., & Elliott, H. (2021). 'It's up to the market to decide': Revealing and concealing power in the sustainable tea supply chain. *Critique of Anthropology*, *41*(3), 227–246. doi:10.1177/0308275X211038607

Bacon, C. (2005). Confronting the coffee crisis: Can Fair Trade, organic, and specialty coffees reduce small-scale farmer vulnerability in Northern Nicaragua? *World Development*, *33*(3), 497–511. doi:10.1016/j.worlddev.2004.10.002

Balineau, G. (2012). *Disentangling the Effects of Fair Trade on the Quality of Malian Cotton*. http://hdl.handle.net/10419/269320

Ballet, J., & Carimentrand, A. (2010). Fair Trade and the Depersonalization of Ethics. *Journal of Business Ethics*, *92*(S2), 317–330. Advance online publication. doi:10.1007/s10551-010-0576-0

Bansal, P., & Song, H. C. (2017). Similar but not the same: Differentiating corporate sustainability from corporate responsibility. *The Academy of Management Annals*, *11*(1), 105–149. doi:10.5465/annals.2015.0095

Barrientos, S., & Dolan, C. (2012). *Ethical Sourcing in the Global Food System*. doi:10.4324/9781849771269

Bartley, T. (2007). Institutional emergence in an era of globalization: The rise of transnational private regulation of labor and environmental conditions. *American Journal of Sociology*, *113*(2), 297–351. doi:10.1086/518871

Bassett, T. J. (2010). Slim pickings: Fairtrade cotton in West Africa. *Geoforum*, *41*(1), 44–55. doi:10.1016/j.geoforum.2009.03.002

Baughman, S. B., & Barclay, S. H. (2011). *Fair Trade and Child Labor*. https://papers.ssrn.com/abstract=1823546

Benson, E., & Connell, K. Y. H. (2014). Fair trade consumption from the perspective of US Baby Boomers. *Social Responsibility Journal*, *10*(2), 364–382. doi:10.1108/SRJ-08-2012-0094

Bigirimana, S., & Masengu, R. (2015). The Contribution of the International Coffee Agreement (ICA) to the Development of Export Markets for Small Scale Coffee Farmers in Manicaland, Zimbabwe. *World Journal of Business and Management*, *1*(1), 57. doi:10.5296/wjbm.v1i1.7931

Bissinger, K. (2019). Price Fairness: Two-Stage Comparison of Conventional and Fairtrade Prices. *Journal of International Consumer Marketing*, *31*(2), 86–97. doi:10.1080/08961530.2018.1482525

Blanchet, V. (2012). *The two faces of Janus: a postcolonial problematization of the fair trade ambivalence*. https://shs.hal.science/halshs-00676060

Bonisoli, L., Galdeano-Gómez, E., Piedra-Muñoz, L., & Pérez-Mesa, J. C. (2019). Benchmarking agri-food sustainability certifications: Evidences from applying SAFA in the Ecuadorian banana agri-system. *Journal of Cleaner Production*, *236*, 117579. Advance online publication. doi:10.1016/j.jclepro.2019.07.054

Brunner, T. A. (2014). Applying neutralization theory to fair trade buying behaviour. *International Journal of Consumer Studies*, *38*(2), 200–206. doi:10.1111/ijcs.12081

Bürgin, D., & Wilken, R. L. (2021). Increasing Consumers' Purchase Intentions Toward Fair-Trade Products Through Partitioned Pricing. *Journal of Business Ethics*. Advance online publication. doi:10.1007/s10551-021-04938-6

Chen, L., Lee, H. L., & Tang, C. S. (2022). Supply Chain Fairness. *Production and Operations Management*, *31*(12), 4304–4318. Advance online publication. doi:10.1111/poms.13849

Coca-Cola. (2023). *Sustainable Agriculture*. https://www.coca-colacompany.com/sustainability/sustainable-agriculture

CraneA.MattenD.SocialC. (2010). https://www.routledge.com/Corporate-Social-Responsibility-Readings-and-Cases-in-a-Global-Context/Crane-Matten-Spence/p/book/9780415683258

Crane, A., Matten, D., & Spence, L. (2014). Chapter 4: Responsibilities to stakeholders. *Corporate Social Responsibility: Readings and Cases in a Global Context*, 133–202. https://ebookcentral-proquest-com.ez.library.latrobe.edu.au/lib/latrobe/reader.action?docID=5725906&amp

Dai, X., Lin, Y. T., Shi, R., & Xu, D. (2020). A manufacturer's responsible sourcing strategy: Going organic or participating in fair trade? *Annals of Operations Research*, *291*(1–2), 195–218. doi:10.1007/s10479-018-3090-x

Das, R. (2011). Gender in Practices of Fair Trade and Decent Work. *Gender, Technology and Development*, *15*(1), 101–125. Advance online publication. doi:10.1177/097185241101500105

Denis, N., Dilda, V., Kalouche, R., & Sabah, R. (2020). *Agriculture supply-chain optimization and value creation.*

Doherty, B. (2011). Resource advantage theory and fair trade social enterprises. *Journal of Strategic Marketing*, *19*(4), 357–380. doi:10.1080/0965254X.2011.581379

Doherty, B., Davies, I. A., & Tranchell, S. (2013). Where now for fair trade? *Business History*, *55*(2), 161–189. doi:10.1080/00076791.2012.692083

Dragusanu, R., Giovannucci, D., & Nunn, N. (2014). The economics of fair trade. *The Journal of Economic Perspectives*, *28*(3), 217–236. doi:10.1257/jep.28.3.217

Dragusanu, R., Montero, E., Nunn, N., Alfaro, L., Atkin, D., Bandiera, O., Donaldson, D., Field, E., Macchiavello, R., Miquel-Florensa, P., Morjaria, A., Muendler, M., Olken, B., Pavcnik, N., Podhorsky, A., & Volpe, C. (2022). *The Effects of Fair Trade Certification: Evidence From Coffee Producers in Costa Rica*. Academic Press.

Durevall, D. (2020). Fairtrade and Market Efficiency: Fairtrade-Labeled Coffee in the Swedish Coffee Market. *Economies, 8*(2), 30. doi:10.3390/economies8020030

Ezedike, E. U. (2013). Globalization, multinationals and business ethics in Nigeria: Ethical imperatives for sustainable development. *Journal of Sustainable Development in Africa, 15*(6).

Fairtrade. (2023). *The Fairtrade system*. https://www.fairtrade.net/about/fairtrade-system

Food and Agriculture Organization of the United Nations. (2008). *Certification in the value chain for fresh fruits: The example of banana industry*. http://www.fao.org

Geysmans, R., & Hustinx, L. (2016). Placing the Distant Other on the Shelf: An Analysis and Comparison of (Fair Trade) Coffee Packages in Relation to Commodity Fetishism. *Sociological Research Online, 21*(1), 35–40. Advance online publication. doi:10.5153/sro.3854

Githui, D. M. (2012). *Responsible Purchasing and Supply Chain Management in Kenya: A Critical Analysis of the Ethical Considerations in Procurement Management*. http://repository.dkut.ac.ke:8080/xmlui/handle/123456789/124

Goff, S. C. (2018). Fair trade: Global problems and individual responsibilities. *Critical Review of International Social and Political Philosophy, 21*(4), 521–543. doi:10.1080/13698230.2016.1252993

Gohary, A., Madani, F., Chan, E. Y., & Tavallaei, S. (2023). Political ideology and fair-trade consumption: A social dominance orientation perspective. *Journal of Business Research, 156*, 113535. doi:10.1016/j.jbusres.2022.113535

Gold, S., Seuring, S., & Beske, P. (2010). Sustainable Supply Chain Management and Inter-Organizational Resources: A Literature Review. *Corporate Social Responsibility and Environmental Management, 17*(4), 230–245. doi:10.1002/csr.207

González-Prieto, A. M. (2018). Conservation of Nearctic Neotropical Migrants: The Coffee Connection Revisited. *Avian Conservation & Ecology, 13*(1), art19. Advance online publication. doi:10.5751/ACE-01223-130119

Gruber, V., & Schlegelmilch, B. B. (2014). How Techniques of Neutralization Legitimize Norm- and Attitude-Inconsistent Consumer Behavior. *Journal of Business Ethics*, *121*(1), 29–45. doi:10.1007/s10551-013-1667-5

Herrell, K. M., Student, G., Tewari, R., & Mehlhorn, J. (2017). *Honduran Coffee Trade: Economic Effects of Fair Trade Certification On Individual Producers*. Academic Press.

Hilson, G., Gillani, A., & Kutaula, S. (2018). Towards Sustainable Pro-Poor Development? A Critical Assessment of Fair Trade Gold. *Journal of Cleaner Production*, *186*, 894–904. doi:10.1016/j.jclepro.2018.03.123

Independent. (2022). *Starbucks cuts back sale of Fairtrade coffee in UK*. https://www.independent.co.uk/news/uk/starbucks-fairtrade-europe-middle-east-fairtrade-foundation-b2017451.html

International Labour Organisation. (2021). *Child labour in cotton a briefing*. ILO.

International Trade Centre (ITC). (2011). *Trends in the trade of certified coffees*. www.intracen.org

Jaffee, D. (2012). Weak coffee: Certification and co-optation in the fair trade movement. *Social Problems*, *59*(1), 94–116. doi:10.1525/sp.2012.59.1.94

Jaffee, D., & Howard, P. H. (2015). Who's the Fairest of Them All? The Fractured Landscape of U.S. Fair Trade Certification. *Agriculture and Human Values*. Advance online publication. doi:10.1007/s10460-015-9663-2

Kashmanian, R. M. (2017). Building Greater Transparency in Supply Chains to Advance Sustainability. *Environmental Quality Management*, *26*(3), 73–104. doi:10.1002/tqem.21495

Kemper, L., & Partzsch, L. (2019). Saving Water While Doing Business: Corporate Agenda-Setting and Water Sustainability. *Water (Basel)*, *11*(2), 297. Advance online publication. doi:10.3390/w11020297

Kim, S., & Chae, S. (2022). Shareholder Value Effects of Ethical Sourcing: Comparing Reactive and Proactive Initiatives. *Journal of Business Ethics*, *179*(3), 887–906. doi:10.1007/s10551-021-04841-0

Kim, S., Colicchia, C., & Menachof, D. (2018). Ethical Sourcing: An Analysis of the Literature and Implications for Future Research. *Journal of Business Ethics*, *152*(4), 1033–1052. doi:10.1007/s10551-016-3266-8

Koos, S. (2021). Moralising Markets, Marketizing Morality. The Fair Trade Movement, Product Labeling and the Emergence of Ethical Consumerism in Europe. *Journal of Nonprofit & Public Sector Marketing*, *33*(2), 168–192. Advance online publication. doi:10.1080/10495142.2020.1865235

Kumar Sharma, N. (2020). *Ethical Dilemma and Marketing Decisions: A Case Analysis of Food Products*. doi:10.26643/think-india.v16i3.7818

Ladhari, R., & Tchetgna, N. M. (2017). Values, Socially Conscious Behaviour and Consumption Emotions as Predictors of Canadians' Intent to Buy Fair Trade Products. *International Journal of Consumer Studies*, *41*(6), 696–705. Advance online publication. doi:10.1111/ijcs.12382

Lawrence, X., Mendoza, D., Tadeo, J. B., Dacanay, J. M., Marla, A. N., Aszhnie, C., & Vergara, B. (2023). Enterprise management strategies in agricultural fairtrade products. *Journal of Enterprise and Development*, *5*(2), 2023.

Le, M. H., & Nguyen, P. M. (2022). Integrating the Theory of Planned Behavior and the Norm Activation Model to Investigate Organic Food Purchase Intention: Evidence from Vietnam. *Sustainability (Basel)*, *14*(2), 816. Advance online publication. doi:10.3390/su14020816

Liu, Y., Sheng, H., Mundorf, N., Redding, C., & Ye, Y. (2017). Integrating norm activation model and theory of planned behavior to understand sustainable transport behavior: Evidence from China. *International Journal of Environmental Research and Public Health*, *14*(12), 1593. Advance online publication. doi:10.3390/ijerph14121593 PMID:29258245

Luna, J. K., Chalit Hernandez, B., & Sawadogo, A. (2021). The paradoxes of purity in organic agriculture in Burkina Faso. *Geoforum*, *127*, 46–56. doi:10.1016/j.geoforum.2021.09.014

Lyon, S., Bezaury, J. A., & Mutersbaugh, T. (2010). Gender equity in fairtrade-organic coffee producer organizations: Cases from Mesoamerica. *Geoforum*, *41*(1), 93–103. doi:10.1016/j.geoforum.2009.04.006

Makita, R. (2011). Fair Trade Certification: The Case of Tea Plantation Workers in India. *Development Policy Review*. Advance online publication. doi:10.1111/j.1467-7679.2012.00561.x

Mark, B., Wendela, H., Elmy, S.-F., & Terya, S. (2018). *Fairtrade Standard for Small Producer Organizations*. www.fairtrade.net/standards.html

Mboga, J. (2017). *Ethical sourcing to ensure sustainability in the food and beverage industry and eliciting millennial perspectives*. Advance online publication. doi:10.5281/zenodo.1009174

Naegele, H. (2020). Where Does the Fair Trade Money Go? How Much Consumers Pay Extra for Fair Trade Coffee and How This Value Is Split Along the Value Chain. *World Development*, *133*, 105006. Advance online publication. doi:10.1016/j.worlddev.2020.105006

Ndiaye, P. G., Diouf, B. L., Tindall, C., Kébé, M., & Holvoet, K. (2009). Evaluating the opportunities, constraints and implications of eco and ethical fish labelling on the octopus value chain in Senegal. *FAO Fisheries and Aquaculture Report*, (904), 165–174.

Neilsno, J., & Pritchard, B. (2010). Fairness and Ethicality in Their Place: The Regional Dynamics of Fair Trade and Ethical Sourcing Agendas in the Plantation Districts of South India. *Environment and Planning a Economy and Space*. doi:10.1068/a4260

Nicholls, A., & Opal, C. (2006). *Fair Trade: Market-Driven Ethical Consumption*. Choice Reviews Online. doi:10.5860/CHOICE.43-2905

Oya, C., Schaefer, F., Skalidou, D., McCosker, C., & Langer, L. (2017). Effects of Certification Schemes for Agricultural Production on Socio-Economic Outcomes in Low- And Middle-Income Countries. *Systematic Reviews*. Advance online publication. doi:10.23846/SR0034

Partzsch, L., & Kemper, L. (2019). Cotton certification in Ethiopia: Can an increasing demand for certified textiles create a 'fashion revolution'? *Geoforum*, *99*, 111–119. doi:10.1016/j.geoforum.2018.11.017

Parvathi, P., Grote, U., & Waibel, H. (2017). *Fair Trade and Organic Agriculture: A Winning Combination?* doi:10.1079/9781786393050.0000

Pazderka, C. (2010). *Procurement for Development: Aligning Social with Commercial Considerations in the Supply Chain*. Academic Press.

Pepsico. (2023). *Sustainable sourcing*. https://www.pepsico.com/our-impact/esg-topics-a-z/sustainable-sourcing

Ribeiro-Duthie, A. C., Gale, F., & Murphy-Gregory, H. (2021a). Fair trade and staple foods: A systematic review. In *Journal of Cleaner Production* (Vol. 279). Elsevier Ltd. doi:10.1016/j.jclepro.2020.123586

Ribeiro-Duthie, A. C., Gale, F., & Murphy-Gregory, H. (2021b). Fair trade governance: Revisiting a framework to analyse challenges and opportunities for sustainable development towards a green economy. *Discover Sustainability*, *2*(1), 58. Advance online publication. doi:10.1007/s43621-021-00063-6 PMID:35425912

Rizqiyanto, S. (2017). Starbucks's Fair Trade in The Edge of Globalization. *ETIKONOMI*, *16*(2), 231–248. doi:10.15408/etk.v16i2.5546

Roy, R., Phau, I., & Lee, T. (2012). Quality and Ethical Sourcing among Smallholder Coffee Producers in Papua New Guinea. *Proceedings of the 12th International Conference of the Society for Global Business & Economic Development*, 1488–1496. https://espace.curtin.edu.au/bitstream/20.500.11937/25122/2/162893_39843_21.2.pdf

Schenk, P. (2019). A Matter of Principle: Comparing Norm-Based Explanations for Fair Trade Consumption. *Journal of Consumer Policy*, *42*(3), 397–423. doi:10.1007/s10603-018-9401-4

Schenk, P., Rössel, J., & Weingartner, S. (2021). It's all about distinction: The lifestyle embeddedness of fair trade consumption. *Sustainability (Basel)*, *13*(19), 10997. Advance online publication. doi:10.3390/su131910997

Sen, D. (2018). Fempreneurs or Organic Tea Farmers? Entrepreneurialism, Resilience and Alternative Agriculture in Darjeeling, India. *Journal of Political Ecology*, *25*(1). Advance online publication. doi:10.2458/v25i1.22386

Sharma, R. K., Abidi, N., & Khan, K. M. (2018). Comparison of conventional and fair trade systems on dimensions of sustainability: A study of basmati rice procurement in India. *International Journal of Innovation and Sustainable Development*, *12*(4), 446. doi:10.1504/IJISD.2018.095066

Shih-Tse Wang, E., & Chen, Y.-C. (2019). Effects of perceived justice of fair trade organizations on consumers' purchase intention toward fair trade products. *Journal of Retailing and Consumer Services*, *50*, 66–72. https://doi.org/https://doi.org/10.1016/j.jretconser.2019.05.004. doi:10.1016/j.jretconser.2019.05.004

Shorette, K. (2014). *Nongovernmental Regulation and Construction of Value in Global Markets*. doi:10.1177/0731121414543027

Simangunsong, E., Hendry, L., & Stevenson, M. (2016). Managing Supply Chain Uncertainty With Emerging Ethical Issues. *International Journal of Operations & Production Management*, *36*(10), 1272–1307. Advance online publication. doi:10.1108/IJOPM-12-2014-0599

Snyder, H. (2019). Literature review as a research methodology: An overview and guidelines. *Journal of Business Research, 104*, 333–339. doi:10.1016/j.jbusres.2019.07.039

Starbucks. (2020). *Starbucks: Ethical Sourcing - Coffee: Starbucks Coffee Company*. https://www.starbucks.com/responsibility/sourcing/coffee/

Talukder, B., vanLoon, G. W., & Hipel, K. W. (2019). Energy efficiency of agricultural systems in the southwest coastal zone of Bangladesh. *Ecological Indicators, 98*, 641–648. doi:10.1016/j.ecolind.2018.11.030

Terstappen, V., Hanson, L., & McLaughlin, D. (2012). Gender, Health, Labor, and Inequities: A Review of the Fair and Alternative Trade Literature. *Agriculture and Human Values*. Advance online publication. doi:10.1007/s10460-012-9377-7

Thao Tran, P., Vettersand Tutor, E., & Blombäck, A. (2012). *Fairtrade-a Competitive Imperative? Title: Fairtrade-a Competitive Imperative?* Author.

Utting, K. (2008). Assessing the Impact of Fair Trade Coffee: Towards an Integrative Framework. *Journal of Business Ethics*. Advance online publication. doi:10.1007/s10551-008-9761-9

Van Baar, A. (n.d.). *A Fair Price for Human Rights Due Diligence*. https://ecco-plan.com/

Van Dam, P. (2015). The limits of a success story: Fair trade and the history of postcolonial globalization. *Comparativ (Leipzig), 25*(1). Advance online publication. doi:10.26014/j.comp.2015.01.05

van der Werff, E., & Steg, L. (2016). The psychology of participation and interest in smart energy systems: Comparing the value-belief-norm theory and the value-identity-personal norm model. *Energy Research & Social Science*, *22*, 107–114. doi:10.1016/j.erss.2016.08.022

WFTO. (2023). *Our 10 Fair Trade Principles - Word Fair Trade Organisation*. https://wfto.com/our-fair-trade-system/our-10-principles-of-fair-trade/

White, G., & Samuel, A. (2016). Fairtrade and Halal Food Certification and Labeling. *Journal of Macromarketing, 36*(4), 388–399. Advance online publication. doi:10.1177/0276146715620236

Zhang, K., Liu, Z., & Wang, P. (2020). Exploring the supply chain management of fair trade business: Case study of a fair trade craft company in China. *Cogent Business and Management*, *7*(1), 1734345. Advance online publication. doi:10.1080/23311975.2020.1734345

Zysk, W. (2020). Fair trade phenomenon and its evolution in Visegrad countries. *International Entrepreneurship Review*, *6*(4), 81–98. doi:10.15678/IER.2020.0604.06

Chapter 5
Cultural Ecology and Agriculture in Rural Areas in Zimbabwe

Blessing Hodzi
https://orcid.org/0000-0001-8599-4275
Durban University of Technology, South Africa

Neil Batsirai Maheve
https://orcid.org/0009-0003-5868-6412
Rhodes University, South Africa

ABSTRACT

This chapter is concerned with indigenous wisdom in agriculture in rural Zimbabwe, with a particular focus Mutoko South and Guruve regions. It argues for the enduring presence of indigenous agricultural systems amidst a predominantly modernized, Western-influenced society. From land preparation to sowing, weed management, harvesting, and yield storage, this study underscores the resilience of indigenous agricultural practices. This chapter shed light on the crucial role of Indigenous knowledge systems (IKS) in fostering sustainable farming practices within Guruve and Mutoko South communities. Passed down through generations and deeply ingrained in daily routines, this knowledge has emerged as a vital tool for predicting weather patterns, managing pests, and enhancing soil fertility. It examines how people in Mutoko South and Guruve have steadfastly embraced traditional methods using qualitative methods. The study primarily involves interviews with 20 subsistence farmers residing in tribal trust lands in 10 villages.

DOI: 10.4018/979-8-3693-2011-2.ch005

INTRODUCTION

This chapter concerns indigenous wisdom in agriculture in rural Zimbabwe, focusing on the Mutoko and Guruve districts. We highlight that the people have continually relied on indigenous agricultural systems in a predominantly modernised, Western-influenced agricultural society. The broader question we sought to address in this chapter has been the role of IKS in farming practices in rural areas of Zimbabwe. We are not bringing a new phenomenon to the discussion; however, we suggest that what has been done as an everyday routine and never termed sustainable farming practice is an essential study area. Traditional, read localised farming practices, from land preparation to sowing, weed management, harvesting, and yield storage, guided by indigenous agricultural practices, should be understood under sustainable farming practices. IKS and sustainable farming in Zimbabwe have not received enough academic attention, not taking away the research that has led to the re-introduction of the Pfumvudza system (by the Government of Zimbabwe), which attests to the importance of IKS. The centrality of IK in sustainable farming practices is vital in the broader discussion of sustainable development goals. The IK on farming has always aimed at providing enough food to the people and hence plays a role in reducing hunger and poverty within the community. It must be highlighted that we are not dismissing the existence of hunger in the areas where research was conducted.

In addition, we established the IK's emphasis on using locally available inputs to help reduce the use of inorganic materials in agriculture. We highlight using organic material, which is arguably good for the soil. Our argument also goes on to show the significance of IK in telling the beginning of the farming season; the shedding of leaves and the emergence of new green foliage serve as a cue for activities like hole-digging and manure application in preparation for the season.

BACKGROUND OF THE STUDY

Despite ongoing efforts to combat climate change, Zimbabwe faces many challenges brought about by these changes, including erratic rainfall patterns, increased temperatures, and extreme weather events. These climatic changes pose significant threats to conventional agricultural practices, necessitating a shift towards more sustainable and resilient farming methods. The failure of conventional systems to address the sustainable use of resources and practices has led to the realisation of the importance of indigenous knowledge. The World Commission on Environment and Development conference, also known as the Brundtland Commission of 1987 and the Convention on Biological Diversity of 1997, underscored the significance of indigenous knowledge in development. While this chapter is not a development paper, it contributes to understanding IKS and sustainable development. Seko et al. (2021) published a compelling chapter assessing the impact of IKS on sustainable agriculture, arguing for its importance. Buthelezi and Hughes (2014) contribute to the limited literature on IKS and agricultural development, focusing on rural areas. Muthee et al. (2019) align closely with our study, noting that while 'modern' education has impacted the full utilisation of IKS, these systems are still in use. Our study aims to address this niche; in Zimbabwe, IKS is particularly crucial in communal areas where subsistence agriculture is practised. While the local communities do not necessarily equate the farming practices influenced by IKS with sustainable farming practices, we have contextualised them as such. In the Guruve and Mutoko districts, indigenous communities possess a wealth of traditional knowledge passed down through generations. This knowledge includes agroecological practices, prudent crop selection, water management techniques, and effective pest control strategies. The focus on the Guruve and Mutoko districts stems from understanding Zimbabwe's unique ecological and cultural contexts. Despite shared challenges, variations in soil composition and agricultural traditions necessitate understanding how IKS facilitates sustainable farming practices. This study fills a gap in academic research by exploring the agricultural practices ingrained in these villages, where indigenous wisdom remains a steadfast anchor amidst a rapidly evolving, Western-influenced society. The central motif that emerges is the irreplaceable role of Indigenous Knowledge in shaping and sustaining farming practices.

Sustainability in Agriculture: A Brief Overview

Sustainability in agriculture is a fundamental concept that addresses the critical need to balance the increasing demand for food production with environmental conservation and socio-economic considerations. Central to this approach is the recognition of the intricate connections between ecological, social, and economic factors within agricultural systems (Velten et al., 2015). Sustainability in agriculture strives to fulfil current food needs without jeopardising future generations' ability, emphasising eco-friendly practices, social responsibility, and economic viability (John et al. 2022). One essential aspect of sustainable agriculture is the preservation and enhancement of soil health. Crop rotation, cover cropping, and reduced tillage improve soil structure, fertility, and water retention. By adopting such measures, farmers aim to ensure long-term productivity while mitigating the risk of soil degradation. Efficient water management is another key focus of sustainable agriculture (Velten,2015). Drip irrigation, rainwater harvesting, and precision farming optimise water usage. These practices reduce agriculture's environmental impact and ensure water resources' availability for future generations. Maintaining biodiversity is critical in sustainable agricultural systems. Through the preservation of natural habitats, integration of diverse crops, and avoidance of overuse of pesticides, farmers contribute to the overall health of ecosystems. Biodiversity enhances resilience to pests and diseases, fosters pollination, and supports essential ecosystem services. As an integral component of sustainable agriculture, agroecology integrates ecological principles into farming practices. Emphasising ecological balance, diversity, and natural processes, agroecology enhances the sustainability of agriculture while reducing dependence on external inputs (John et al., 2022). Responsible resource management is underlined in sustainable agriculture, encompassing energy, nutrients, and land. Engaging local communities in their knowledge contributes to the social sustainability of agricultural systems. Given the challenges posed by climate change, sustainable agriculture focuses on building climate-resilient farming systems, enhancing resilience and food security in a changing climate(Carlisle et al., 2019). Sustainability in agriculture is a comprehensive approach that seeks to align the imperatives of food production with environmental and social considerations.

Sustainable agriculture is vital to achieving the United Nations Sustainable Development Goals (SDGs), balancing food production with environmental and socio-economic considerations. Sustainable agriculture is central to SDG 2 (Zero hunger) by promoting diverse farming systems for stable food production. SDG 13 (Climate Action) mitigates greenhouse gas emissions and enhances farming systems' resilience. Sustainable agriculture connects to SDG 15 (Life on Land) by emphasising biodiversity to prevent soil degradation and promote ecosystem health (sdgs.un.org). Despite the positive impacts, challenges persist in the widespread adoption of sustainable practices, economic constraints, and the need for supportive policy frameworks (Ladha et al., 2020). However, there are ample opportunities for innovation, collaboration, and knowledge-sharing to overcome these challenges and advance sustainable agriculture. This chapter explains sustainable agriculture as a linchpin in achieving food security and environmental conservation.

OBJECTIVES OF THE STUDY

The primary aim of this study is to examine the Indigenous Knowledge Systems that form the basis of sustainable farming practices in the districts of Guruve and Mutoko. The specific objectives of this research are:

- To examine how the villagers deal with soil fertility, pest control problems, and other factors affecting agricultural productivity.
- To explore and identify the specific components of IKS that contribute to achieving optimal yields.

Framework

Recently, there has been a global shift towards alternatives to conventional farming systems, with particular attention to the global south. This increased interest stems from a growing acknowledgement of the limitations and adverse impacts of conventional agricultural practices. Our investigation zooms in on the agricultural practices in villages in Guruve and Mutoko districts, where integrating Indigenous Knowledge Systems (IKS) and a deep understanding of the ecology of culture provides valuable insights for understanding sustainable farming systems. To comprehend these practices, we utilise two closely connected frameworks: Indigenous Knowledge Systems (IKS) and the ecology of cultures. Indigenous knowledge systems and ecology of culture as the framework help understand sustainable farming systems in Guruve and Mutoko villages.

IKS is the accumulated knowledge, traditions, and practices local communities shape through their extensive engagement with the environment. This localised and orally transmitted knowledge system, as emphasised by Semali and Kincheloe (1999), embodies the distinct cultural identities and experiences of the Guruve and Mutoko communities (Mkabela, 2006). The ecology of culture investigates how humans adapt to their social and physical environment (Sutton and Anderson, 2020). The ecology of culture framework, highlighting the reciprocal relationship between cultural and ecological factors (Chertkovskaya, 2019), underscores the significance of cultural diversity and environmental stewardship in sustainable farming practices.

As Posey (2002) noted, gaining a comprehensive understanding of IKS begins by establishing the meaning of 'indigenous' and addressing key questions such as 'What does indigenous mean?' and 'Who are considered indigenous people? To avoid these questions, this chapter contextualises indigenous people as those residing in the study areas. To contextualise 'indigenous', this chapter defines it as the people residing in the areas of study, avoiding broader questions of identity. The definition of IKS is varied, with Semali and Kincheloe (1999) elaborating on the concept that indigenous knowledge comprises the knowledge produced by individuals within a specific community through their ongoing engagement with the surrounding environment. Following this perspective, Maurial (1999) and Reynar (1999) assert that this knowledge system is localised, holistic, and not documented; instead, it is transmitted orally from one generation to the next. The definition of Indigenous Knowledge Systems here encompasses indigenous peoples' unique cultures, languages, values, histories, governance, and legal systems, representing everyday life experiences. This chapter resonates with Mkabela (2006: v), who views indigenous knowledge as culture-specific, embedded in communities' oral traditions and part of their everyday lives. This chapter defines Indigenous knowledge systems as reflecting Indigenous peoples' unique cultures, languages, values, histories, governance and legal systems. The body of knowledge is the total of everyday life experiences. While acknowledging the diverse nature of Indigenous Knowledge Systems (IKS), our exploration reveals more similarities than differences in the knowledge systems of Guruve and Mutoko. This perspective aligns with our departure from the strict localisation emphasis proposed by Ngulube and Lwonga (2007). We emphasise that IKS is inseparable from the communities that create and uphold it, encompassing customs, values, traditions, and beliefs intricately woven into their farming practices.

METHODOLOGY

This qualitative research was conducted in two distinct geographic areas: Guruve District in Mashonaland Central Province and Mutoko in Mashonaland East Province, Zimbabwe. Despite the distance between the two, these regions exhibited unique, universally interchangeable cultural nuances. Guruve is divided into communal and resettlement (former commercial farming areas). This demarcation is pertinent in this study, providing the reader with context. Community lands were mainly for subsistence farming, whereas the resettlement areas were meant to be commercial areas for farming. The land area in communal areas is at an average of one hectare per farmer. Whilst the communal areas have such small areas in former commercial farms, the landholding size starts from 6-hectare plots known as A1 farms to the most significant plots known as A2. The clarification is essential because most individuals at the large farming scale have taken up commercial farming systems. The focus of this chapter is on subsistence communal areas.

In Guruve, the study was conducted in Date, Kanounama, Gondo, Muzika and Gezi villages. The villages in Guruve were selected because they provided easy access to participants and made it easier for the respondents to associate and relate with one of the authors. In Mutoko, the study was conducted in Mutoko South villages four, five, six, seven and ten. The primary data collection phase transpired between March and September 2022. The protracted period was because we conducted interviews whenever our schedule allowed. Sometimes, we would return to the respondents once we realised the information provided was insufficient. Various tools were employed for data collection, including questionnaires, voice recorders, and notebooks. The purposive sampling method through convenience sampling was elected because the participants were readily available and accessible to the researcher, making them convenient to include in the study (Flick 2018,1998).

Interviews are purposeful conversations between two or more people (Bremborg,2011; Saunders et al., 2007). The main objective of interviews is to collect reliable and meaningful data. This study used semi-structured and unstructured interviews because they offer more flexibility and adaptability in questioning. Semi-structured interviews are exploratory interviews employed in qualitative research (Frances et al. 2006). Saunder et al. (2007) assert that semi-structured interviews enable participants to share their thoughts, opinions, and experiences in more detail, leading to a richer understanding of the research topic. The semi-structured approach allowed us to change or adapt questions based on the direction and flow of the conversation, allowing for a more natural and open-ended exchange of information. The semi-structured interviews enabled us to gather rich information about individuals' attitudes, beliefs, and experiences. Moreover, follow-up questions, or probes, were formulated based on the information shared by the interviewees (Bremborg, 2011; Roulston & Choi, 2018). The semi-structured approach captured the interplay between environmental factors, cultural distinctions, and the intricate processes of knowledge transmission, contributing to a nuanced exploration of indigenous farming systems in Guruve and Mutoko. The interviews conducted allowed us to think with and alongside the respondents. This chapter comes out of a study exploring Indigenous knowledge systems in broader sense. This study uses 48 interviews which directly resonate with IKS and sustainable farming practices. While the number may be considered small to deduce themes, we argue that we reached a point saturation on IKS. Saturation is a criterion employed in qualitative research to discontinue data collection based on the notion of not getting anything new (Saunders et al, 2018).

RESULTS AND DISCUSSION

Inquiries directed at respondents sought to elucidate the origins of indigenous knowledge within their respective villages. The respondents conveyed that this knowledge is predominantly transmitted intergenerationally and partly derived from keen observations of natural phenomena accumulated over time. This echoes findings in comparable research conducted by Borthakur and Singh (2021) and Buthelezi and Hughes (2014), focusing on indigenous agricultural knowledge. These studies underscore indigenous agricultural knowledge emanates from farmers' experiences and systematic experimentation with nature. A gradual refinement marks the evolutionary trajectory of this knowledge fostered through sustained interaction with natural processes. Moreover, this knowledge seamlessly traverses generations, embodying a continuum that intricately weaves together the agro-climatic factors specific to a given geographical area. Indigenous knowledge is used to plan and make decisions in the communities under study.

Indigenous Knowledge and Weather Forecast

According to Ajani, Magbenka, and Okeke (2013) and Materechera (2021), farmers across Sub-Saharan Africa have devised indigenous early warning systems for weather forecasting, prediction interpretation, and decision-making. This indigenous knowledge, particularly in the agricultural practices of Guruve and Mutoko South villages, is pivotal in managing farmers' vulnerability by informing decisions on cropping patterns and planting dates. The indicators are varied and treated as signs of the impending rainy season. The indicators do not happen in any form or pattern but usually follow each other. Farmers in selected villages of the Guruve district rely on discernible patterns observed over time to anticipate the upcoming farming season. An illustrative example is the emergence of new leaves on trees, a tangible indicator of the onset of a new season. All trees shed leaves in spring, and the locals will begin the farming season when the new leaves start protruding. The local communities in Guruve refer to that as Pfumvudza. This prompts farmers to initiate preparations for the impending agricultural activities. Willard, a respondent from Date Village, underlined this reliance, stating, "I rely on 'pfumvudza' (when I see trees with new leaves); it is an indication of a new season, and I start to prepare".

In addition to Pfumvudza, farmers tend to be assured of the advent of the rainy season when they receive the *Gukurahundi* or *bumharutsva*. This is rain that usually precipitates between August and September. *Gukurahundi/ Bumharutsva* holds significant cultural and agricultural importance. The name Gukurahundi loosely translates to the rain that washes away the chaff, and the name Bumharutsva translates to the rain that cleans the residues of burnt material. The rain is considered a cleansing rain, symbolising the initiation of a new season. Mbuya Jack, in an interview with her, asserted, "When *Gukurahundi/bumharutsva* arrives, it heralds the beginning of the new season". Farmers use these indigenous indicators to prepare for the forthcoming farming season.

Farmers in Mutoko South rely on keen observations of nature to predict rain and adapt to changing seasons. They examine trees, birds, and wild fruits like loquats (mazhanje) for clues. An abundance of loquats, for example, signals potential drought, as it indicates scarcity for baboons, who might raid crops. Mbuya Moffat and Gift) concurred and asserted that "*Gore rine mazhanje emusango akawanda kunenge kusina mvura, kana ari amshoma mvura inenge yakawanda* (when there is plenty of loquats (*Eriobotrya japonica)*, there is drought)". This belief is rooted in the notion that a surplus of wild fruits implies a lack of food for baboons in farmers' fields, indicating drought. Peasant farmers in these communities, characterised by their keen observations of surrounding activities, adeptly use these indigenous indicators to predict seasons, enabling them to adapt to environmental changes. Similarly, Gezi, Muzika, and Kanounama villagers use bird movements, like the Abdim's stork (rain bird), to predict rain. Willard and Nometah alleged that "*ukaona shuramurove moziva kuti mvura yauya*, (when you see Abdim's stock bird {rain bird}it is a sign that the rains are about to come. Mbuya Zuze, Sarudzai, and Maria agreed with the duo from Guruve by saying that when they see the rain birds, they will know that the rainy season has begun. This aligns with Mapfumo's (2015) observations of traditional rain prediction practices, but Jiri et al. (2016) argue that climate change affects this knowledge. However, we contend that indigenous knowledge is evolving based on the changes in the environment of its proponents, highlighting its resilience.

Indigenous Methods of Controlling Weeds and Pests

Weeds and pests pose a significant threat to crop production, as they extract essential nutrients, water, and moisture from the soil while competing for light and air, adversely impacting crops (World Economic Forum, 2021). The term 'weed' commonly refers to a plant considered undesirable in agricultural settings (Gavali and Kulkarni, 2014), causing diminished production and negatively affecting overall yield. To address this issue, farmers employ various strategies, with manual weeding using hoes emerging as the predominant method across the ten villages under investigation. Fungai, among other respondents, affirmed that *'tosakura nemapadza'* (we use hoes to weed). In concordance with traditional practices, Sirmour (2016) reported that weeding traditionally uses indigenous hand tools. Additionally, participants highlighted a cultivation process between seed rows aimed at weeding, elucidating that "we use cultivators (plough) drawn by oxen, typically in the early stages of plant growth" (Mbuya Zuze). Rabo Usha (2022) observed that this traditional animal traction ploughing method breaks down soil crust, mitigates runoff, enhances infiltration, and effectively controls weeds.

Furthermore, indigenous methods, rooted in local agricultural practices, have been employed over time for weed control. These include intercropping, crop rotation, early planting, and mulching. In Guruve, farmers utilise various approaches to manage weeds. One commonly employed method is known as *kutema mukuruware*, which can be understood as a modified or adapted form of zero tillage. Albert walked us through the process: "In August, we started to dig planting holes in a field that would have been burnt soon after harvesting. The seeds of the weeds are usually burnt, and there are low chances of the other weeds cropping up. We dig these holes and plant the seeds *(kuwarika)*. The plants would germinate because, in the areas outside of the holes, the grass and weeds cannot." He explained that this method is essential in mitigating weed growth as tilling the land tends to bring weed seeds to the surface, exposing them to sunlight. This method, however, does not mean that the field would be free of weeds; the rate and percentage of the weeds are drastically different.

Additionally, in the Guruve district, another noteworthy practice called '*kuwarika*' is employed, involving planting seeds a month before the onset of the first rains. A participant expounded on the process, stating that '*kuwarik*a' is conducted in September, wherein individuals dig holes and allow them to dry for a week, depleting the soil's moisture content. Subsequently, seeds are planted, poised to germinate with the advent of the first rains in October. Kuwarika is particularly suitable for long-season varieties of seeds like *hakirikingi* (Alufeyo). We could not ascertain the scientific name for the variety continually referred to as *hakirikingi*. Kuwarika process works closely with the process of *kutema mukuruware*.

Furthermore, Kanounama farmers employ intercropping for pest and weed control. John and Mbuya Jack highlight maize intercropped with "mona wega" beans, which climb maize stalks. They described the intercropping that typically includes maize and a type of bush bean colloquially known as '*mona wega'*. This name comes from the character of the leguminous plant. Betina and Alufeyo from Kanonama and Gondo villages interplant African horned cucumber with maize, as locusts avoid it but attract pest-eating ants, which will feed on worms in the field. The rationale behind this method lies in the ability of certain crops to suppress weeds and repel specific pests. This aligns with Muthee et al. (2019) findings on crops like black nightshade, amaranth, pumpkin leaves, pigweed, and cowpeas, enhancing soil fertility and suppressing weeds and pests.

Farmers in Mutoko South, villages 7, 10, and 5, employ intercropping, increased plant population, and mulching to deprive weeds of sunlight effectively. Masara elucidated their methods, stating, "We engage in early planting to ensure that our crops outgrow weeds". Concerning what Masara said, Pauline said, "We enhance plant population to cover the ground, thereby shading out weeds from sunlight". Sarudzai highlighted that they use maize plant residue from the previous season or dried grass to control weeds. This process, which can be understood as mulching, involves using plant residues to cover unused ground in the field, inhibiting the germination and growth of weeds. According to Ajani, Mgbenka, and Okeke (2013), mulching serves various purposes, including moderating soil temperatures, suppressing diseases and harmful pests, and conserving soil moisture. Moreover, it plays a pivotal role in reducing wind and water erosion by controlling surface runoff and mitigating the impact of raindrops on the soil surface.

In addition to these practices, an indigenous pest control method observed in Mutoko South villages 7 and 5 involves intentionally releasing natural enemies into the fields. Participants noted, "We release chickens into the field to consume worms after cultivation (Nhamo and Sarudzai). This release of natural enemies typically occurs post-weeding or animal-drawn cultivation, coinciding with when worms are brought to the surface.

Indigenous Methods of Maintaining Soil Fertility

The research established various indigenous methods to sustain and enhance soil fertility in the researched areas. These practices include fallowing, digging basins, mixed cropping, intercropping, mulching, composting (utilising dead leaves), and the application of animal manure, including cow dung, goat droppings, and chicken waste. Respondents from Date and Gezi villages highlighted the significance of basin digging, expressing that it aids in moisture preservation and soil fertility. Participants claimed that they dig basins to preserve moisture and soil fertility. They dug small holes where seedlings are planted, representing a non-tillage farming technique that minimises soil disturbance. Our study corroborates with Negi and Verma's (2015) assertion that non-tillage is a conservative method where seeds are sown directly (into undisturbed soil), making land use more sustainable.

Participants in both Guruve and Mutoko South villages underscored using intercropping and crop rotation as integral strategies to enhance soil fertility. Expressing this agricultural approach, Betina and Nometah said, 'We sometimes plant maize seeds together with groundnuts'. The same sentiments were echoed by Mbuya Moffat, who claimed that "we engage in intercropping by planting maize with cowpeas (nyemba)'. Intercropping, the simultaneous cultivation of multiple crop species in the same field (Muthee et al., 2019), is an efficient method for optimising natural resources such as land, light, water, and nutrients. This practice augments biodiversity and contributes to heightened agro-ecosystem productivity, resilience, and stability (Ning et al., 2017). The intentional combination of cereals with leguminous plants in intercropping serves a dual purpose of enriching the soil with nitrogen and other essential nutrients, thus elevating soil fertility. This symbiotic relationship is facilitated by the legumes' association with rhizobium bacteria, which aids in nitrogen fixation (Duchene et al., 2017). This symbiotic relationship elevates fertility, boosting yields (Muthee et al., 2019). Additionally, legume intercropping combats erosion by strengthening soil structure (Forte et al., 2017).

Participants also highlighted practising crop rotation as a crucial mechanism for enhancing soil fertility. Crop rotation involves the sequential planting of different crops in the same field over successive seasons, a time-tested practice renowned for its positive impact on soil health. This practice, rooted in ancient agricultural traditions (Tanveer, Ikram & Ali, 2019), has been consistently recognised for its efficacy in maintaining soil fertility. Most participants in the Guruve villages commented on the reliance on crop rotations. John highlighted that "we also implement crop rotation, and for me, the sequence typically includes maize seeds, beans, sorghum, and sweet potatoes". Similarly, Jeniffer observed a tangible improvement in soil quality with the implementation of crop rotation, stating, "There is a noticeable increase in yield when we rotate crops. When we grow soybeans this year and next year we put maize, our maize yield increases". Beyond its historical significance, crop rotation emerges as an effective strategy for carbon sequestration when compared to continuous monoculture (Triberti et al., 2016).

Shifting cultivation is another sustainable agricultural method employed in Mutoko villages to enhance soil fertility. Faulkner and Mackie (2013) describe shifting as a method wherein farmers utilise a piece of land until it becomes less fertile and then leave it fallow for two to four years, contingent upon land availability. This approach is designed to allow the land to rejuvenate its fertility. Mbuya Moffat, Maria and Jenifer from different Mutoko South villages emphasised this practice. They claimed they refrain from using a field for two or more farming seasons to facilitate nutrient replenishment. This process of refraining is called *kusiya munda uchirarira*, which means letting the soil recuperate. This deliberate hiatus from cultivation permits the land to regenerate its nutrient content naturally.

Due to the high costs associated with inorganic fertiliser, rural farmers traditionally now rely on natural manures derived from animals and decomposed leaves. Jairos and Mbuya Zuze underlined the utilisation of manure from domestic animals, such as *ndowe yemombe* (cow dung), *matosi ehuku* (chicken waste), *nhoko dzembudzi* (goat droppings), and *murakwani* (leaves compost for vegetables in the garden). These forms of manure continue to be a pivotal strategy in fortifying soil fertility. Additionally, mulching using mufudze is employed to enhance soil fertility. Pauline averred that "after harvesting maize, beans, cowpeas, sorghum, and groundnuts, we spread the residues in the field where they will decompose". In the long run, our findings on using residue to serve as mulching and manure align with Briggs and Moyo's (2012) study in Malawi and the current study's findings regarding the use of crop residues in mulching. This research affirms that incorporating crop residues contributes to maintaining soil fertility, leading to the vigorous growth of maize with larger cobs compared to those grown without decomposing dry matter in the soil.

Farmers, drawing on their knowledge and experience, demonstrated an ability to classify soils based on texture, colour, and fertility. This informed classification guides their plant selection and adaptation strategies. In the context of Guruve, for instance, farmers grappling with sandy soils articulated a strategic approach; as Edith elucidated, "We dig anthills commonly found in the fields and distribute the soil across the expanse of the field. This intentional practice, as revealed by Alufeyo, is undertaken with the explicit aim of augmenting soil fertility and enhancing moisture retention. If the anthills are limited in the fields, the owners dig up them and use them to cover the seeds sown. This deliberate combination of sandy and clay soils is conducted to conserve water and enrich nutrient content. This synergistic blending of different soil types underscores the farmers' commitment to optimising agricultural productivity. Drawing on the insights of the Food and Agriculture Organization (FAO), as documented in the FAO soils portal (2017), preventing soil degradation, enhancing water retention ability, and improving humus content through suitable land use practices are identified as pivotal components in achieving food security. The farmers' practices in Guruve align with these principles, emphasising their proactive stance towards sustainable soil management for agricultural productivity and broader community well-being.

This finding adds to the assertion made by Singh and Sureja (2008: 643), who affirm that farmers can broadly categorise soils into agricultural and non-agricultural types, delineating their specific utility for farming and domestic purposes. Moreover, participants elucidated that the nature of the soil plays a pivotal role in dictating its designated use. Specifically, the consensus among participants underscored that sandy soils are conducive to cultivating groundnuts, round nuts, and cowpeas. In contrast, clay soils are earmarked for the cultivation of maize, with wetlands left for pastoral activities.

Indigenous Methods of Post-Harvesting Preservation and Storage

People have developed post-harvest preservation and storage methods, leveraging the knowledge passed from generation to generation. Among these methods, using ashes is a unique approach to preserving sweet potatoes *(kupfimbika)* in the areas under study. Masara, in an interview, explained the process, 'We dig a hole, then put ashes first, followed by sweet potatoes. We ensure the sweet potatoes have no cuts so they do not rot'. This preservation technique, although applied explicitly to sweet potatoes in this context, resonates with practices in Kenya, where ashes derived from burned beanstalks are utilised to preserve grains and cereals (Muthee et al., 2019).

Additionally, the preservation of grain involves a careful process of harvesting and sun-drying. Gathered evidence shows that the farmer would ensure the grain dries appropriately. Mbuya Zuze, Nhamo, Jenifer, and Pauline consistently underscored the imperative of meticulously drying maize grains before storage, expressing the sentiment that 'chibage chofanira kunyatsooma (maize grains should be thoroughly dried) before we place it in granaries. The dried grain is stored in airtight granaries plastered with cow dung. Plastering the granaries with cow dung serves a dual purpose: it not only prevents the intrusion of weevils and other insects but also points to the significance of cultural knowledge in sustainable agricultural storage. It ensured that they had food throughout the dry seasons.

Seed Preservation and Planting

Farmers in Mutoko and Guruve face economic constraints and often forego purchasing new seeds each season, relying instead on indigenous knowledge and preservation methods. Insights gathered from interviews with farmers unveiled a common practice of selecting seeds for the upcoming season during harvest time. John, aJairos, Mbuya Moffat and Maria articulated how they, using experience, chose the best maize cob with more lines and bigger grains, big and fat bean seeds, and sweet sorghum. This process involves an examination of the harvested produce, with the farmers choosing the best cobs and seeds based on grain size and germination capacity. The selected seeds are then carefully preserved for the subsequent planting season. According to ISSD Africa (2022), smallholder farmers in developing countries predominantly rely on seeds from previous seasons or informal distribution channels like farmer exchanges, community-sharing systems, and local markets. Sarudzai and Mbuya Zuze have used smoke soot as the best preservative approach. They argue that the weevils, rats, and even moisture do not destroy the grains preserved using soot. The maize cobs and sweet sorghum fingers hang over the fireplace, and the soot accumulates on them. Sekai and Betina shared a peculiar way of preserving seeds using ashes (*hundi*). Hundi is attained by burning the maize cobs and using the ashes as preservatives. The seed is preserved until the time they sow the seed. Before planting, these seeds are soaked in water to remove the soot and the ashes and quicken germination.

Furthermore, farmers engage in seed germination enhancement practices when planting. Farmers in Mutoko South Villages rely on a method referred to as "kuuchika" (breaking seed dominance/inducing germination) is employed. Mbuya Moffat and Masara Participants detailed the intricacies of this process, stating that the process involves soaking and scarification of seeds. Emphasising the rationale behind this practice, Fungai noted that it weakens the outer seed coat, especially for seeds with a resilient outer layer that could impede prompt germination. In parallel, farmers employ soaking to expedite the germination process in the villages under study in the Guruve district. As elucidated by Albert, "I soak the maize seed overnight and plant them in the morning". Thus, *kuuchika* and *kunyika* mbeu (soaking) enhance water absorption and initiate germination. This indigenous approach resonates with traditional agricultural wisdom, reflecting a resourceful adaptation to ensure a heightened germination rate, mainly when dealing with seeds that may exhibit delayed or inconsistent sprouting.

Limitations of the Findings

This study is vital in understanding the place of IKS in sustainable farming practices in the Guruve and Mutoko. However, it is pertinent to recognise that utilising these findings has limitations. While providing valuable insights, the Guruve and Mutoko South villages represent a microcosm of rural farming communities in Zimbabwe. The study's findings may not universally apply to people with different beliefs. The association of the rain bird with the absence of rain and farming is a perspective many interested farmers hold. However, some may view these birds simply as migratory birds. Consequently, the arguments presented in this chapter should be understood within the context of the specific communities studied and may not have universal relevance. An additional limitation of this chapter is the non-empirical nature of the findings. The assertion that the abundance or scarcity of wild fruit signifies an excellent or poor harvest could not be empirically validated. While qualitative methods offer rich insights into indigenous knowledge's cultural and contextual aspects, the lack of extensive quantitative data may limit the ability to make statistically robust generalisations.

Furthermore, the perception of sustainable farming practices can be subjective, varying among different stakeholders. The study acknowledges that the residents perceive the practices as routine agricultural procedures, but outsiders may interpret sustainability differently. Finally, the study does not account for external factors such as economic, political, or environmental changes and their impact on the sustainability of indigenous farming practices. These external factors could disrupt the continuity of traditional methods in unforeseen ways. While these limitations should be considered, they also offer opportunities for future research into specific aspects, expand the geographical scope, and employ a more diverse research method to understand indigenous agricultural wisdom comprehensively.

CONCLUSION

This chapter sheds light on the role of Indigenous Knowledge Systems in fostering sustainable farming practices within Guruve and Mutoko South communities. Indigenous knowledge emerges as a vital tool for predicting weather patterns, managing pests, and enhancing soil fertility. The evidence indicates that these communities leverage their traditional wisdom to combat crop-destroying pests and weed proliferation. In the villages under study, intercropping serves a dual purpose for the residents—enhancing soil fertility and curbing weed invasion. This chapter underscores the value of indigenous knowledge and its significant role in sustainable farming, underscoring the importance of recognising, preserving, and incorporating it into modern agricultural practices. We paint a vision of a sustainable agricultural future where environmental respect and innovation work hand in hand. Looking ahead, researchers, policymakers, and farmers can find inspiration in these longstanding practices.

REFERENCES

Aditya, S. (2016). *Design and development of single-row power weeder for rice* [Master's thesis].

Africa, I. S. S. D. (2022). *Policies, laws, and regulations in support of farmer seed systems*. Available: https://issdafrica.org/2022/03/16/policies-laws-and-regulations-in-support-of-farmer-seed-systems/

Borthakur, A., & Singh, P. (2021). Indigenous Agricultural Knowledge Towards Achieving Sustainable Agriculture. In V. Kumar Singh, R. Singh, & E. Lichtfouse (Eds.), *Sustainable Agriculture Reviews 50. Emerging contaminants in agriculture*. Springer. doi:10.1007/978-3-030-63249-6_15

Bremborg, A. D. (2011). Interviewing. In M. Stausberg & S. Engler (Eds.), *The Routledge Handbook of Research Methods in the Study of Religion*. Routledge.

Briggs, J., & Moyo, B. (2012). The Resilience of Indigenous Knowledge in Small-scale African Agriculture: Key Drivers. *Scottish Geographical Journal, 128*(1), 64–80. doi:10.1080/14702541.2012.694703

Buthelezi, N. N., & Hughes, J. C. (2014). Indigenous knowledge systems and agricultural rural development in South Africa: Past and present perspectives. *Indilinga, 13*, 231–250.

Carlisle, L., DeLonge, M. S., Iles, A., Calo, A., Getz, C., Ory, J., Galt, R., Melone, B., & Knox, R. (2019). Transitioning to Sustainable Agriculture Requires Growing and Sustaining an Ecologically Skilled Workforce. *Frontiers in Sustainable Food Systems, 3*, 3. doi:10.3389/fsufs.2019.00096

Duchene, O., Vian, J., & Celette, F. (2017). Intercropping with legume for agroecological cropping systems: Complementarity and facilitation processes and the importance of soil microorganisms. A review. *Agriculture, Ecosystems & Environment, 240*, 148–161. doi:10.1016/j.agee.2017.02.019

Egeru, A. (2012). Role of Indigenous Knowledge in Climate Change Adaptation: A Case Study of the Teso Sub-Region, Eastern Uganda. *Indian Journal of Traditional Knowledge, 11*(2), 217–224.

Faulkner, O. T., & Mackie, J. R. (2013). *West African Agriculture*. Cambridge University Press. doi:10.1017/CBO9781316530122

Flick, U. (1998). *An introduction to qualitative research: Theory, method, and applications*. Sage.

Flick, U. (2018). Doing qualitative data collection – charting the routes. In *The SAGE Handbook of Qualitative Data Collection*. SAGE Publications Ltd. doi:10.4135/9781526416070.n1

Food and Agriculture Organisation (FAO). (2017). *Degradation/Restoration.* Available: https://www.fao.org/soils-portal/soil-degradation-restoration/en/

Forte, A., Fagnano, M., & Fierro, A. (2017). Potential role of compost and green manure amendment to mitigate soil GHG emissions in Mediterranean drip-irrigated maize production systems. *Journal of Environmental Management*, *192*, 68–78. doi:10.1016/j.jenvman.2017.01.037 PMID:28142125

Frances, R., Coughlan, M., & Cronin, P. (2006). Interviewing in qualitative research. *International Journal of Therapy and Rehabilitation*, *16*(16).

Gavali & Kulkarni. (2014). Comparative Analysis of Portable Weeders. *International Journal of Innovative Research in Science, Engineering and Technology*, *3*(4), 11004–11013.

Jacob-John, J., D'Souza, C., Marjoribanks, T., & Singaraju, S. (2022). Sustainable Development Practices for SDGs: A Systematic Review of Food Supply Chains in Developing Economies. In N. Nguyen, H. V. Nguyen, C. D'Souza, & C. Strong (Eds.), *Environmental Sustainability in Emerging Markets Consumer, Organisation and Policy Perspective*. Springer. doi:10.1007/978-981-19-2408-8_10

Jiri, O., Mafongoya, P. L., Mubaya, C., & Mafongoya, O. (2016). Seasonal climate prediction and adaptation using indigenous knowledge systems in agriculture systems in Southern Africa: A review. *Journal of Agricultural Science (Toronto)*, *8*(5), 156–172. doi:10.5539/jas.v8n5p156

Ladha, J. K., Jat, M. L., Stirling, C. M., Chakraborty, D., Pradhan, P., Krupnik, T. J., Sapkota, T. B., Pathak, H., Rana, D. S., Tesfaye, K., & Gerard, B. (2019). Achieving the Sustainable Development Goals in Agriculture: The Crucial Role of Nitrogen in Cereal-based Systems. *Advances in Agronomy*, *163*, 39–116. doi:10.1016/bs.agron.2020.05.006

Mapfumo, P., Mtambanengwe, F., & Chikowo, R. (2015). Building on indigenous knowledge to strengthen the capacity of smallholder farming communities to adapt to climate change and variability in southern Africa. *Climate and Development*.

Maurial, M. (1999). Indigenous knowledge and schooling: A continuum between conflict and dialogue. In L. M. Semali & J. L. Kincheloe (Eds.), *What is indigenous knowledge system: Voices from the academy* (pp. 59–77). Falmer.

Mishra, P. K., & Rai, S. C. (2013). Use of Indigenous Soil and Water Conservation Practices among Farmers in Sikkim Himalaya. *Indian Journal of Traditional Knowledge, 12*(3), 554–564.

Muthee, D. W., Gwademba, G. K., & Masinde, J. M. (2019). The role of indigenous knowledge systems in enhancing agricultural productivity in Kenya. *Eastern Africa Journal of Contemporary Research, 1*(1).

Negi, M. S., & Verma, H. (2015). Prospects of Conservation Agriculture in Hill Situation". Training Course Series 32, Compendium of lectures. *Tillage and Nutrient Dynamics for Better Crop Production (October 10-30, 2015), organized by CAFT, Department of Agronomy, GBPUA&T, Pantnagar, Sponsored by ICAR, New Delhi*, 66-72.

Ning, C., Qu, J., He, L., Yang, R., Chen, Q., Luo, S., & Cai, K. (2017). Improvement of yield, pest control and Si nutrition of rice by rice-water spinach intercropping. *Field Crops Research, 208*, 34–43. doi:10.1016/j.fcr.2017.04.005

Posey, D. (2002). Upsetting the sacred balance: Can the study of indigenous knowledge reflect cosmic connectedness? In P. Sillitoe, A. Bicker, & J. Pottier (Eds.), *Participating in development: Approaches to indigenous knowledge* (Vol. 39, pp. 24–42). Routledge. doi:10.4324/9780203428603_chapter_2

Rabo Usha, M. (2022). *Impact of tillage practices on runoff, soil and nutrient loss and wheat yield at kulumsa, central Ethiopia* [Doctoral dissertation]. Hawassa University.

Reynar, R. (1999). Indigenous people's knowledge and education: A tool for development? In L. M. Semali & J. L. Kicheloe (Eds.), *What is indigenous knowledge system: Voices from the academy* (pp. 285–304). Falmer.

Roulston, K., & Choi, M. (2018). Qualitative Interviews. In U. Flick (Ed.), *The SAGE Handbook of Qualitative Data Collection*. SAGE Publications. doi:10.4135/9781526416070.n15

Saunders, B., Sim, J., Kingstone, T., Baker, S., Waterfield, J., Bartlam, B., Burroughs, H., & Jinks, C. (2018). Saturation in qualitative research: Exploring its conceptualization and operationalization. *Quality & Quantity, 52*(4), 1893–1907. doi:10.1007/s11135-017-0574-8 PMID:29937585

Saunders, M., Lewis, P., & Thornhill, A. (2007). *Research methods for business students*. Pearson Education.

Semali, M. J., & Kincheloe, J. L. (1999). What is indigenous knowledge, and why we should study it? In M. L. Semali & J. L. Kincheloe (Eds.), *What is indigenous knowledge system: Voices from the academy* (pp. 3–57). Falmer.

Singh, R. K., & Sureja, A. K. (2008). Indigenous Knowledge and Sustainable Agricultural Resources Management under Rainfed Agro-Ecosystems. *Indian Journal of Traditional Knowledge*, *7*(4), 642–654.

Tanveer, A., Ikram, R. M., & Ali, H. H. (2019). Crop Rotation: Principles and Practices. In M. Hasanuzzaman (Ed.), *Agronomic Crops*. Springer. doi:10.1007/978-981-32-9783-8_1

Triberti, L., Nastri, A., & Baldoni, G. (2016). Long-term effects of crop rotation, manure and mineral fertilisation on carbon sequestration and soil fertility. *European Journal of Agronomy*, *74*, 47–55. doi:10.1016/j.eja.2015.11.024

Velten, S., Leventon, J., Jager, N., & Newig, J. (2015). What Is Sustainable Agriculture? A Systematic Review. *Sustainability (Basel)*, *7*(6), 7833–7865. doi:10.3390/su7067833

World Economic Forum. (2021). *What effect do insects and pests have on crop production.* Available at https://www.weforum.org./agenda/2021/06/

Chapter 6
Incorporating Indigenous Knowledge Systems (IKS) in Enhancing Sustainable Agriculture in Selected Rural Areas of Zimbabwe

Joshua Risiro
Great Zimbabwe University, Zimbabwe

Divaries Cosmas Jaravaza
https://orcid.org/0000-0002-8930-1242
Bindura University of Science Education, Zimbabwe

Paul Mukucha
Bindura University of Science Education, Zimbabwe

ABSTRACT

The purpose of the study was to explore how Indigenous knowledge systems (IKS) could be integrated with modern farming techniques in order to enhance sustainable agriculture in selected rural areas. The sustainable livelihoods approach and post-colonial theory were used as the theoretical framework for the study. The interpretive approach was adopted for data generation by the researchers. Data were generated using individual interviews, focus group discussions (FGD), and field observations. Data were presented and analyzed using thematic approach. The study found that sustainable indigenous farming strategies including indigenous land management strategies, indigenous water conservation methods, and biodiversity management strategies could be integrated with modern farming techniques to reduce land degradation, loss of biodiversity, and lessen the impacts of climatic change in rural communities. The study recommends that farmers incorporate indigenous strategies with modern forms of farming in enhancing sustainable agriculture.

DOI: 10.4018/979-8-3693-2011-2.ch006

INTRODUCTION AND BACKGROUND TO THE STUDY

This paper argues that disregard of indigenous methods of land conservation and management particularly through modern agriculture and agricultural commercialization has resulted in massive deforestation, land degradation and loss of biodiversity in Zimbabwe. The situation has been worsened by excessive emission of greenhouse gases into the atmosphere from the use of agricultural machinery, excessive use of chemicals and fertilizers which is causing global warming and climate change (Risiro et al., 2013). In the Eastern parts of Zimbabwe stretching from Nyanga to Chimanimani, tracts of land have been cleared to give way to timber, tea and coffee plantations. The drier South East Lowveld and the Zambezi valley have not been spared for commercial activities such as growing of cotton. Commercial agriculture has destroyed natural forests and displaced some animals while others died due to lack of proper habitats. This chapter therefore proposes the integration of indigenous knowledge with modern methods of farming in order to promote sustainable agriculture. Modernisation in agriculture disregarded existing sacred places, taboos and cultural beliefs on environmental management within indigenous communities (Risiro et al., 2013). Worse so, the colonial British government in Zimbabwe did little to integrate IK on environmental practices as observed by Masaka (2016), Ogunniyi (2016) and Pedzisai (2013There is need to give due importance to indigenous knowledge systems to safeguard the environment.

Indigenous Knowledge

Indigenous/local knowledge is knowledge that a particular community possess which determine their decisions (Hewson & Ogunniyi, 2011;Tefflo, 2013; UNESCO, 2017;). It has been defined as knowledge of people of a geographical area that has survived for a long period of time (Mawere, 2015; Mapara, 2009; UNESCO, 2017). Several authors define IK in different ways. It has been referred to as "ways of knowing" (Nyota & Mapara, 2008, p.190) "local knowledge" (Graham & Ireland, 2008,p. 32). Mawere (2015) and Shizha (2010) perceive IK as constituting people's culture, the activities they practice and their beliefs. Pedzisai (2013, p.245) views IK as "community knowledge unique for a given culture that solves societal problems."

The current study is guided by the definitions given by Shizha's (2010) and Mapara (2009). Shizha (2010, p. 28) views IK as "African indigenous science which is culturally specific including their knowledge on culture, their lives, spiritual world, natural environment and the activities they carry out such as farming."

International Conventions and Sustainable Development Goals (SDGs)

A number of Conventions and Sustainable Development Goals (SDG) have been developed as measure to manage environmental degradation and the impacts of rapid expansion and commercialization of agricultural activities. The Stockholm Convention held in1972 focused in the preservation of the environment. Further, The Brundtland Commission was constituted in 1983 in order to outline the link between the environment and economic development. The idea of sustainable development was formalized at the forum. Sustainable development refers to the development that allows the present population to meet their needs without disadvantaging future generations to meet their own wants [World Commission on Environment and Development (WCED), 1987]. The Earth Summit in Rio de Janeiro in 1992 produced the Rio Principles Agenda 21 which gave an outline of actions to be carried out in order to achieve sustainable development at various scales (Chenje et al., 1998).

Linked to International conventions are SDGs which were adopted in New York at the United Nations Summit. There are seventeen (17) goals to be achieved by 2030 (Filho at al.,2018). Of particular to the study are SDG 14 that aim at the conservation and sustainable use of life under water; SDG 15 (conservation and sustainable use of terrestrial ecosystems); food security and sustainable Agriculture (SDG2); sustainable use of water and sanitation (SDG 6) and climate change (SDG 13) (Filho at al., 2018).

THEORETICAL FRAMEWORK FOR THE STUDY

The study is guided by Edward Said;s post colonial theory and The Sustainable Livelihoods Approach

Edward Said's Post-Colonial Theory

The theory explains how the colonized people resisted the dominance of Eurocentric knowledge (Subedi &Daza, 2008). The effect of colonial education on the local people was to make them despise their own indigenous practices. The colonial education system in Zimbabwe was tailored to serve the interest of the colonial power through provision of cheap labour in the industry (Mavhunga, 2006). The post-colonial theory is suitable for the study as the research pursued to integrate indigenous and modern methods of farming. Studies by Ogunniyi (2016), Pedzisai (2013) and Risiro, (2021) has shown that very little Indigenous Knowledge (IK) has been integrated in environmental management practices.

Sustainable Livelihood Approach (SLA)

Natarajan (2022, p. 1) provide a detailed introduction to the Sustainable Livelihoods Approach. It is a method of analyzing and changing the lives of people especially in rural areas. Su et al., (2021, p. 2) reveal, that the Sustainable Livelihoods Approach, "is being widely used in poverty reduction and development projects around the World." The researchers chose the Sustainable Livelihoods Approach because it "seeks to understand changing combinations of modes of livelihood in a dynamic and historical context" (Serrat, 2017, p. 25).

STATEMENT OF THE PROBLEM

Modern methods of farming characterized by use of advanced farm technology, chemicals and herbicides have contributed to massive deforestation, loss of biodiversity and environmental degradation. The situation has been worsened by emission of greenhouse gases by agricultural machinery into the atmosphere which has caused global warming and climate change. Further, disregard of indigenous methods of farming has resulted in un-sustainable agricultural practices. It against this background, that the study explores the integration of indigenous and modern methods of farming as a strategy to promote sustainable agriculture.

KNOWLEDGE GAP

Indigenous methods of farming are crucial in promoting sustainable farming yet its practice has been over-shadowed by modern methods of farming. Further very little studies have focused on the integration of indigenous and modern methods of farming in order to promote sustainable farming practices. Most of the studies on sustainable agriculture have focused indigenous and modern methods of farming separately. This study therefore aims at integrating indigenous and modern practices of farming in order to promote sustainable farming practices.

RESEARCH QUESTIONS THAT GUIDED THE STUDY

1. What indigenous knowledge farming practices are available in rural communities that promote sustainable farming?
2. How can indigenous knowledge practices be integrated with modern farming methods in order to promote sustainable agriculture?

3. What challenges can be faced by farmers in integrating indigenous and modern methods of farming?

METHODOLOGY OF THE STUDY

Research Design

The interpretive approach was used for the study as it accommodates various perspectives on the same phenomenon (Addae & Quan- Baffour, 2015; Pham, 2018). In this study the views of rural communities on the integration of indigenous and modern methods of farming were explored in the rural areas in the Eastern Highlands of Zimbabwe. A case study was relevant as it provided detailed information about a phenomenon as supported by Patton (2015). Convenience sampling was adopted to select the rural areas basing on accessibility and a need to reduce costs (Houser, 2015; Risiro, 2020).

Data Collection

Data was collected using open-ended interviews and Focus Group Discussions (FGD). Thirty (30) individual interviews were conducted to generate data using open ended questions. Patton (2015, p.14) observed that, "open ended questions and probes yield in-depth responses about people's experiences, feelings and knowledge." Four (4) FGD that consisted of 3-6 participants were facilitated. According to Remler and Van Ryzin (2015) at least 2-3 FGD can provide credible findings. Data was captured through audio-video recordings using smart phones. The data was stored on computers in folders and external drive and kept by the corresponding author.

Access to Participants

The Ministry of Rural Development, Promotion and Preservation of National Culture and Heritage is responsible for the administration of rural areas' Permission was granted from this Ministry to gather data in the study area. Purposive sampling aided by the village heads was used to identify community elders and farmers knowledgeable in indigenous methods of farming. The Village Heads assisted the researchers to select the participants since they have the demographic information and registers of the rural communities.

Data Analysis and Presentation

Data was analyzed using the thematic approach in line with Creswell (2014). Data generated was transcribed, coded and grouped into main and sub-themes. Data was then presented according to themes as well as direct quotes from the participants.

Ethical Issues

Research participants willing to participate in the research were required to sign informed consent forms after the researchers explained to the participants the objectives of the study and issues of confidentiality and anonymity. In addition the researchers sought verbal permission from the participants. In order to ensure participants' confidentiality, the responses were assigned alpha-numeric codes, R1, R2, and R3 and so on. R stands for Respondent whereas 1, 2, 3—represents respondent number.

FINDINGS AND DISCUSSION

Thematic approach based on the three research questions is used to present research findings and discussion.

Research Question 1: Sustainable Indigenous Knowledge Farming Practices in Rural Communities

Research Question 1 is presented and discussed under three main themes namely: 1. Indigenous land use and land planning practices; 2: Sustainable indigenous food preservation strategies and 3: Sustainable indigenous water conservation and management.

Theme 1: Indigenous Land Use and Land Planning Practices

Table 1 shows indigenous land use and land planning practices that contributes to sustainable farming practices by communal farmers. The indigenous farming practices include multi-cropping, indigenous food crops and small livestock, planting basins and organic farming, fallowing and stone ridging. These farming practices are presented in a tabular form followed by discussion of these farming practices.

Multi-cropping. It was evident from the interviews with R1 and R4 that small-scale farmers (communal farmers) in rural communities of Zimbabwe practice multi-cropping to ensure food security and to preserve moisture and reduce soil erosion.

Table 1. Indigenous land use and land planning practices

Main Theme	Sub-Themes	Verbatim Quotations From the Respondents
Indigenous land use and Land planning practices	Multi-cropping	*-We plant different crops in the same field like maize, beans, pumpkins* **(R4)** *-Planting different on the same field protects the soil from erosion, preserves moisture and improves soil fertility* **(R1)**
	Indigenous food crops, early maturing varieties and small livestock	*-In Bikita District we keep small livestock and plant finger millet, sorghum, round nuts and groundnuts in early October so that the crops make use of first rains in view of conditions caused by climate change* **(R12)** *-Planting early and growing short maturing varieties ensures that the soil is protected from erosion. The early maturing varieties also endure dry conditions and resist most of the pests and diseases* **(R5)**
	Planting basins (*Pfumvudza*) and Organic farming	*-We dig planting basins as early as in September and put organic manure in the basins so that water collects in the planting basins* **(R7).** *-Use of organic manure minimizes the use of fertilizer that may be washed away and pollute water* **(R2).**
	Fallowing	*-In Chivi, some farming portions are left uncultivated for at least 3-5 years so that the soil regains is fertility* **(R1)** *-As grass and plant grows and finally decompose they add fertility to the soil* **(R20)**
	Stone ridging	-In Chimanimani where there are steep slopes, ridges are constructed to break the flow of water **(P3).** - Stone ridging also allow deposition of sediments thus adding soil fertility **(P17).**

The soil is always covered by crop cover that matures at various times. This protects the farming land from raindrop splash erosion and surface runoff that causes soil erosion and loss of nutrients from the soil. The crop cover further provides shade to the soil thus reducing potential and actual evapo-transpiration. Multi-cropping is the practice of growing two or more crops in the same piece of land during one year, instead of one crop (Adams et al., 1990). Burton (2000) observed that multi-cropping preserves moisture as well as boosting soil fertility.

Indigenous food crops, early maturing varieties and small livestock. Indigenous small grains, legumes and keeping small livestock were some of the strategies adopted by rural farmers to ensure sustainable agriculture (R12). Small grains were planted as early as October to ensure maximum utilization of all rains that falls in the season particularly in view of frequent droughts induced by climate change. The rearing of small livestock such as goats ensures that there is less overgrazing. Small animals require small amounts of vegetative cover to sustain them. Indigenous small grains endure dry conditions, pests and diseases (R5) thereby minimizing usage of chemicals that are harmful to the environment, a view supported by Shava et al., (2009).

Planting basins (*Pfumvudza*) and Organic farming. The use of planting basins and application of organic manure was viewed by the respondents (R7; R2) as sustainable form of agriculture as the practice make use of minimal chemical fertilizers through the use of organic manure that is derived from animal manure and crop residues (*murakwani*). The communal farmers viewed *pfumvudza* as a strategy to mitigate the consequences of climate change impacts that result from reduced rainfall and high temperatures. *Pfumvudza* involves farmers digging some basins and adding organic manure and little fertilizer in the planting basins. The basins collect water and retain moisture for a long period of time. The moisture is utilized by the plants during periods of extended drought. A study by Deen and Katari (2003) found that the use of planting basins provide ground water storage and improves soil fertility.

Fallowing. Communal farmers practice fallowing as a natural way of retaining soil fertility The vegetation (grass, shrubs and trees) that grows over the land that has been fallowed eventually dies and decompose and add humus to the soil. This increases crop yields and food security for the households. The type of farming is sustainable and cause minimal soil poisoning and water pollution as there is no use or little application of fertilisers and chemicals. However with population increase fallowing is no longer practiced.

Stone Ridging. In mountainous areas such as Chimanimni where there are steep slopes, stones are arranged across the slope to reduce the velocity of overland flow and soil erosion. Stone ridging allow sediments to deposit thus adding to soil depth and fertility. More so, moisture is retained due to accumulation of flowing water along the stone ridges. In the North East of Zimbabwe among the Tangwena people, Mapara (2009) found that stone ridging was practiced to reduce soil erosion on steep slopes.

Theme 2: Sustainable Indigenous Food Preservation Strategies

This section presents finding on sustainable food preservation strategies used by communal farmers in implementing sustainable agriculture using sub themes as shown in Table 2.

Boiling and Sun-Drying Vegetables. Rural communities boil and sun-dry indigenous vegetables (muboora, nyevhe,black jack)for consumption during off season (R7). This ensures the community achieves food security in a sustainable manner without overusing the land. Sun-drying vegetables reduces wastage and loss of vegetables. The removal of excess water from vegetables through sun-drying (R8) preserves the vegetables from going bad. Sun-drying extend the shelf life of vegetables. Fellows (2009) acknowledge that drying removes excess water from the vegetables thereby extending the shelf life of the vegetables.

Sun Drying of Sweet Sorghum (*ipwa*). Sweet sorghum is sun dried to get rid of excess water that can allow microbes to thrive and result in the rotting of the sweet

Table 2. Sustainable indigenous food preservation strategies

Main Theme	Sub-Themes	Verbatim Quotations From the Respondents (R)
Food Preservation	Boiling and Sun-Drying vegetables	*- Indigenous vegetables such as nyevhe/rudhe and black jack are picked up and sun-dried (mufushwa) for use during off season* **(R7)** *-Sun-drying vegeatables get rid of excessive water content in the vegetables* **(R8)**
	Sun-drying sweet sorghum	*-Sweet sorghum is peeled and cut into small pieces whilst fresh and then sun-dried. It is then tied into small bundles and kept in a sack* **(R2)**
	Shade drying of vegetables	*-Indigenous vegetables like muboora, nyevhe, black jack are dried under shade of big trees to preserve their nutritional value and green pigment* **(R3)**
	Smoking	*Meat products are preserved through smoking. We cut the meat into long strips and hang the meat above a fire pace in a hut* **(R8)** *-We preserve and store seeds in a kitchen hut through smoking after selecting the best seeds from the previous harvest. This prevents the seeds from pest and diseases attack* **(R15)**
	Underground trenches (*Pfimbi*)	*-Sweet potatoes and ground nuts are preserved in underground trenches (pfimbi). Sand is poured at the bottom of the trench before sprinkling some ash to avoid attack from pests and diseases. We then cover either the sweet potatoes or round nuts with another layer of sand. The sweet potatoes or round nuts are stored in this cool dry dark place up to six months* **(R19)**.
	Crop Granary (*Tsapi/Dura*)	*-Un-thrashed crop like finger millet and sorghum are kept in a tsapi where as thrashed crop is stored and preserved in a dura* **(R14)** *-The tsapi/dura are built with specific materials and design. It is built on raised ground by laying logs on large rock boulders. The tsapi/dura (storage hut) is constructed from pole and dagga and grass thatched. It is smeared with cow dung and ash is sprinkled before the crop is stored there* **(R20)**

sorghum. The peeled sweet sorghum is tied into small bundles to make sure certain quantities of sweet sorghum are consumed up to the next harvesting season. The household are guaranteed of availability of the crop throughout the year. It is also imperative to note that the dried sweet sorghum was kept in a sack to allow highest ventilation of the stored crop in order to reduce losses from bacterial infection. The notion of drying vegetables is widespread as shown by studies carried out by Roday (2014) and Prasad (2011), However, the preservation of sweet sorghum is unfamiliar to many researchers and rural communities.

Shade Drying of Vegetables. Communal farmers ensured that there is food security in terms of relish by drying the vegetables under tree shade. The shade-dried vegetables keeps the vegetables green, tasty as well as retaining their nutritional value. The vegetables dry off through indirect heat and evaporation of moisture from air movement under a tree shade. The preservation of vegetables through sun-drying has been recorded by various scholars (Mashoko, 2018; Fellows, Lean, 2006), however, the concept of drying vegetables under shade is not common. This study found that drying vegetables under shade allow the dried vegetables to retain its nutritional value, green colour and tastiness. Air movement under shade wipes off moisture over the drying vegetables thereby extending the life shelf of the dried vegetables. Fellows (2009) argue that the practice of drying discourages any microbial growth and extends the shelf span of vegetables.

Smoking. Meat is preserved through smoking. The meat is cut into strips and hung above the fire place in the kitchen. The heat from the fire removes water that causes the meat to rot. Smoking reduces microbial activity as well as scaring away insects like flies that can cause the meat to go bad. The communal farmers smoke selected seeds from different crops hanging them in the kitchen hut where there is smoke and soot from the fire. Smoking deters seeds from pest attack.

Underground Trenches (*Pfimbi*).The study established that communal farmers store and preserve sweet potatoes and round nuts in underground trenches for a period of up to six (6) months. This allows families to have food during offseason. The sweet potatoes and the round nuts are preserved underground where there is darkness. This has an effect of reducing bacterial activity. Secondly, the women explained that they spread sand at the bottom and after placing the sweet potatoes and the round nuts. Sand is a good insulator that allows free movement of air. This lowers soil temperatures and microbial activity thereby extending the shelf span of the stored food (sweet potatoes and round nuts) up to six (6) months. The findings on the use of *pfimbi* to preserve and store sweet potatoes concur with a study by Mashoko (2018). However, the new idea coming up from this study is the application of sand in the trench (*pfimbi*) and the sprinkling of ash in the trench to protect the sweet potatoes and round nuts from pests. Ash is alkaline and discourages the proliferation of diseases (Amoah et al., 2011). The use of underground trenches in food preservation allows rural communities to consume round nuts while fresh well after harvesting period.

Crop Granary (Tsapi/Dura). *Tsapi/dura* is a grass thatched hut made of dagga and wooden poles built on raised ground which is used to store and preserve un-thrashed crop. *Dura* is constructed in a similar way as tsapi but is used to preserve thrashed crop like finger millet and sorghum. The *dura* may have several partitions known as *tsapana* (granary with sub-divisions) which are used to store different types of crops. The *tsapi/dura* is constructed on logs laid across large rock boulders to allow ventilation as well as protecting the harvested crop from flooding or moisture. The thatched roof, pole and dagga materials used to construct the *tsapi/dura* are good insulators which allow cooling inside thereby avoiding the crop from getting spoiled or rot. Brown (2008) observed that insulators keep temperatures of the environment constant and this has an effect of slowing down the action of bacteria and fungi on food. The interior of the granary is smeared with cow dung to protect the crop from pest attack.

Theme3: Sustainable Indigenous Water Conservation and Management

This section discusses sustainable indigenous water-related strategies adopted by communal farmers to achieve sustainable agriculture. The strategies include use of *mufuku* (shallow sand water holes), rain making ceremonies, water taboos and myths as well as land use and land use planning.

Use of Mufuku (Shallow Sand Water Holes). The study found that due to shortage of water in rivers the community is collecting water from shallow sand water holes (*mufuku*). P1 explained:

We collect water from shallow sand holes (Mufuka) to irrigate vegetables and to provide drinking water to our domestic animals. We are resorting to keeping small livestock like goats and donkeys that requires small amounts of water

In dry regions communal farmers have resorted to keeping small livestock that can requires small amounts of water for their survival. The livestock are fed with water from shallow dug wells along the river beds. The low rainfall and high temperature induced by warming due to climate change has caused most streams and rivers to dry off. The community has therefore resorted to use of *mufuku* to obtain water for growing vegetables and to supply water for the small livestock and donkeys.

Infiltration Pits and Sustainable Agriculture. The study found that some of the farmers in Bikita and Chivi make use of infiltration pits to harvest overland flow that is utilized by field crops as explained by one of the small holder farmer:

Some farmers are digging infiltration pits on the upslope of the field in order to trap running water after some rains. The water collected in the infiltration pits infiltrates into the ground and recharge ground water which is then taken up by plant roots **(R6)**.

Another farmer (R9) concurred with R6 and explained the nature of the pits:

The pits measure two meters long, one meter wide and one meter deep and interspaced at about twenty meter.

Infiltration pits stores water after the rains. This stored water recharges the ground water and increase soil moisture which is useful for plant growth especially during dry. This reduces moisture stress in plants and keeps the crops turgid and green. Crop yields improve from increased ground water storage. However, a study carried out in Rushing in Zimbabwe by Nyakudya et al., (2014) on infiltration pits showed that infiltration pits did not improve crop yields. The variation from these findings may be that the study was done when the seasons were wet such that the effect of infiltration pits could not be realized.

Rain Making Ceremonies in Sustainable Agriculture. The community elders believed performing rain making ceremonies (*mukwerere/musoso*) would allow ancestral spirits to provide some rains and crop protection from environmental hazards. One elderly villager (R16) proclaimed:

During the rainmaking ceremony the ancestral spirits (mhondoro) requests the ancestors to provide rain for a good harvest and to protect the community from environmental hazards like drought pests and diseases.

The local Chief (R3) echoed the views from R16:

If rain making ceremony is done properly, you may not reach your homesteads after the ceremony before heavy rains falls.

Participant C12 revealed the importance of performing rainmaking ceremonies as mitigation against the negative impacts of climate change. The elderly women suggest that when people perform rainmaking ceremony the ancestors would respond by providing some rains, fruits and protection from natural hazards like droughts, pests and diseases. However, the local Chief (C3) seem to suggest that these rain making ceremonies are no longer being performed properly resulting in drought conditions. The findings are consistent with Risiro (2020) who found that indigenous people practiced rain making ceremonies in sacred forests and mountains as well as under trees related to the ancestors such as fig tree (Muonde) in return for rains and protection from natural hazards. The brewing of beer of which some of it is poured onto the ground is viewed as a way of communicating with the ancestors. Dei (2013) and Khupe (2014) observed that the ancestors could be linked through objects such as beer. Arunotai (2006) found similar practices of rainmaking ceremony in Thailand and in Malawi.

Water Taboos and Myths in the Preservation of Springs and Wetlands. Water taboos and myths ensured there is sustainable use of water for irrigation farming and for animals.

There are wetlands such as Bikita Wetlands which provide water for irrigation of crops and drinking water for animals. These are protected by mermaids and anyone who dares pollute them risky disappearing (R11).

R5 echoed R11 sentiments:

No one is allowed to fetch water from springs and wetlands from Ucherere wetlands in Bikita with utensils coated with soot. Only gourds are allowed to fetch water from the springs.

The research participants (R5; R11) strongly believed in observing water taboos in the preservation and sustainability of spring water and wetlands for use in agricultural practices. The community is not allowed to fetch water with smoke coated containers from Uchere wetlands and springs in Bikita. The use of a gourde is a strategy to reduce wastage and gurantees continued supply of water for domestic animals and water to irrigate vegetable gardens. Restricting the community from polluting the springs and wetlands ensures that there is continued supply of clean water for domestic animals as well as crop farming. Dirwai (2007) observed that myths and taboos help to conserve endangered plants and animal species within an ecosystem. In a similar study of Igbo community in Nigeria, Obiora and Emeka (2015) found that the community sustainably preserved natural resources through observing their beliefs, taboos totems and myths. In the Cape Coast of Ghana the respect of sacred places contributed in the conservation of flora and fauna (Emmanuel, 2020).

Indigenous weather forecasting as a climate change adaptation. Interviews and FGD with communal farmers revealed that they used various indigenous methods to predict weather which help them to make decisions on farming and land use practices. The community uses primarily biological indicators to forecast seasonal patterns of weather conditions. Participant R16 proclaimed on the use of indigenous methods of weather forecasting in determining land use practices:

If indigenous fruit trees like muhacha and mazhanje produce a lot of fruits there is pending drought as such we grow drought resistant crops like sorghum and finger millet and if indigenous fruit trees produce very little fruits there is bound to be more rains and as such we can grow maize crops that requires high rainfall.

It was evident from R16 that fruit trees such as *muhacha* and *muzhanje* were used by communal farmers to forecast seasonal weather patterns. Indigenous weather forecasting complements conventional weather forecasting in providing accurate information on weather patterns. The community believed that the ancestors provide indigenous fruits for the survival of the community during extreme weather conditions like drought caused by climate change. Mapara (2009) in a study in Nyanga area in Zimbabwe found that fruit trees like *mazhanje* were plenty when there was pending drought. Indigenous weather forecasting allows farmers to plan farming activities such as time to plant and type of crop depending on the weather forecast. In a study in Mozambique, Salite (2019) found that small farmers use indigenous weather forecasting to make decisions on farm land use. Mapara (2009) further observed that it was not allowed to harvest unripe fruits. This ensures that there is sustainable use of fruit resources by the community. Further, Risiro et al., (2013) in Zaka District of Masvingo observed that the community was not allowed to harvest indigenous fruits and insects from sacred mountains of Bvuma and Chikona hills in Zaka District of Masvingo until a harvesting ceremony (*mutoro*) was performed. Indigenous weather forecasting thus promotes sustainable land use planning in view of predicted weather pattern as well as reducing wastage from the wild fruits.

Soil moisture conservation strategies. The study unpacked that communal farmers use various strategies to retain moisture in the soil as illustrated by the following exerpts:

Mulching using crop residues is used to cover the soil in the field to reduce evaporation as well as adding fertility to the soil (R3)

Terraces and stone ridges are constructed across the slopes to break the velocity of overland flow. This promotes infiltration and percolation thus soil moisture is recharged for use by field crops (R15).

Crop rotation and fallowing are used to improve soil fertility and structure. This enhances the soil ability to retain nutrients and soil moisture (R19).

Mulching is water conservation method that can be easily applied by small-scale farmers. As mulch slowly decomposes it provides organic matter to the soil which improves root growth and increases the water-holding capacity of the soil. Mulching further improves infiltration of water into the soil as well as reducing soil moisture loss through evaporation. It protects the soil from erosion, reduces compaction from heavy rains and prevents weed growth. Terraces and stone ridges reduce the flowing velocity of overland flow and allow water to sink and recharge ground water storage as well as curtailing soil erosion. A study by Mapara (2009) among the Tangwena people in Nyanga and findings by Risiro (2021) in Chimanimani have revealed that stone ridging was a common practice in mountainous areas to reduce soil erosion and allow water to sink into the ground. Crop rotation and fallowing is practiced by the farmers retain soil fertility; reduce the spread of pests and diseases and to enhance crop yields. Magni (2017) in a study in Thailand among the Karen community observed that farmers practice shifting cultivation and fallowing in order to allow soil to regain soil fertility.

Research Question 2: Integrating Indigenous Knowledge Farming Practices With Modern Farming Methods for Sustainable Agriculture

The research question shows how indigenous knowledge farming practices could be incorporated with modern agricultural practices to achieve sustainable agriculture.

Table 3. Integrating indigenous knowledge with modern agricultural practices

Agricultural Practice	Modern Agricultural Practices	Indigenous Knowledge Agricultural Practices	Contribution of IKS to Sustainable Agriculture
Land use and land use planning	-Mixed farming on a large scale - Commercial ranches -Pivot, sprinkler flooding and basin -Contour ridges -Geographical Information System (GIS) and Remote sensing on land use planning	-Multi-cropping -Planting basins -Fallowing -Stone ridging	-Multi-cropping ensures continuous soil cover and reduce erosion. -Stone ridges preserves moisture and reduce erosion -Planting basins retains soil moisture and improves crop productivity -Multi-cropping Improves food security
Food preservation	-Refrigeration -Cold rooms -Preservatives	-Sun drying -Shade drying -Smoking -Underground trenches (*Pfimbi*)	-Sun drying and shade drying. causes no pollution to the environment -Sustained food security during drought periods -Underground trenches ensures food is preserve fresh for a long period of time
Water conservation and management	-Large dams -Large tanks -Cloud seeding -Modern weather forecasting (Use of satellites, radar, GIS)	-Hand dug shallow sand holes -Infiltration pits -Rainmaking ceremonies -Water taboos ad myths -Indigenous weather forecasting	-Ensures continued supply of vegetables, food crops and drinking water for the animals -Continued availability of water for crops and animals -Water taboos reduces wastage and pollution -Indigenous weather forecasting result in Informed decisions on land use and time of planting
Crops	-Commercial crops (e.g. Cotton, tobacco, citrus fruits	Short term grain crops (Sorghum; Finger millet) -Indigenous fruits	-Grain crops are more resistant to drought as well as more resistant to pests and diseases -Indigenous fruits supplement diet and reduces hunger and starvation -Less fertilizer and chemicals are used on grain crops thus reducing water, soil and land pollution.
Livestock	-Hybrids -Big livestock	-Small livestock -Hard *Mashona* (indigenous) breeds	-Small livestock requires less grass and supplements thereby reducing overgrazing and land degradation. -Less chemicals, vaccination are used on small livestock

Modern methods of farming uses technology such GIS in land use planning. The communal farmers determine the type of land use (multi-cropping, fallowing, stone ridging). The integration of indigenous and modern methods of farming result in sustainable farming practices and effective land use planning. Multi-cropping involve growing different types of crops on the same piece of land. The farming practice is sustainable since the crops mature at different times of the year. This is important for household food security. There is guaranteed availability of food supply. Planting basins and stone ridges break the flow of running water and allow more to infiltrate and recharge soil moisture storage. The findings resonate with studies carried out in Uganda by Nuwategeka and Nyeko (2019) who have shown that modeling land suitability based on IK and Geographic Information System (GIS) tools generate agricultural land management models that are more effective in realizing sustainable farming practices and environmental management.

Indigenous food preservation methods are environmental friendly as they do not use preservatives or refrigeration unlike modern methods of farming. Indigenous preservation methods are sustainable methods since they use natural resources like solar energy and trees that provide shade for preserving crop and meat produce from the farm. Preserving food like sweet potatoes underground trenches ensures that communal farmers have food throughout the year. A study by Wane (2014) in Sub-Saharan Africa revealed that indigenous methods of food preservation are sustainable and provide the majority of households with affordable, adequate and nutritionally balanced diet over long periods of time.

.The integration of indigenous methods in food preservation is a sustainable farming practice especially in rural communities of Zimbabwe where some rural areas do not have electricity or experience frequent power cuts.

Modern methods of farming entail conserving water for farming in artificial structures such as dams and large tanks. More so, cloud seeding and modern methods of weather forecasting can be practiced. This is costly and not sustainable particularly for small scale farmers. Indigenous methods of conserving water include use of infiltration pits, hand-dug shallow holes in river beds, which are sustainable as they are natural, environmental friendly and less costly. Critchley et al., (1994) has shown that in Sub-Saharan Africa smallholder farmers in semi arid regions and in the Sahel region have adopted a variety of innovative traditional rainwater harvesting techniques such as pits to produce water for crops including millet and sorghum and livestock.

In addition, indigenous ways of water management are embedded in indigenous beliefs (water taboos and myths). Risiro (2021) in a study in Eastern Highlands of Zimbabwe observed water taboos and myths ensured that water was preserved clean without use of water purifying detergents which may pollute land resources. The belief in water taboos and myths by communal farmers in sync with United Nations

Declaration on the Rights of Indigenous Peoples (UNDRIP, 2007). Article 12 that gave rights to the indigenous people to practice their beliefs and spiritual traditions. These indigenous beliefs support the achievement of Sustainable Development Goals (SDG) 14 and 15 whose focus is the sustainable use of resources under water and on land surfaces respectively. The integration of indigenous methods of water conservation and management in farming activities is sustainable especially for small scale farmers with limited financial resources to purify water and to construct storage facilities to store water for irrigating crops and for animal use.

Growing grain crops or short maturing varieties such as sorghum and finger millet are sustainable forms of farming practices in drought prone areas and during seasons characterized by prolonged droughts caused by climate change. Small grains are hard and require limited use of chemicals since they are more resistant to pests and diseases in comparison to cash crops like tobacco and cotton.

Modern farming involves use of biotechnology, cross-breeding and rearing of big animal breeds. The use of high technology in farming is not sustainable for small scale farmers with limited financial resources. The integration of indigenous forms of farming involving rearing small livestock like goats and drought resistant animals like hard *mashona* breeds is more sustainable form of farming for communal farmers in view of climate change which has increased the frequency of droughts. Indigenous and small animal breeds require less food (grass and food supplements) for their sustenance as such communal farmers who rely largely on natural farming inputs with limited financial resources can easily rear these small animal breeds. Further the indigenous animal breeds are more resistant to dry conditions, pests and diseases. Rearing indigenous animals does not cause much environmental damage since the farmer uses less pesticides and chemicals to treat the animals.

Research Question 3: Challenges Faced by Farmers in Integrating Indigenous and Modern Methods of Farming

The challenges of incorporating IK to enhance sustainable agriculture are discussed in this section in three categories namely: Environmental; Economic and Social challenges.

Environmental Challenges

Climate change and variability. Climate change and variability is making it difficult for communal farmers to make decisions on planning land use for a particular season. One communal farmer (R2) explained:

It is difficult to determine the type of crop to grow in a particular season. In one season the rains can come early enough to grow crops like groundnuts and sorghum and in another season like this year, 2023/24 season the rains had delayed too much such that most crops may not mature.

Another communal farmer (R12) added:

Prolonged droughts causes moisture stress in crops resulting in low yields for the sustenance of our families. At times cyclones damage our crops and wash away the soil due to flooding.

It was apparent from communal farmers (R2; R12) that sustainable agriculture is hindered by climate change and variability. Weather extremes are making it difficult for farmers to plan what to grow or the type of livestock to keep. IPCC (2023:46) reported that since 1950 droughts and tropical cyclones have become more frequent. Prolonged droughts caused by climate change results in poor crop productivity and inadequate grass for livestock grazing. Occasional cyclones have also caused land degradation as soil has been washed away. Farming land along flood plains has been reduced as some parts flood plain got washed away by flooded rivers thereby compromising food security. The findings are in agreement with IPCC (2023) that found that nearly 50% of coastal wetlands have been lost over the past 100 years.

Physiographic factors. Physiographic factors including topography and poor soils are a hindrance to sustainable farming by communal farmers. A communal farmer (R18) proclaimed on the effect of topography on sustainable agriculture in Chimanimani District of Manicaland:

Our area is mountainous and people farm on steep slopes such that when heavy rain falls most of the soil and the crops are washed away leaving behind poor soils that is less productive to crops.

Another communal farmer (R13) added:

In Rusitu area of Chimanimani it is difficult to practice viable livestock farming because there is limited grazing land as a result of steep slopes. We keep small livestock like goats that can move around on slopes searching grass to feed on.

In Chivi District of Masvingo sustainable farming was limited due to poor soils that mainly sustain small grains and drought resistant livestock like donkeys and hard *mashona* cattle as revealed by the following excerpt:

Chivi District is dry and has poor soils that require irrigation. The soils sustain little grass for animal grazing. We therefore mainly grow drought resistant crops such as sorghum and keep small livestock like goats as well as donkeys and hard mashona animal breeds (R17)

Respondent (R13) revealed topographic factors as a hindrance in achieving sustainable agriculture in mountainous areas such as Rusitu in the Eastern Highlands of Zimbabwe. Steep slopes reduces soil infiltration and increase surface runoff which causes soil erosion and washing away of crops. Soil erosion decreases soil fertility and crop production. Topography also limits the type of farming practiced in an area. In steep terrain it is difficult for cattle to move around in search of grass. Poor soils and dry conditions have been found to be a limiting factor to sustainable crop production and rearing of big cattle breeds that require a lot of fodder. The communal farmers in Chivi therefore mainly grow small grains and keep drought resistant animals.

Biotic conditions. Pests and diseases are a deterrent to sustainable farming in communal areas. Pests and diseases attack crops and animals resulting in low productivity and food insecurity as illustrated by the following excerpts:

Pests and diseases attack crops at various stages of farming. Pests like arm worm and locusts attack crops resulting in low crop production. Food and mouth diseases can kill many livestock within a short period of time (R19).

We have a challenge of storing our grains since we do not have money to buy chemicals to preserve our crop yields (R2).

The participants (R19; R2) revealed that sustainable agriculture in communal areas is threatened by pests and diseases that attack crops and livestock at various stages. The IPCC (2023) reported an increase in pests and diseases in some regions due to global warming. The situation by communal farmers is worsened by lack of financial resources to buy chemicals to preserve the harvested crop or to spray/vaccinate livestock. The prevalence of pests and diseases lowers crop and animal production especially to small holder farmers.

Economic Factors

The study revealed that lack of adequate financial resources to buy farming inputs, farm machinery and to fund farming operations were found to be set back in achieving sustainable agriculture by communal farmers:

I do not have enough money to buy farming equipment and to put up farm infrastructure (R5).

For effective farm operations I need to develop roads, storage facilities and dams for irrigation but I cannot afford to do so (R10).

The study unpacked that lack of financial resources by communal farmers were an obstacle to sustainable agriculture. The communal farmers lack capital to buy farming inputs as well as money to develop infrastructure like farming equipment and storage facilities. Farming requires money to buy inputs like organic fertilizer, certified seeds and to construct crop granaries to preserve harvested crop. Financial resources are also needed to develop paddocks for livestock and to buy food supplements. Lack of financial resources contributes to low crop and animal productivity from communal farmers.

Social Factors

Lack of Knowledge and Skills. The lack of knowledge and skills in sustainable farming practices by the some communal farmers are a deterrent to practicing sustainable agriculture. One elderly participant (R2) proclaimed:

We now have young generation in the community who lacks knowledge on sustainable indigenous farming practices for example the young generation does not know natural means to preserve food stuffs or how to construct granaries to store harvested crop, some even despise this indigenous knowledge.

Rural communities now have few elderly people who can teach/apprentice the young people on sustainable indigenous farming practices and methods of preserving food or crops. Rural urban migration has resulted in many elderly people knowledgeable in IK now living in urban areas. In a similar study in South Africa Seehawer (2018) observed that integrating IK in the school curriculum was hampered by lack of indigenous knowledge among the teachers. Further,, modernization and western education received by the young people lacks indigenous education on farming practices. Eurocentric education has despised indigenous education such that the young farmers perceive indigenous farming methods as backward (Shizha, 2010).

Religious beliefs. Religious beliefs were viewed as a hindrance to the achievement of sustainable indigenous farming practices in communal areas. One elderly participant (R1) expressed the following sentiments:

We have our traditional beliefs on preservation of natural resources like forests, water, soil and land management but those who believe in Christian values are not prepared to respect these indigenous natural resources management techniques.

Some are using sacred water pools for baptism where as some have turned sacred places used to perform rain making ceremonies to worshipping places (shrines).

Christian values are disregarding sustainable indigenous ways of conserving natural resources such as forests, water and soil. Sacred places like water pools and sacred forests and mountains have been turned into worshipping places as observed by Risiro (2020). This could result in water pollution and degradation of forest resources which are important for livestock grazing. Christian beliefs have ignored and disregarded sustainable indigenous knowledge on farming and this has affected food security for local communities. Seehawer (2018) in a study in South Africa observed that, some religious denominations discouraged practicing indigenous practices as they viewed these IK practices as evil and against their ideology of spreading Christianity.

Population and demographic characteristics. The study established that the migration or relocation of people from Chimanimani West to Chimanimani East has resulted in many farming land abandoned or underutilized. One Village Head (R10) had this to say:

There is underutilization of land as many people in the community have migrated to Chimanimani to occupy and farm former European farms which they perceive as more productive.

Another participant (R3) commented on the migration of people to Chimanimani East:

This part of Chimanimani West (Mhakwe and Chikwakwa) is left with largely elderly people and school going children who cannot productively work in the fields. Most pieces of land are left uncultivated.

The participants (R10 and R3) revealed that migration from Chimanimani West to Chimanimani East has contributed low agricultural productivity in parts of Chimanimani West. Most of the farming land has been abandoned or underutilized due to shortage of productive labour force. He sending region (Chimanimani West) is left behind with elderly and school going age group who cannot maximize land usage. The elderly no longer have the strength to work in the fields whereas the young children spent most of their time attending to school work. United Nations, Department of Economic and Social Affairs, Population Division (2019) has reported that urbanization was increasing at a rate of 43%. Rapid urbanization coupled with climate change has aggravated the situation of food security (Ehui et al. 2020). Unlike some areas whose challenge on sustainable farming is overpopulation, parts of Chimanimani West has got a challenge of under population to utilize available farming land thus contributing to food insecurity.

CONCLUSION

The study came out with indigenous knowledge farming strategies that could be incorporated in farming in order to achieve sustainable farming in selected rural areas of Zimbabwe. The suggested farming practices include multi-cropping, growing indigenous food crops and rearing indigenous livestock, the use of planting basins, fallowing and stone ridging. These indigenous farming practices have the effect of reducing soil erosion, improve moisture storage and soil fertility. The research participants suggested sustainable indigenous food preservation that makes use of natural methods including sun drying, shade drying, smoking use of underground trenches and crop granaries. The natural methods of food preservation are environmentally friendly and are meant to increase the shelf life of food products at the same time maintaining the nutritional value and tastiness of the preserved food. The communal farmers have indigenous ways in which conserve water as well as ensuring its quality. The communal farmers believed in the spiritual world (ancestors) to provide rain and in water taboos to ensure water is conserved and minimize pollution. Infiltration pits conserve water and recharge soil moisture for the sustenance of crops. Farming practices like crop rotation and fallowing ensures that soil fertility is maintained while pests and diseases are kept under control. The study explored ways in which modern methods of farming could be integrated with indigenous practices in order to achieve sustainable agriculture. Finally, the study focused on the various challenges (physical, physiographic, economic, social0 which are faced by communal farmers in implementing sustainable agriculture. The study proposes the incorporation of modern methods of farming with indigenous practices in order to achieve sustainable agriculture.

RECOMMENDATIONS

Based on the research findings the study recommends the following:

Climate change is heavily impacting upon the implementation of sustainable agriculture by communal farmers. The communal farmers should therefore work in partnership with the government and private sector to develop and adapt more climate change adaptations to combat the negative impacts of climate change on sustainable agriculture.

The communal farmers are facing financial challenges on buying farming inputs and to develop farm infrastructure in order to improve on food security. The government through the Ministry of Agriculture and Climate Change can act as a guarantor for communal farmers so that get farming loans from banks.

There is lack of sufficient knowledge on sustainable indigenous farming practices among some communal farmers especially the young and middle aged group. The elderly who are knowledgeable on Indigenous farming practices could be invited as resource persons on organized workshops. Most of these indigenous practices are not documented. Funding for more research on indigenous knowledge and documentation of IK is needed.

In Chimanimani West in particular, it was observed that there was underutilization of farming land due to shortage of labour caused by rural to urban migration and migration to former colonizer's farms in Chimanimani East. The challenge could be alleviated through cooperatives whereby farmers pool their resources and buy agricultural machinery such as tractors to counter the shortage of labour. The District Rural Council can also lend tractors to the communal farmers at a small fee.

Some religious groups regarded indigenous knowledge farming practices as backward and retrogressive. The government in partnership with local leadership of various religion can work together to educate people on the importance of indigenous knowledge practices in food security.

Recommendations for Future Research

Indigenous knowledge is locally based and geographic specific such that what was studied in the three selected districts many not apply to other regions due to variations geographically, beliefs and the level of modernization. There is therefore need to carry out extensive research in other districts of Zimbabwe and find sustainable strategies suitable for each locality.

REFERENCES

Adams, R. M., Rosenzweig, C., Peart, R. M., Ritchie, J. T., McCarl, B. A., Glyer, J. D., Curry, R. B., Jones, J. W., Boote, K. J., & Allen, L. H. Jr. (1990). Global climate change and U.S. agriculture. *Nature, 345*(6272), 219–224. doi:10.1038/345219a0

Addae, D., & Quan-Baffour, K. P. (2015). The place of mixed methods research in the field of adult education: Design options, prospects and challenges. *International Journal of Education and Research, 3*(7), 151–162.

Amoah, R. S., Teye, E., Abano, E. E., & Tetteh, J. P. (2011). The storage performance of sweet potatoes with different pre-stage treatments in an evaporative cooling barn. *Asian Journal of Agricultural Research, 5*(2), 137–145. doi:10.3923/ajar.2011.137.145

Arunotai, N. (2006). Moken traditional knowledge: An unrecognised form of natural resources management. *International Social Science Journal, 58*(187), 139–150. doi:10.1111/j.1468-2451.2006.00599.x

Brown, A. (2008). *Understanding food: Principles and preparation* (3rd ed.). Thomson Wadsworth.

Burton, I. (2000). Adaptation to climate change: Advancing the agenda for collective global benefit. *BRIDGES-Between Trade and Sustainable Development, 4*(8), 9.

Chenje, M., Sola, L., & Paleczny, D. (1998). The State of Zimbabwe. Harare: Ministry of Mines, Environment and Tourism.

Critchley, W., Reij, C., & Wilcocks, T. (1994). Indigenous soil and water conservation: A review of the state of knowledge and prospects for building on traditions. *Land Degradation & Development, 5*(4), 293–314. doi:10.1002/ldr.3400050406

Deen, W., & Katari, P. K. (2000). Carbon sequestration in a long- term convention vernus conservation tillage experience. *Soil & Tillage Research, 74*(2), 27–142.

Dei, G. J. S. (2013). Indigenizing the school curriculum: The case of the African University. *Proceedings of the Fourth International Conference of the Science and Indigenous Knowledge Systems Project/ South African-Mozambican Collabourative Research Programmes*, 162-177.

Dirwai, C. (2007). Sustainable environmental management: An ethno-based approach: The case of totems, spirostachys africana and acacia nigrescens in conserving Zimbabwean fauna and flora. *Journal of Sustainable Development in Africa, 9*(4), 48–80.

Ehui, S., Kray, H., & Mghenyi, E. (2020). Policy priorities for achieving food and nutrition security by 2030. In B. S. Coulibaly (Ed.), *Foresight Africa: top priorities for the continent 2020 to 2030* (pp. 56–59). Africa Growth Initiatives at Brookings.

Fellows, P. J. (2009). Food Processing and Practice (2nd ed.). Woodhead Publishing Limited.

Filho, W. L., Azeiteiro, U., Alves, F., Pace, P., & Mifsud, M. (2018). Reinvigorating the sustainable development research agenda: The role of the sustainable development goals (SDG). *International Journal of Sustainable Development and World Ecology, 25*(2), 131–142. doi:10.1080/13504509.2017.1342103

Graham, S., & Ireland, B. (2008). *Integrating St'at'imc Knowledge Systems into Lillooet Area K-12 School curricula and Pedagogy.* Power of place (PoP), Final research report, 2008, Canadian Council on Learning.

Hewson, M. G., & Ogunniyi, M. B. (2011). Argumentation teaching as a method to introduce indigenous knowledge into science classrooms: Opportunities and challenges. *Cultural Studies of Science Education*, *6*(3), 679–692. doi:10.1007/s11422-010-9303-5

Houser, R. A. (2015). *Counselling and Educational Research: Evaluation and Application* (3rd ed.). Sage Publications.

IPCC. (2023). Sections. In *Climate Change 2023: Synthesis Report Contributions of Working Groups i, ii and iii to the Sixth Assessment Report of the Intergovernmental on Climate Change* (pp. 35–115). IPCC. doi:10.59327/IPCC/AR6-9789291691647

Khupe, C. (2014). *Indigenous knowledge and school science: Possibilities for integration* [PhD Thesis]. Faculty of Science, University of Witwatersrand, Johannesburg.

Magni, G. (2017). Indigenous knowledge and implications for the sustainable development agenda. *European Journal of Education*, *52*(4), 437–447. doi:10.1111/ejed.12238

Mapara, J. (2009). Indigenous knowledge systems in Zimbabwe: Juxtaposing Post-Colonial Theory. *The Journal of Pan African Studies*, *3*(1), 139–155.

Mashoko, D. (2018). *Integrating indigenous knowledge of food preservation with school science* [PhD Thesis]. University of Witwatersrand, Johannesburg.

Mavhunga, F. (2006). Africanising the school curriculum. A case for Zimbabwe. The Zimbabwe Journal of Educational Research, 18(3).

Mawere, M. (2015). Indigenous knowledge and public education in Sub-Saharan Africa. *Africa Spectrum*, *50*(2), 57–71. doi:10.1177/000203971505000203

Nyota, S., & Mapara, J. (2008). Shona traditional children's games and play songs as indigenous ways of knowing. Journal of Pan African Studies, 2(4), 184-202.

Obiora, A. C., & Emeka, E. E. (2015). African indigenous knowledge system and environmental sustainability. *Int J Environ Prot Policy*, *3*(4), 88–96. doi:10.11648/j.ijepp.20150304.12

Ogunniyi, M. B. (2016). Explicating the philosophy of *ubuntu* into science education: A project Experience. *Proceedings of SAARMSTE 24th Annual Conference at Tshwane University of Technology,* 417-431.

Patton, M. (2015). *Qualitative Research and Evaluation Methods* (4th ed.). Sage Publications.

Pedzisai, C. (2013). Teachers' perceptions on inclusion of agricultural indigenous knowledge systems in crop production: A case study of Zimbabwe's Ordinary Level Agriculture Syllabus (5035). *Journal of Biology, Agriculture and Healthcare, 3*(16), 37–44.

PhamL. (2018). *Qualitative Approach to Research, A review of Advantages and Disadvantages of three paradigms: Positivism, Interpretivism and Critical Inquiry.* University of Adelaide: ResearchGate. doi:10.13140/RG.2.2.13995.54569

Prasad, U. (2011). *Food Science and Nutrition.* Sonali Publications.

Remler, D. K., & Van Ryzin, G. G. (2015). Research Methods in Practice, 2nd ed: Strategies for Description and Causation. SAGE Publications.

Risiro, J. (2020). *Integrating indigenous knowledge into the teaching of weather and climate in the geography curriculum in secondary schools: The case of Manicaland in Zimbabwe* [PhD Thesis]. University of KwaZulu-Natal.

Risiro, J. (2021). Integrating indigenous knowledge on environmental practices in the teaching of Geography in secondary schools: A case of Chimanimani Community in Zimbabwe. In Sustainable Development in Africa, Fostering Sustainability in one of the World's Most Promising Continents. Springer Nature. doi:10.1007/978-3-030-74693-3_7

Risiro, J., Tshuma, D.T. & Bhasikiti, A. (2013). Indigenous knowledge systems and environmental management: A case study of Zaka District, Masvingo Province, Zimbabwe. *International Journal of Academic Research in Progressive Education and Development 2*(1).

Roday, S. (2014). *Food Science and Nutrition* (2nd ed.). Oxford University Press.

Salite, D. (2019). Traditional prediction of drought under weather and climate uncertainty:analyzing the challenges and opportunities for small-scale farmers in Gaza Province SouthernRegion of Mozambique. *Natural Hazards, 96*(3), 1289–1309. doi:10.1007/s11069-019-03613-4

Seehawer, M. (2018). South African science teachers' strategies for integrating indigenous and western knowledges in their classes: Practical lessons in decolonisation. *Educational Research for Social Change*, *7*(0), 91–110. doi:10.17159/2221-4070/2018/v7i0a7

Serrat, O. (2017). Proposition 5: The sustainable livelihoods approach. In O. Serrat (Ed.), *Knowledge Solutions: Tools, Methods, and Approaches to Drive Organisational Performance*. SpringerOpen. doi:10.1007/978-981-10-0983-9_5

Shizha, E. (2010). The interface of neoliberal globalization science education and indigenous african knowledge in Africa. *Journal of Alternative Perspectives in the Social Sciences*, *2*(1), 27-58.

Su, Song, Ma, Sultanaliev, Ma, Xue, & Fahad. (2021). An assessment of poverty alleviation measures and sustainable livelihood capability of farm households in rural China: A Sustainable livelihood approach. *Agriculture, 11*(12). doi:10.3390/agriculture11121230

Subedi, B., & Daza, S. L. (2008). The possibilities of post-colonial praxis in education. *Race, Ethnicity and Education*, *11*(1), 1–10. doi:10.1080/13613320701845731

Tefflo, L. (2013). Rural communities as sites of knowledge: A case for african epistemologies. *Indilinga*, *12*(2), 188–202.

UNESCO. (2017). *What is local and indigenous knowledge*. Available at: https://www.unesco.org/new/en/naturalsciences/priority-areas/links/related-information/what-islocal-and-indigenous-knowledge/

United Nations. (2007). *Declaration on the Rights of Indigenous Peoples (UNDRIP)*. https://www.un.org/esa/socdev/unpfii/documents/DRIPS_en.pdf

United Nations, Department of Economic and Social Affairs, Population Division. (2019). *World urbanization prospects: the 2018 revision* (ST/ESA/SER.A/420). United Nations. https://www.un-documents.net/wced-ocf.htm

Chapter 7
Digitizing Marketing in Agriculture:
Leveraging Information Communication Technologies for Success in Zimbabwe

Option Takunda Chiwaridzo
https://orcid.org/0000-0001-9755-9123
University of Science and Technology Beijing, China

Rodwell Musiiwa
Midlands State University, Zimbabwe

Tariro Hlasi
Midlands State University, Zimbabwe

ABSTRACT

This chapter explores the transformative role of ICTs in revolutionizing agricultural sector of developing nations like Zimbabwe. By scrutinizing the current state of traditional and inefficient agricultural marketing in Zimbabwe, the study advocates for the potential of ICTs to enhance competitiveness and profitability. The benefits of digitizing marketing include minimizing information asymmetry, facilitating price discovery, expanding market linkages, improving financial access, and enabling traceability. Simultaneously, the paper delves into key challenges such as infrastructure limitations, skills gaps, affordability issues, gender disparities, and cultural factors that must be addressed. To create an enabling environment for digitizing agricultural marketing, the study recommends targeted policy and regulatory interventions. It concludes that strategic investments in ICT infrastructure and the capacity-building of farmers are crucial to fully unlock the potentials of ICTs, leading to improved incomes and enhanced rural livelihoods in Zimbabwe.

DOI: 10.4018/979-8-3693-2011-2.ch007

INTRODUCTION

The global digital revolution in agriculture, marked by innovations like digital advisory services, e-commerce platforms, precision techniques, and blockchain traceability, is reshaping how smallholders engage with information, markets, finance, and services. Developing countries, including Zimbabwe, have the opportunity to leverage digitization for enhanced transparency, commercialization, productivity, sustainability, and inclusion in agricultural systems.

Despite the potential benefits, Zimbabwe lags behind regional and global peers in technology adoption. Barriers such as inadequate connectivity infrastructure, high data costs, limited local content, low digital literacy, unsupportive policies, and a lack of specialized capacity hinder progress. Presently, less than 27% of farming households have internet access, exacerbating the existing "digital divide" in agriculture, which is crucial for rural development.

This paper underscores the importance of coordinated efforts among public, private, and development partners to create an enabling environment for digitization tailored to Zimbabwe's smallholder context. The approach necessitates integrated investments, policies, innovations, and skills building across the agricultural ecosystem. Solutions must align with farmers' priorities, addressing concerns related to affordability, access, transparency, and value.

Information and Communication Technologies (ICTs) refer to technologies that provide access to information through telecommunications. This includes the Internet, wireless networks, cell phones, computers, software applications, and other communication mediums(Mapiye et al., 2023). ICTs enable the rapid exchange of information and support the transition to digital economies. They have transformed how individuals, organizations, and societies create, collect, store, and share information, enabling innovation, productivity, and social and economic benefits(Tiwari, 2022). The study emphasizes the role of ICTs in digitizing agricultural marketing systems in Zimbabwe. It explores how ICTs can facilitate commercialization, enhance climate resilience, empower farmers, and contribute to systemic competitiveness. However, realizing these benefits requires the establishment of 21st-century digital infrastructure as the foundation for a successful digital transformation in Zimbabwe's agricultural sector.

BACKGROUND OF THE STUDY

Historically, Zimbabwe held the title of the "breadbasket of Africa" due to a robust large-scale commercial farming sector, complemented by significant smallholder production for domestic markets (Chipuriro, 2021). Agriculture played a pivotal role, contributing over 45% of all exports in the 1980s and supporting livelihoods for more than 66% of the population (Moyo, 2016). However, a confluence of factors, including fast-track land reform, hyperinflation, declining productivity, climate change, and deteriorating infrastructure, resulted in a substantial decline in agricultural output(Chihambakwe, 2018; Raftopoulos et al., 2021). Between 2000 and 2008, maize production, a staple crop, contracted by 76%, impacting food self-sufficiency in Zimbabwe (Matema, 2021).

Zimbabwe underwent a transformation from a net agricultural exporter to relying heavily on food and raw material imports over the past two decades(Gwanongodza, 2020). Challenges contributing to this shift include constraints in fertilizer, machinery, and irrigation; inconsistent trade policies; climate vulnerabilities; water scarcity; financial exclusion limiting investments; weak rural infrastructure; and fragmented, inefficient value chains (Mujeyi et al., 2022; Sachikonye, 2021). Smallholder farmers, in particular, face formidable barriers in accessing inputs, credit, extension services, technologies, and formal markets, impeding viability, commercialization, and competitiveness (Mango et al., 2017).

Addressing these barriers and modernizing agriculture is imperative for rural employment, food security, export revenues, and overall economic growth(Ba, 2016; Norton et al., 2021). Digitization of farming operations and agribusiness models provides avenues for innovation, efficiency, and systemic transparency(MacPherson et al., 2022). However, there is a significant knowledge gap in research on digital agriculture adoption determinants, usage patterns, and socioeconomic impacts in the Zimbabwean context (Mayoyo et al., 2023; Pedzisa, 2016). Questions persist regarding the factors influencing the uptake of digital advisory, e-commerce, precision techniques, and data analytics across gender, demographic, and geographic lines (El Bilali & Allahyari, 2018; Siankwilimba et al., 2023). Infrastructure access, affordability, trust, awareness, and farmer-centric solution design also require deeper investigation through user experience assessments and field evaluations (Adewale, 2022).

While emerging studies analyze barriers, opportunities, and early impacts of digitizing Zimbabwe's agricultural systems (Gwaka, 2022; Mapiye et al., 2021), research gaps persist around adoption determinants, usage patterns, and localization requirements to drive widespread impact (Kimathi, 2023; Kuipa, 2019; Mandipaza, 2022). User assessments highlight limited awareness and content relevance as key access barriers, emphasizing the need for localized awareness campaigns and languages (Mugwisi, 2015; Pedzisai, 2020). Rural-urban digital divides exacerbate through uneven literacy, electricity, and internet connectivity (Kim et al., 2020). Representative panel data remains limited, hindering disaggregated analysis of socioeconomic impacts across gender, youth, and geographic lines (Chineka, 2016; Mayoyo et al., 2023).

Digitizing agricultural marketing systems through information and communication technologies (ICTs) offers immense potential to enhance market linkages, transparency, traceability, and incomes for farmers(Duncan et al., 2021; Liu et al., 2021). However, current marketing systems remain largely traditional, inefficient, and have limited reach. Adoption of digital platforms like mobile applications, e-commerce, blockchain, satellite imaging, and data analytics can modernize marketing but requires assessing farmer awareness, affordability, infrastructure availability, and trust. Targeted policies and investments into ICT infrastructure and capacity building are vital to harness ICTs for competitive, profitable, and inclusive agricultural marketing systems in Zimbabwe.

This book chapter aims to analyze the potential role of information and communication technologies (ICTs) in transforming and modernizing agricultural marketing systems in Zimbabwe. It advocates for digitizing marketing channels to enhance transparency, efficiency, competitiveness, and inclusivity across agricultural value chains. The study scrutinizes limitations of current traditional marketing methods and makes the case for ICT platforms to facilitate commercialization, price discovery, market linkages, and traceability. It also examines persistent barriers related to infrastructure, affordability, skills, policies, and socio-cultural factors that must be addressed alongside technology interventions. The chapter provides targeted recommendations on investments, innovations, capacity building, and regulatory reforms needed to create an enabling environment for ICT adoption in agricultural marketing. The overarching goal is to highlight how strategic deployment of digital technologies, combined with farmer-centric solutions, can unlock new opportunities to improve smallholder incomes and rural livelihoods in Zimbabwe.

RESEARCH OBJECTIVES

The overarching aim of this research is to analyze opportunities and barriers to accelerating adoption of digital technologies across Zimbabwe's agricultural marketing ecosystems to drive transparency, efficiency, competitiveness and inclusivity. Specific objectives include:

1. Assessing the current state of traditional agricultural marketing systems and their limitations for farmers in Zimbabwe.
2. Identifying relevant ICT platforms and tools with potential to transform and modernize agricultural marketing.
3. Analyzing the benefits of digitizing agricultural marketing for improving farmer incomes, competitiveness, and rural livelihoods.

4. Examine the key challenges and barriers to adopting ICT-enabled agricultural marketing systems in the Zimbabwean context.
5. Develop targeted policy and regulatory recommendations for creating an enabling environment for successful ICT adoption in agricultural marketing.

LITERATURE REVIEW

Information and Communication Technologies (ICTs) in Agriculture

(Mapiye et al., 2023; Okediran, 2019)provides a comprehensive review of how ICTs can support inclusive agricultural innovation and value chain development. The paper analyses case studies across Africa and Asia, finding that ICT platforms have enabled smallholder farmers to access information, financing, and output markets, thereby enhancing productivity and incomes. However, the author notes that ICT interventions often focus on commercialization and exports rather than local food security needs. Tailored, context-specific approaches are needed to make ICTs more inclusive. (Ayesiga, 2023; Njoh, 2018)present an in-depth analysis of an ICT coordination platform for smallholder farmers. This integrated mobile and web-based system connects farmers to extension, veterinary services, inputs and markets. The authors foundthat it has increased productivity and quality, commercialization, and farmer incomes. However, challenges remain around low literacy and lack of trust in using the platform. Ongoing training and support is key to successful adoption.

(El Bilali & Allahyari, 2018; Lajoie-O'Malley et al., 2020; Lezoche et al., 2020) provide a broad review of how various digital technologies can promote sustainability across the agriculture and food sector. This ranges from farm management software, remote sensing, Internet of Things sensors, robotics, through to blockchain for traceability and mobile platforms linking actors. The authors highlight the potential of these technologies but note adoption barriers like costs, skills and infrastructure must be addressed, especially for smallholders. (Anantha et al., 2021; Makate, 2019)specifically evaluates smart farming innovations for enhancing productivity, resilience, and environmental sustainability of smallholder farming systems in Africa and Asia. The papers examined technologies like drones, smartphone applications, farm management information systems and sentinel land imaging. Theyconcluded that ICTs can provide benefits but often lack localization for smallholders' contexts and capacity. The author recommends participatory approaches to tailor and target solutions. (Kikuchi et al., 2021)conduct an insightful comparative analysis of agricultural drone awareness, perceptions of benefits, and adoption factors among farmers in Thailand, Myanmar and Laos based on survey data. Adoption was highest in Thailand, followed by Myanmar then Laos. Key farmer concerns around drones included costs, technical skills, and relevance. Targeted training programs are needed to increase adoption.

Literature put forward that ICTs increase the impact of young entrepreneurship and facilitate new avenues of addressing systemic barriers, such as skills acquisition, financing, marketing and business networks (Ratinho et al., 2020). According to (AlBar & Hoque, 2019) ICT has got a greater influence as a medium of promotion, communication and information. Utilization is much more significant very influential on the effectiveness and efficiency of work processes, if it is carried out intensely and maximally. The benefits felt by business people directly and indirectly have a positive influence on the farming community involved, especially from the increasingly broad marketing channels of business people that increase production demand and encourage production procurement among the farming community, which is always expected to increase production with standards determined quality, with more integrated farming activities to produce more and more quality, then the opportunity to improve the lives of farmers and their families is wide open.

According to (Kiel et al., 2017), the use of ICT allows companies to sell products directly to other companies or to end consumers. The potential to bypass distribution channels brings support for expanding the market by setting lower prices when it comes to final consumers, this is because it does not involve many levels or processes in distribution that want to get more profit by increasing selling prices to consumers, but directly from farmers to end consumers decision for the farmers which has positive implication on food security status of the farming households. (Lele & Goswami, 2017)studied the working pattern of technological services to the agricultural marketing. The study observed that the proper and timely implementation of ICT projects can bring revolutionary change in the Indian agricultural sector of our country. Due to the unavailability of records in the farms studied, the cost and return particulars were collected verbally from the farmers in the study. (El Bilali & Allahyari, 2018)in their study Role of ICT Based Services in Agricultural Marketing found out that ICT has maximum impact on agricultural marketing. It has become the most important weapon in the agricultural reform process in India, and through the various scheme, the price related information and the market-related information is made available to common people and the farmers easily and even free of cost.

(Majumdar & Singh, 2019)in their study of the impact of information and communication technology on marketing of rice in Uttar Pradesh found out that education and land holding type have a positive and significant impact on the ICT use. However, although age and land ownership type have a positive impact, but they have an insignificant, impact on the use of ICT. A significant positive impact of ICT use was also observed on the net price received by the farmers of Uttar Pradesh. In a similar study (Bala, 2021)found similar results that education and training factors have a positive bearing on the adoption of ICTs as a source of information, highlighting the relevance of capacity building initiatives for enhancing the use of ICTs in Indian agriculture.

(Janssen et al., 2017) pointed out that farmers who have a high level of information technology knowledge have a higher chance to use information technology in marketing agricultural products than farmers who are less informed about information technology and use of information technology. Thus the need for information and communication technology training related to the use of technology so that not only uses smartphones for communication, but can be used for marketing media. Therefore, farmers can increase sales volume and product price levels which have an impact on improving farmers' welfare

In most developing countries, rural farmers are far from the market and also they have no access to appropriate market information. Thus, the farmers are exploited by traders and get low price for their products (Ahmed et al., 2017). In the region the farmers used traditional means of communication and hence they are deterred from the market access. Furthermore, information asymmetry, lack of information transparency and mistrust between farmers and traders, obliged farmers to sell their product in the local market which makes the trades to benefit more at the expense of farmers (Quayson et al., 2020). Therefore, applying ICTs in agriculture is not a matter of choice if poverty reduction of the farming community is aimed at large (Mapiye et al., 2023; Maulu et al., 2021).

FACTORS INFLUENCING ICTS ADOPTION IN AGRICULTURE SECTOR

Economic Factors

The costs associated with acquiring, implementing, and maintaining ICTs can deter adoption, especially for farmers with limited resources (Gavai et al., 2018). However, studies show that perceptions of return on investment and long-term economic benefits encourage adoption, even when initial costs are high (AlBar & Hoque, 2019). Providing affordable and flexible financing options can help overcome capital constraints (Yadav et al., 2018). Impact assessments also build confidence in economic returns, driving adoption (Havemann et al., 2022). Quantitative analyses by (Akinyi et al., 2022; Mutenje et al., 2019)demonstrate that access to financing and clear cost-benefit analyses strongly influence farmers' decisions to adopt new agricultural technologies.

Infrastructure and Connectivity

Lack of electricity, broadband internet, cellular networks, and technical support impede ICT use in rural areas (Owolabi et al., 2021). Recent initiatives to address these gaps through public-private partnerships, universal service funds, and development projects have expanded rural access (Gwaka et al., 2018). But reliability issues remain, highlighting the need for policies and investments in robust, sustainable digital infrastructure (Kamutuezu et al., 2021). Study by (Blimpo & Cosgrove-Davies, 2019) found connectivity and electricity access to be top predictors of ICT adoption across sites in Africa and Asia.In developing countries, even if ICTs are believed to have a potential to bring rapid change in the agricultural sector in general, and in agricultural marketing section in particular, inadequate ICT facilities, lack of personnel, insufficient infrastructure, harmonization of knowledge and language, power supply and farmers' perception are challenges that obstruct successful implementation of ICT in the sector (Akmal, 2021; Girma & Abebe, 2019)

Knowledge and Awareness

Unfamiliarity with ICTs and limited digital literacy among farmers often hinder adoption (Ireri et al., 2021). Hands-on training and educational programs to improve tech skills and demonstrate benefits are essential (Chen et al., 2020). Partnerships with extension services, agribusinesses, and NGOs can facilitate knowledge dissemination (Kaur & Kaur, 2018). Research by (Mapiye et al., 2021) underscores the importance of sustained, community-based ICT training programs for increasing adoption.(Shah, 2022)indicated that education and training factors have a positive bearing on the adoption of ICTs as a source of information, highlighting the relevance of capacity building initiatives for enhancing the use of ICTs in Indian agriculture

Institutional Support and Policies

Clear governance frameworks, coordination across public agencies, and partnerships with private tech providers facilitate ICT initiatives in agriculture (Tiwari, 2022; Wolfert et al., 2023). Supportive policies on data systems, digital transactions, and network expansion also drive adoption (Smidt & Jokonya, 2022). Ongoing engagement between policymakers, companies, and farmers is needed (Fabregas et al., 2019). Study by (Makate, 2019)highlight that institutional capacity and government commitment are preconditions for scaling agricultural technology adoption in agriculture.

Socio-Cultural Factors

Designing technologies that align with farmers' values and needs enhances adoption (Lee et al., 2021). Community participation in tool development also counters misperceptions in agriculture sector (Wong & Guggenheim, 2018). Gender dynamics, caste issues, and other social factors should inform inclusive ICT strategies (Hansda, 2017). As shown by (Rola-Rubzen et al., 2020), addressing sociocultural barriers through participatory approaches leads to more equitable ICT adoption in agriculture. (Alemu et al., 2021)pointed out that to use market information for long-term decisions, farmers should be aware of storage costs, to decide whether to store or not; in addition, they should understand production costs, so that they can use MIS to plan whether to grow new crops or to move to off-season production

Perceived Utility and User Experience

Farmers are more likely to adopt user-friendly, localized ICT tools that offer tangible benefits like increased productivity, market access, and streamlined operations (Lamsal et al., 2023). Ongoing engagement for design improvements as well as support services improve user experience and sustained adoption (Piñeiro et al., 2020). Recent user experience assessments by (Rose et al., 2019) reiterate the primacy of usefulness, ease-of-use, and continued training in boosting ICT adoption. Different ICTs innovations have come to application and are changing the lives of farmers in developing countries. Among ICTs mobile phone is widely accessible to billions of people in developing countries. Innovations taking in the agriculture sector include commodity and stock market price information, meteorological data collection, advisory services to farmers for agricultural extension, early warning systems for disaster prevention and control, and also financial services are crucial to make informed.

THEORETICAL FRAMEWORK

Technology Acceptance Model (TAM)

The Technology Acceptance Model (TAM), originally proposed by Fred Davis in 1989, serves as a useful theoretical framework for examining ICT adoption by farmers in the Zimbabwean context. TAM posits that perceived ease of use and perceived usefulness determine user attitudes and acceptance of new technologies. This model has been widely applied to analyze adoption of innovations from mobile learning to enterprise systems across various contexts (Barrio-García et al., 2015; Ibrahim et al., 2017; Sánchez Prieto et al., 2016). In this book chapter's literature review, applying TAM can provide insights into key factors driving farmers' intentions to accept or reject agricultural ICT platforms based on their perceptions of usability and value-add. However, TAM's focus on individual technology perceptions risks overlooking broader socioeconomic, cultural, and contextual factors that influence adoption, such as infrastructure availability or gender disparities, which are crucial considerations in the Zimbabwean smallholder setting. Thus, while integrating TAM within the literature review offers a starting point to assess ICT acceptance determinants, its application must be supplemented by qualitative approaches and adapted to the local context to gain a holistic understanding of barriers, incentives, and impacts shaping technology adoption across varying farmer demographics in Zimbabwe.

RESEARCH METHODOLOGY

Data Type and Sources

This study utilizes primary data obtained through questionnaires. Data was collected from a total of 271 individuals that were selected randomly from 30 ground nut farming Wards in Gokwe South District. A total of 9 people was sampled in each Ward. Meanwhile, 10 people were sampled in Ward 22 as it had a relatively larger population as compared to all other Wards. Statistics that were supplied were on the basis of 2022-2023 farming season.

Variables

Income of the Farmer (Inc)

This represents net income received by the farmer after deducting all associated fixed and variable costs per tin (20 litre bucket)of ground nuts sold. Some of the costs include, among others, labour, marketing, harvesting and wedding costs. This is one of the dependent variables in this study.

Information Communication Technology (ICT)

This represents various forms of old and new technologies used to market products by farmers. Bolarinwa et al. 2014 and Das, 2014 asserted that the mode of ICT use is bifurcated into three categories which are very old, old and new.

Accordingly, they categorized books, posters, newspapers and drama as very old ICTs. Radio and television were classified as old ICTs while New ICT ATM, E-banking, E-commerce, Cash credit card and mobile devices were classified as new technologies. The general expectation is that adoption of ICT in marketing should positively impact incomes on farmers. Therefore, a positive coefficient is expected on this variable.

Variables That Influence Use of ICT in Marketing

Age (Age)

This represents actual number of years of the farmer being interviewed. Age will be in full years attained by the responded. Generally, due to dynamism of technologies, it is likely that old people tend to ignore rapid changes in technology and hence tend not to use technologies in marketing of their product. Therefore, a negative relation is expected.

Education (Educ)

This variable captures the level of education of the responded. To capture different levels of education a value of one (1) is recorded for primary level, value of two (2) represents a responded who has secondary level of education while three (3) represents those with tertiary levels and above. Generally, the higher the level of education, the more likely that the person embraces technologies in marketing. However, due to the everchanging nature of modern technologies even those with higher levels of education especially the elderly may not fully utilize the modern (new) technologies. Therefore, this variable can take either positive or negative sign.

Land Ownership Type (OwnTyp)

This variable represents the status of ownership of the land being utilize to produce ground nuts. A value of one (1) is used to represent owned land and a value of zero (0) is representing leased land.

Land Size (LanSiz)

This represents the actual size of land (hectors) being planted ground nuts.

THE MODELS

Model 1: OLS Regression to Capture the Impact of ICT Usage on Farmer Income

$$Inc_{fi} = \propto + \beta ICTu_1 + \gamma_i \tag{1}$$

Where:

Inc_{fi} represents net income received by farmer *I;*

$\propto$ is the constant term;

β is the coefficient of ICT use;

$ICTu_1$ is a dummy variable which takes the value of one (1) if a responded uses ICT and zero (0) for otherwise; and

γ_i is the error term or disturbance term.

Model 2: Multiple Dummy Regression Model to Analyze the Impact of Various Forms of ICT on Farmer INCOME

$$Inc_{fi} = \rho + aICTuN + bICTuO + \vartheta_i \tag{2}$$

Where: $\rho, a \ and \ b$ are parameters;

$ICTuN$ represents use of new ICTs;

$ICTuO$ represents use of old ICTs and

ϑ_i is the disturbance term.

$ICTuN$ and $ICTuO$ are dummy variables. A value of 1 represents their use and 0 for otherwise.

Model 3: Multinomial Logit Model to Analyze the Factors That Influence the Use of ICT in Marketing of Ground Nuts in Gokwe South

$$ICTu = \pi + \varnothing_1 Age_{1i} + \varnothing_2 Educ_{2i} + \varnothing_3 LanSiz_{3i} + \varnothing_4 OwnTyp_{4i} + \epsilon_i \tag{3}$$

Where: $\pi, \varnothing_1, \varnothing_2, \varnothing_3 \ and \ \varnothing_4$ are parameters while ϵ_i is the disturbance term;

Age_{1i} is the age of individual *i*in years;

$Educ_{2i}$ represents the level of education of the responded. It ranges from 1-3 where 1 is primary level, 2 is secondary level and 3 is tertiary level and above.

$LanSiz_{3i}$ is the size of land (hectares) being used to produce ground nuts.

$OwnTyp_{4i}$ shows whether the land is owned or leased by the respondent.

PRESENTATION OF RESULTS

Descriptive Statistics

As indicated in Table 1 below, a total of 271 observations were made in this analysis. Results show that on average farmers in 30 ground nuts farming wards of Gokwe South District of Zimbabwe receive a net income of \$3.60 per bucket of ground nuts. Meanwhile, minimum and maximum net incomes received for this product stood at \$2.50 and \$5.00 respectively. Average usage of ICT in these wards was found to be 0.47 meaning that majority of the farmers as represented by the sample are not utilizing technologies in marketing their produce and particularly ground nuts. Results further revealed that most of those who are using ICT in marketing their products are using old technologies compared to new technologies like internet and e-commerce. On socio-demographic factors that influence the adoption of ICT, a mean of 2 was recorded in level of education of the sampled farmers. This indicates that most sampled farmers in the district under study attained secondary level education. In terms of age, average age was recorded to be 51 with the oldest being 85-year-old. Farm characteristic variable land size and ownership type were also utilized to analyze their impact on usage of ICT. Average land planted ground nuts was 3 hectares and majority of the sampled people own land.

Impact of ICT on Farmer's Income

To evaluate analyze the impact of ICT usage on farmer's income (Inc), variable income was regressed against ICT usage (ICTu). Income represented net income received the farmer per bucket (20 litre tin) of ground nuts after deducting all associated average fixed and variable costs from planting to marketing stage. Some of the variable and fixed costs include, land preparation, gypsum fertilizers, weeding, harvesting and transportation costs. This variable was in monetary value. Meanwhile, ICT usage was utilized as the sole regressor. This variable was in dummy form where the farmer was recording one (1) if ever the farmer was utilizing any form of ICT in marketing of their ground nuts. A value of zero (0) was recorded

Table 1. Summary statistics

Variable	Obs.	Mean	Std.Dev	Min	Max
Income	271	3.60	0.68	2.50	5.00
ICTu	271	0.47	0.50	0.00	1.00
ICTuO	271	0.49	0.50	0.00	1.00
ICTuN	271	0.46	0.50	0.00	1.00
ICTtyp	271	1.90	0.78	1.00	3.00
Edu	271	1.98	0.52	1	3
Age	271	51	17	25	85
LanSiz	271	3	1.8	1	7
OwnTyp	271	0.7	0.46	0	1

Source: researchers' own computation

for an otherwise response by the responded. This research utilized E-views 12 as econometric package for this evaluation.

$$Income = 3.6396 + 0.6314 ICTu + \varepsilon \,. \tag{4}$$

Results from Table 4 above shows that a one (1) percent increase in ICT usage by farmers will increase the farmer's income by 0.63%. The coefficient of ICT usage on farmers income is also positive indicating that ICT adoption in marketing by farmers in Gokwe South leads to increased incomes for ground farmers. The goodness of fit as shown by R^2 was recorded to be 0.57. this implies that 57% of the factors that influence farmers income are captured under ICT variable while

Table 2. Regression results on the impact of ICT on farmer's income

Regress Inc ICTu				
Method: Least Squares				
Sample: 1271				
Observations: 271				
Variable	**Coefficient**	**Std. Error**	**t.stat**	**p-value**
c	3.6396	0.0569	63.937	0.6341
ICTu	0.6314	0.0832	-3.0252	0.0216
R^2	0.5719			
Adj. R^2	0.5698			

Source: researchers' own computation

0.43% of the factors are captured by the error term ε . Furthermore, a p-value of 0.0216 for ICTu indicates that the explanatory variable significantly impacts farmer's income at 5% level of significancy. Findings of this study also goes in line withPartovifar, P. (2016) in Israel and Majumdar K and Singh R. K. (2019) in Uttah Pradesh Districts of India.

Impact of Various Types of ICT on Farmer's Income

$$Income = 3.6933 - 0.0562ICTuO + 0.1437ICTuN + \mu \tag{5}$$

The purpose of this regression was to assess the impact of various forms of ICT of farmer's income. ICTuO and ICTuN were utilized as explanatory variables against farmers income where ICTuO represented old ICTs like news papers and posters while ICTuN represented new and modern ICT techniques like Social Media Platforms, Internet and E-Commerce. As indicated by equation five (5) above, Old ICT usage has a negative and insignificant relationship with farmer's income while new ICT usage has a positive and significant (at 10% level of significancy) relationship with incomes of farmers. A coefficient of 0.1437 for New ICTs in marketing indicates that a unit increase in the level of adoption of new digital marketing of ground nuts by farmers in Gokwe South will lead to a 14% increase in their incomes. Findings of this study also concur with findings by Lashgarara, F., Mohammadi, R. and Najafabadi, M.O. (2017b) in Iran.

Table 3. Results of the impact of ICT type of farmer's incomes

Regress Inc ICTuOICTUn				
Method: Least Squares				
Sample: 1271				
Observations: 271				
Variable	**Coefficient**	**Std. Error**	**t.stat**	**p-value**
C	3.6933	0.0683	54.0564	0.0000
ICTuO	-0.0562	0.0829	-3.0252	0.4984
ICTuN	0.1437	0.0831	-1.7296	0.0848
R^2	0.6347			
Adj. R^2	0.6344			
Prob(F)	0.0701			

Source: researchers' own computation

Table 4. Results of the factors that influence the use of ICT in marketing of ground nuts in Gokwe South

Dependent Variable: ICTu				
Method: ML-Binary Logit (Newto-Raphson/Marquardt steps)				
Sample: 1271				
Observations: 271				
Variable	**Coefficient**	**Std. Error**	**z.Stat**	**p-value**
C	-0.0508	0.6305	-0.0805	0.0000
Age	-0.1894	0.0072	1.7429	0.0984
Educ	0.1644	0.2351	2.1463	0.0848
LanSiz	0.0606	0.0605	0.8755	0.3813
OwnTyp	0.2177	0.2682	0.8118	0.4169
Prob(LR statistic)	0.0743			
Obs with Dep = 0	144			
Obs with Dep = 1	127			

Source: researchers' own computation

Factors Affecting Use of ICT

$$ICTu = -0.0508 - 0.1894Age + 0.1644Educ + 0.0606LanSiz + 0.2177OwnTyp + \tau \qquad (6)$$

Equation six (6) gives a summary of the variables that impact usage of ICT in marketing of ground nuts in Gokwe South. Age was found to have a negative and significant (10%) relationship with ICT usage in marketing. This implies that as people grow old, their chances of utilizing technologies be it new or old diminishes. Thus, young farmers are the ones who are likely to adapt to fast and dynamic trends of technologies. Education was also found to be statistically significant (10%) in explaining the adoption of technologies in marketing of ground nuts in Gokwe South. Findings on age and education also concur with results from a study by Majumdar K and Singh R. K. (2019).

DISCUSSION OF RESULTS AND RECOMMENDATIONS

The study findings confirm the well-documented inverse relationship between age and ICT adoption in agriculture, highlighting the necessity for targeted interventions to bolster digital literacy among older farmers. To address this disparity effectively, focused digital literacy programs tailored to the unique needs of older individuals should be implemented. Additionally, the study underscores the positive correlation between education and ICT adoption, consistent with previous research. It suggests that farmers with higher educational qualifications are more inclined to embrace ICTs in agricultural marketing, aligning with findings by Oladele (2011). This emphasizes the importance of educational attainment in facilitating access to agricultural information through ICT platforms. However, the study also emphasizes that formal education alone does not guarantee tech proficiency, advocating for localized, in-language training content to accommodate diverse educational backgrounds. The overarching recommendation emphasizes a comprehensive approach, integrating user-centric design principles and sustained community engagement to surmount demographic barriers and promote widespread ICT adoption in Zimbabwe's agricultural sector.

Furthermore, based on the analysis, it is recommended that agricultural extension officers and organizations such as the Agricultural Marketing Authority (AMA) and Zimbabwe Trade (ZIMTRADE) enhance the availability of information to farmers through educational initiatives, such as workshops, aimed at elucidating the benefits of ICT adoption for improving marketing both locally and internationally. These efforts aim to empower farmers with the necessary knowledge and skills to leverage ICT tools effectively in their agricultural practices and market engagement.

In conclusion, the proposed multifaceted strategy encompasses intuitive tool development, targeted digital literacy initiatives, localized training content, and inclusivity in design to bridge the technology gap among different demographic groups within Zimbabwe's agricultural landscape. Policymakers and stakeholders are urged to recognize and address the diverse needs of both younger and older farmers, as well as those with varying educational backgrounds, to ensure the successful integration of ICTs into agricultural practices. Additionally, policy interventions focused on improving infrastructure, access to credit, education, and training, along with the facilitation of farmers' organizations, can further enhance technology adoption and market participation among farmers in Zimbabwe.

LIMITATIONS OF THE STUDY

Key limitations include the geographic focus on selected districts which constrains generalizability across Zimbabwe. The small sample amongst digital technology users also inhibits statistical representativeness and cross-tabulation analysis on demographic factors. Being a cross-sectional study, it cannot assess changes over time regarding evolving technology awareness, adoption decisions and usage patterns.

Suggestions for Further Research

Follow up studies can address these limitations through expanded, longitudinal panel surveys tracking usage dynamics by gender, income, education and age groups over time. In-depth qualitative assessments should probe user barriers, enablers and experiences especially for women smallholders. Detailed ecosystem analyses can map network infrastructure gaps, solutions landscape and institutional capabilities to inform regulatory reforms. Overall, structured datasets tracking awareness-to-adoption journeys across farmer profiles are imperative.

REFERENCES

Adewale, H. A. (2022). *A Community Knowledge-Based Assessment of Smallholder Farmers' Perception of Cluster Farming and its Potential Role in Sustainable Accessibility to Farmer-Centric Support in Nigeria*. Polytechnic University.

Ahmed, U. I., Ying, L., Bashir, M. K., Abid, M., & Zulfiqar, F. (2017). Status and determinants of small farming households' food security and role of market access in enhancing food security in rural Pakistan. *PLoS One*, *12*(10), e0185466. doi:10.1371/journal.pone.0185466 PMID:29077719

Akinyi, D. P., Karanja Ng'ang'a, S., Ngigi, M., Mathenge, M., & Girvetz, E. (2022). Cost-benefit analysis of prioritized climate-smart agricultural practices among smallholder farmers: Evidence from selected value chains across sub-Saharan Africa. *Heliyon*, *8*(4), e09228. doi:10.1016/j.heliyon.2022.e09228 PMID:35399383

Akmal, N. (2021). *Information Communication Technology (ICT) and Livelihood Improvement in Rural Pakistan: A Comparative Study of Small-and Large-Holder Citrus Farming Households in the Sargodha District*. University of Canberra.

AlBar, A. M., & Hoque, M. R. (2019). Factors affecting the adoption of information and communication technology in small and medium enterprises: A perspective from rural Saudi Arabia. *Information Technology for Development*, *25*(4), 715–738. doi:10.1080/02681102.2017.1390437

Alemu, G. T., Nigussie, Z., Haregeweyn, N., Berhanie, Z., Wondimagegnehu, B. A., Ayalew, Z., Molla, D., Okoyo, E. N., & Baributsa, D. (2021). Cost-benefit analysis of on-farm grain storage hermetic bags among small-scale maize growers in northwestern Ethiopia. *Crop Protection (Guildford, Surrey)*, *143*, 105478. doi:10.1016/j.cropro.2020.105478 PMID:33941995

Anantha, K., Garg, K. K., Barron, J., Dixit, S., Venkataradha, A., Singh, R., & Whitbread, A. M. (2021). Impact of best management practices on sustainable crop production and climate resilience in smallholder farming systems of South Asia. *Agricultural Systems*, *194*, 103276. doi:10.1016/j.agsy.2021.103276

Ayesiga, C. (2023). *Digitalising African agricultural value chains: the role of ICT platforms in smallholders' uptake of agricultural inputs in Uganda*. Wageningen University.

Ba, M. N. (2016). Strategic agricultural commodity value chains in Africa for increased food: The regional approach for food security. *Agricultural Sciences*, *7*(09), 549–585. doi:10.4236/as.2016.79055

Bala, M. (2021). Role of Information, Communication and Technology (ICT) in Rural Development. *National Research Journal of Social Sciences*, *6*, 57–63.

Blimpo, M. P., & Cosgrove-Davies, M. (2019). *Electricity access in Sub-Saharan Africa: Uptake, reliability, and complementary factors for economic impact*. World Bank Publications.

Bolarinwa, K. K. L., Irene, S. E., & Apantaku, S. O. (2014). Utilization of information communication technology's components for coordination of marketing, agricultural commodities in oyo State Nigeria. *European Journal of Business and Management*, *6*(6), 12–16.

Chen, J. C., Huang, Y., Lin, K. Y., Chang, Y. S., Lin, H. C., Lin, C. Y., & Hsiao, H. S. (2020). Developing a hands-on activity using virtual reality to help students learn by doing. *Journal of Computer Assisted Learning*, *36*(1), 46–60. doi:10.1111/jcal.12389

Chihambakwe, M. T. (2018). *Urban and peri-urban agriculture: an analysis of the perceptions and use of food sovereignty among low-income dwellers in Harare*.

Chineka, J. (2016). Analysis of drought incidence, gendered vulnerability and adaptation in Chivi South. Academic Press.

Chipuriro, R. M. (2021). *Land reform and local economic development: elderly women farmers' narratives in Shamva District, Zimbabwe*. University of Johannesburg.

Das, B. (2014). Adoption for accessing agricultural information: Evidence from Indian agriculture§. *Agricultural Economics Research Review*, *27*(2), 199–208. doi:10.5958/0974-0279.2014.00024.X

Davis, F. D. (1989). Perceived usefulness, perceived ease of use, and user acceptance of information technology. *Management Information Systems Quarterly*, *13*(3), 319–340. doi:10.2307/249008

del Barrio-García, S., Arquero, J. L., & Romero-Frías, E. (2015). Personal learning environments acceptance model: The role of need for cognition, e-learning satisfaction and students' perceptions. *Journal of Educational Technology & Society*, *18*(3), 129–141.

Duncan, E., Abdulai, A.-R., & Fraser, E. D. (2021). Modernizing agriculture through digital technologies: Prospects and challenges. *Handbook on the human impact of agriculture*, 138-161.

El Bilali, H., & Allahyari, M. S. (2018). Transition towards sustainability in agriculture and food systems: Role of information and communication technologies. *Information Processing in Agriculture*, *5*(4), 456–464. doi:10.1016/j.inpa.2018.06.006

Fabregas, R., Kremer, M., & Schilbach, F. (2019). Realizing the potential of digital development: The case of agricultural advice. *Science*, *366*(6471), eaay3038. doi:10.1126/science.aay3038 PMID:31831641

Gavai, P., Musungwini, S., & Mugoniwa, B. (2018). A Model for the Adoption and Effective Utilization of ICTs in Commercial Agriculture in Zimbabwe. *Journal of Systems Integration*, *9*(4), 40.

Girma, Y., & Abebe, A. (2019). ICTs and Agricultural Marketing in Africa: A Review. *Marketing, 9*(10).

Granic, A., & Marangunic, N. (2019). Technology acceptance model in educational context: A systematic literature review. *British Journal of Educational Technology*, *50*(5), 1–40. doi:10.1111/bjet.12864

Gwaka, L. T. (2022). Computer Supported Livestock Systems: The Potential of Digital Platforms to Revitalize a Livestock System in Rural Zimbabwe. *Proceedings of the ACM on Human-Computer Interaction, 6*(CSCW2), 1-28. 10.1145/3555085

Gwaka, L. T., May, J., & Tucker, W. (2018). Towards low-cost community networks in rural communities: The impact of context using the case study of Beitbridge, Zimbabwe. *The Electronic Journal on Information Systems in Developing Countries*, *84*(3), e12029. doi:10.1002/isd2.12029

Gwanongodza, T. (2020). *Agricultural exports and economic growth in Zimbabwe*. Stellenbosch University.

Hansda, R. (2017). Small-scale farming and gender-friendly agricultural technologies: The interplay between gender, labour, caste, policy and practice. *Gender, Technology and Development, 21*(3), 189–205. doi:10.1080/09718524.2018.1434990

Havemann, T., Negra, C., & Werneck, F. (2022). Blended finance for agriculture: exploring the constraints and possibilities of combining fi nancial instruments for sustainable transitions. In *Social innovation and sustainability transition* (pp. 347–358). Springer. doi:10.1007/978-3-031-18560-1_23

Ibrahim, R., Leng, N. S., Yusoff, R. C. M., Samy, G. N., Masrom, S., & Rizman, Z. I. (2017). E-learning acceptance based on technology acceptance model (TAM). *Revue des Sciences Fondamentales et Appliquées, 9*(4S), 871–889. doi:10.4314/jfas.v9i4S.50

Ireri, D. M., Awuor, M., Ogalo, J., & Nzuki, D. (2021). Role of ICT in the dissemination and access of agricultural information by smallholder farmers in South Eastern Kenya. *Acta Informatica Malaysia, 5*(1), 31–41. doi:10.26480/aim.01.2021.31.41

Janssen, S. J., Porter, C. H., Moore, A. D., Athanasiadis, I. N., Foster, I., Jones, J. W., & Antle, J. M. (2017). Towards a new generation of agricultural system data, models and knowledge products: Information and communication technology. *Agricultural Systems, 155*, 200–212. doi:10.1016/j.agsy.2016.09.017 PMID:28701813

Kamutuezu, E. U., Winschiers-Theophilus, H., & Peters, A. (2021). An exploration of factors influencing the adoption of ICT enabled entrepreneurship applications in Namibian rural communities. *arXiv preprint arXiv:2108.09789.*

Kaur, K., & Kaur, P. (2018). Agricultural extension approaches to enhance the knowledge of farmers. *International Journal of Current Microbiology and Applied Sciences, 7*(2), 2367–2376. doi:10.20546/ijcmas.2018.702.289

Kiel, D., Arnold, C., & Voigt, K.-I. (2017). The influence of the Industrial Internet of Things on business models of established manufacturing companies–A business level perspective. *Technovation, 68*, 4–19. doi:10.1016/j.technovation.2017.09.003

Kikuchi, M., Mano, Y., Njagi, T. N., Merrey, D., & Otsuka, K. (2021). Economic viability of large-scale irrigation construction in sub-Saharan Africa: What if mwea irrigation scheme were constructed as a brand-new scheme? *The Journal of Development Studies, 57*(5), 772–789. doi:10.1080/00220388.2020.1826443

Kim, J., Shah, P., Gaskell, J. C., & Prasann, A. (2020). *Scaling up disruptive agricultural technologies in Africa*. World Bank Publications. doi:10.1596/978-1-4648-1522-5

Kimathi, L. M. (2023). *Impact of Conservation Agriculture on Smallholder Farmer Livelihoods in Kenya and Zimbabwe*. University of Pretoria.

Kuipa, P. J. T. (2019). *The agricultural financing gap in Zimbabwe: rationing, sustainable credit access and participation in rural financial markets*. University of the Free State.

Lajoie-O'Malley, A., Bronson, K., van der Burg, S., & Klerkx, L. (2020). The future (s) of digital agriculture and sustainable food systems: An analysis of high-level policy documents. *Ecosystem Services*, *45*, 101183. doi:10.1016/j.ecoser.2020.101183

Lamsal, R. R., Karthikeyan, P., Otero, P., & Ariza, A. (2023). Design and Implementation of Internet of Things (IoT) Platform Targeted for Smallholder Farmers: From Nepal Perspective. *Agriculture*, *13*(10), 1900. doi:10.3390/agriculture13101900

Lashgarara, F., Mohammadi, R., & Najafabadi, M. O. (2017b). Identifying appropriate information and communication technology (ICT) in improving marketing of agricultural products in Garmsar City, Iran. *African Journal of Biotechnology*, *10*(5), 11537–11540.

Lee, C.-L., Strong, R., & Dooley, K. E. (2021). Analyzing precision agriculture adoption across the globe: A systematic review of scholarship from 1999–2020. *Sustainability (Basel)*, *13*(18), 10295. doi:10.3390/su131810295

Lele, U., & Goswami, S. (2017). The fourth industrial revolution, agricultural and rural innovation, and implications for public policy and investments: A case of India. *Agricultural Economics*, *48*(S1), 87–100. doi:10.1111/agec.12388

Lezoche, M., Hernandez, J. E., Díaz, M. M. E. A., Panetto, H., & Kacprzyk, J. (2020). Agri-food 4.0: A survey of the supply chains and technologies for the future agriculture. *Computers in Industry*, *117*, 103187. doi:10.1016/j.compind.2020.103187

Liu, W., Shao, X.-F., Wu, C.-H., & Qiao, P. (2021). A systematic literature review on applications of information and communication technologies and blockchain technologies for precision agriculture development. *Journal of Cleaner Production*, *298*, 126763. doi:10.1016/j.jclepro.2021.126763

MacPherson, J., Voglhuber-Slavinsky, A., Olbrisch, M., Schöbel, P., Dönitz, E., Mouratiadou, I., & Helming, K. (2022). Future agricultural systems and the role of digitalization for achieving sustainability goals. A review. *Agronomy for Sustainable Development*, *42*(4), 70. doi:10.1007/s13593-022-00792-6 PMID:35818482

Majumdar, K., & Singh, R. K. (2019). Impact of information and communication technology on marketing of rice: A study of Uttar Pradesh. *International Journal of Social Economics, 46*(9), 1061–1080. doi:10.1108/IJSE-02-2019-0105

Makate, C. (2019). Effective scaling of climate smart agriculture innovations in African smallholder agriculture: A review of approaches, policy and institutional strategy needs. *Environmental Science & Policy, 96*, 37–51. doi:10.1016/j.envsci.2019.01.014

Mandipaza, B. F. (2022). *Socio-Cultural Dimensions of Farming, Small Farm Households and Conservation Agriculture in Nyanga District, Zimbabwe*. University of Pretoria.

Mapiye, O., Makombe, G., Molotsi, A., Dzama, K., & Mapiye, C. (2021). Towards a revolutionized agricultural extension system for the sustainability of smallholder livestock production in developing countries: The potential role of icts. *Sustainability (Basel), 13*(11), 5868. doi:10.3390/su13115868

Mapiye, O., Makombe, G., Molotsi, A., Dzama, K., & Mapiye, C. (2023). Information and communication technologies (ICTs): The potential for enhancing the dissemination of agricultural information and services to smallholder farmers in sub-Saharan Africa. *Information Development, 39*(3), 638–658. doi:10.1177/02666669211064847

Matema, T. (2021). *State-led Agricultural Reform and Food Security in Zimbabwe: Command Agriculture and Food Production in Makoni Rural District, Manicaland*. University of Witwatersrand Johannesburg.

Maulu, S., Hasimuna, O. J., Mutale, B., Mphande, J., & Siankwilimba, E. (2021). Enhancing the role of rural agricultural extension programs in poverty alleviation: A review. *Cogent Food & Agriculture, 7*(1), 1886663. doi:10.1080/23311932.2021.1886663

Mayoyo, A., Chapoto, A., Matchaya, G., Aheeyar, M., Chiwunze, G., Ebrahim, G., Ajayi, O. C., Afun-Ogidan, K., Fakudze, B., & Kasoma-Pele, W. (2023). *Digital climate adaptation in agriculture profile for Zimbabwe*. Academic Press.

Moyo, S. (2016). *Family farming in sub-Saharan Africa: its contribution to agriculture, food security and rural development*. Academic Press.

Mugwisi, T. (2015). Communicating agricultural information for development: The role of the media in Zimbabwe. *Libri, 65*(4), 281–299. doi:10.1515/libri-2015-0094

Mutenje, M. J., Farnworth, C. R., Stirling, C., Thierfelder, C., Mupangwa, W., & Nyagumbo, I. (2019). A cost-benefit analysis of climate-smart agriculture options in Southern Africa: Balancing gender and technology. *Ecological Economics*, *163*, 126–137. doi:10.1016/j.ecolecon.2019.05.013

Njoh, A. J. (2018). The relationship between modern information and communications technologies (ICTs) and development in Africa. *Utilities Policy*, *50*, 83–90. doi:10.1016/j.jup.2017.10.005

Norton, G. W., Alwang, J., & Masters, W. A. (2021). *Economics of agricultural development: world food systems and resource use*. Routledge. doi:10.4324/9780429316999

Okediran, O. O. (2019). An e-agriculture framework for inclusive agricultural value chains in Nigeria. Annals Computer Science Series, Tibiscus University of Timisoara, Romania, 17(2), 209-219.

Owolabi, O., Oku, A.-R. O., Alejo, A., Ogunbiyi, T., & Ubah, J. I. (2021). Access to electricity, information and communications technology (ICT), and financial development: Evidence from West Africa. *International Journal of Energy Economics and Policy*, *11*(2), 247–259. doi:10.32479/ijeep.10344

Partovifar, P. (2016). *Role of ICT in marketing of garden products* [MSc thesis]. Islamic Azad University.

Pedzisa, T. (2016). *Determinants of yield impact and adoption of conservation agriculture among smallholder farmers in Zimbabwe*. University of Pretoria.

Pedzisai, C. (2020). *An analysis of indigenous agricultural knowledge systems in the Zimbabwe secondary school agriculture curriculum: Prospects and opportunities*. Academic Press.

Piñeiro, V., Arias, J., Dürr, J., Elverdin, P., Ibáñez, A. M., Kinengyere, A., Opazo, C. M., Owoo, N., Page, J. R., Prager, S. D., & Torero, M. (2020). A scoping review on incentives for adoption of sustainable agricultural practices and their outcomes. *Nature Sustainability*, *3*(10), 809–820. doi:10.1038/s41893-020-00617-y

Quayson, M., Bai, C., & Osei, V. (2020). Digital inclusion for resilient post-COVID-19 supply chains: Smallholder farmer perspectives. *IEEE Engineering Management Review*, *48*(3), 104–110. doi:10.1109/EMR.2020.3006259

Ratinho, T., Amezcua, A., Honig, B., & Zeng, Z. (2020). Supporting entrepreneurs: A systematic review of literature and an agenda for research. *Technological Forecasting and Social Change*, *154*, 119956. doi:10.1016/j.techfore.2020.119956

Rola-Rubzen, M. F., Paris, T., Hawkins, J., & Sapkota, B. (2020). Improving gender participation in agricultural technology adoption in Asia: From rhetoric to practical action. *Applied Economic Perspectives and Policy*, *42*(1), 113–125. doi:10.1002/aepp.13011

Rose, D. C., Sutherland, W. J., Barnes, A. P., Borthwick, F., Ffoulkes, C., Hall, C., Moorby, J. M., Nicholas-Davies, P., Twining, S., & Dicks, L. V. (2019). Integrated farm management for sustainable agriculture: Lessons for knowledge exchange and policy. *Land Use Policy*, *81*, 834–842. doi:10.1016/j.landusepol.2018.11.001

Sánchez Prieto, J. C., Olmos Migueláñez, S., & García-Peñalvo, F. J. (2016). Informal toolsin formal contexts: Development of a model to assess the acceptance of mobile technologies among teachers. *Computers in Human Behavior*, *55A*, 519–528. doi:10.1016/j.chb.2015.07.002

Shah, D. (2022). Role of Information and Communication Technology in Agricultural Development of India. *Review of Market Integration*, *14*(2-3), 113–132. doi:10.1177/09749292221119286

Siankwilimba, E., Hiddlestone-Mumford, J., Hoque, M. E., Hang'ombe, B. M., Mumba, C., Hasimuna, O. J., Maulu, S., Mphande, J., Chibesa, M., Moono, M. B., Muhala, V., Cavaliere, L. P. L., Faccia, A., & Prayitno, G. (2023). Sustainability of agriculture extension services in the face of COVID-19: A study on gender-specific market systems. *Cogent Food & Agriculture*, *9*(2), 2284231. doi:10.1080/23311932.2023.2284231

Singh, A. K. (2013). Income and livelihood issues of farmers: A field study in Uttar Pradesh. *Agricultural Economics Research Review Conference No*, *26*, 89–96.

Smidt, H. J., & Jokonya, O. (2022). Factors affecting digital technology adoption by small-scale farmers in agriculture value chains (AVCs) in South Africa. *Information Technology for Development*, *28*(3), 558–584. doi:10.1080/02681102.2021.1975256

Tiwari, S. P. (2022). Information and communication technology initiatives for knowledge sharing in agriculture. *arXiv preprint arXiv:2202.08649*.

Wolfert, S., Verdouw, C., van Wassenaer, L., Dolfsma, W., & Klerkx, L. (2023). Digital innovation ecosystems in agri-food: Design principles and organizational framework. *Agricultural Systems*, *204*, 103558. doi:10.1016/j.agsy.2022.103558

Wong, S., & Guggenheim, S. (2018). Community-driven development: myths and realities. *World Bank Policy Research Working Paper* (8435).

Yadav, P., Davies, P. J., & Abdullah, S. (2018). Reforming capital subsidy scheme to finance energy transition for the below poverty line communities in rural India. *Energy for Sustainable Development*, *45*, 11–27. doi:10.1016/j.esd.2018.04.001

Chapter 8
Side Marketing:
A Threat to Contract Farming Viability in Zimbabwe

Paul Mukucha
Bindura University of Science Education, Zimbabwe

Divaries Cosmas Jaravaza
https://orcid.org/0000-0002-8930-1242
Bindura University of Science Education, Zimbabwe

Joshua Risiro
Great Zimbabwe, Zimbabwe

Trymore Chingwaru
Bindura University of Science Education, Zimbabwe

Phillip Dangaiso
Zimbabwe Ezekiel Guti University, Zimbabwe

Fungai Ngoma Mauchi
Bindura University of Science Education, Zimbabwe

ABSTRACT

Contract farming is increasingly becoming a major source of funding for most agricultural commodities on the basis of its ability to ensure a consistent supply of raw materials to agro processing firms. Drawing from the market imperfection theory and the information asymmetry theory, extant literature has documented a number of factors that contribute to side marketing, but is very limited on profiling the common demographic factors attributed to that practice. Therefore, this study sought to develop a comprehensive profile of contract farmers that are likely to side market. A sample of 400 tobacco contract farmers drawn from Bindura District was surveyed. The data was analysed using logistic regression analysis, and the results indicated that the men, youth, single, primary level educated, and communal farmers were more likely to side market than their counterparts in the same demographic groups. The study recommended that agro-processing firms should be cautious when they enter into farming contracts with the flagged demographic groups.

DOI: 10.4018/979-8-3693-2011-2.ch008

SIDE MARKETING: A THREAT TO CONTRACT FARMING VIABILITY IN ZIMBABWE

Contract farming is one of the means of financing agricultural activities, and in Zimbabwe it dates back to the colonial era (Shonhe & Scoones, 2021; Mazwi et al., 2018). However, it gained prominence after the controversial Fast Track Land Reform Programme (FTLRP), when the financial institutions drastically scaled back on financing agricultural activities (Mazwi et al., 2020). The FTLRP leveraged on the Land Reform Act which vested all the land rights to the state through revoking all the existing title deeds of commercial farmers and redistributed it to the majority of the landless peasants (Scoones et al., 2018). The peasants were not awarded the accompanying title deeds and as a result they could not possess the necessary surety needed by the financial institutions in order to qualify for accessing liquidity for financing their farming activities (Ngarava, 2020).

The beneficiaries of the land reform programme, were not only limited in terms of farming resources, but they also lacked agronomic skills needed for conducting lucrative farming activities (Mazwi et al., 2018). This lack of technical skills further alienated the new farmers from the financial institutions who could not fathom how the farmers would be able to repay their loans without the ability to viably conduct the farming activities (Mazwi et al., 2018). This left many agro processing merchants at the brink of insolvency due to operational challenges emanating from the less constant supply of raw materials from the upstream agricultural industry (Mukucha & Chari, 2022). As a response to this unfavourable development, the agro processing industry took upon itself to fund the agricultural industry through contract farming (Shonhe & Scoones, 2021). Contract farming covered the supply chain for several crops such as cotton (Mutambara and Mujeyi, 2021), tobacco (Pangapanga-Phiri et al., 2024; Mango & Kugedera., 2022), watermelon (White & Wijaya, 2022), palm oil (Hambloch, 2022; Pratama., et al., 2021; Vamuloh et al., 2020), cassava (Obi-Egbedi et al., 2022), wheat (Li et al., 2022), and small livestock such as poultry (Mukucha et al., 2024).

Contract farming did not ensure supply chain resilience for agro-processing firms (Mukucha & Chari, 2022), but also improved the livelihoods of small-scale farmers. The use of mainly small-scale suppliers for contract farming, although inevitable due to unavailability of and to large scale farmers (Mukucha et al., 2023), ensured that the agro-processing forms comply with social pillar of the sustainable supply chain management (Lysons & Farrington, 2020). More specifically the improvements on the livelihoods of the small-scale farmers were noticed in the form of increased income (Chen & Chen, 2021), employment generation among the less educated members of the community (Kaur et al., 2021; Meemken & Bellemare, 2020) and good agronomic practices (Shonhe & Scoones, 2021; Mazwi et al., 2018).

The viability of contract farming in Zimbabwe has been threatened by the emerging retrogressive practices among farmers such as side marketing (Shonhe & Scoones, 2021). Side-marketing, also known as side-selling (Otsuka et al., 2016), occurs when a farmer fails to comply with contract terms and chooses to sell his or her produce to another merchant that did not finance the harvest (Alemu et al., 2021). This challenge is faced by most agro-processing industries for cash crops such as cotton (Mutambara & Mujeyi, 2021), maize (Shonhe & Scoones, 2021), and more worrisome tobacco (Shonhe & Scoones, 2021). Several reasons for side marketing have been advanced in the extant literature such as exploitative nature of most of the contracts (Shonhe, 2019), and farmers' failure to meet the stringent quality demands of the merchants (Mazwi et al., 2018). However, missing in the extant literature are the profiles of the farmers that are inclined to side marketing. Therefore, this study seeks to develop a demographic profile of the people that are likely to practice side marketing.

The rest of the study is organized as follows; literature review of the predictive theories that account for side marketing in identified demographic groups such as the market imperfection theory, the opportunistic behaviour theory, and the information asymmetry theory. Literature review shall be followed by an outline of the methodology adopted for empirically evaluating the collected data, the presentation and discussion of the results, and finally the policy implications of the study.

LITERATURE REVIEW

Contract Farming

The Contract farming is a forward agreement between a farmer and a merchant to supply a specified agricultural commodity at a predetermined price (Eaton & Shepherd, 2001). Contract farming gained popularity in the agricultural industry against the background of underfunding of agricultural activities and market failures in most developing countries (Shonhe & Scoones, 2022). The necessity for contract farming arose as a result of the agrarian changes that took place during the Fast-Track Land Reform Program (FTLRP) that disposed most commercial farmers and left most of the peasants as the beneficiaries of the arable land (Mukucha et al., 2023). The new land ownership structure left most of the agro-processing firms with very limited supplies of agricultural commodities due to the undercapitalization of the new land owners (Mazwi et al., 2020). It therefore followed that contract farming emerged as the most viable funding alternative (Shonhe & Scoones, 2021).

Contract farming is generally necessitated by the unavailability of land for merchants to conduct estate farming for the provision of their own resources (Ruml

& Qaim, 2020). In some cases, contract farming is politically acceptable (Vicol, 2018) making it the only practical way to obtain raw materials (Ruml & Qaim, 2021). Furthermore, contract farming is necessitated by the unavailability of quality and consistent supply of raw materials from the spot markets (Mukucha et al., 2024; Mukucha & Chari, 2021).

Contract farming offers several benefits to the agro-processing merchants. These befits include uninterrupted supply of high-quality raw materials (Mukucha & Chari, 2022), reduction in transaction costs on the basis of sourcing from a carefully selected limited number of suppliers (Mazwi, 2020), and the ability to control the quality of the production outputs from the suppliers (Ruml & Qaim, 2020). Moreover, farming contracts reduces agro-processing merchants' procurement risks (Ruml & Qaim, 2020). This is achieved through transferring to the farmers all production risks such as droughts, diseases and pest outbreaks, and price volatilities (Mazwi et al., 2018). Thus, contract farming has been embraced in Zimbabwe as the main financing scheme for most agricultural produces notably tobacco (Mukucha & Chari, 2021), cotton (Mutambara & Mujeyi, 2021), sugarcane (Mazwi, 2020), horticultural products and maize (Mazwi *et al.*, 2019).

However, contract farming has been associated with negative outcomes that affected farmers such as turning farmers into quasi-employees and paupers (Ruml & Qaim, 2021), increased indebtedness (Shiringo et al., 2022), and diversion of resources from farming of food products leading to food insecurity (Mazwi et al., 2018). The agro processing firms also faced challenges with contract farming chief among them being inputs diversion (Mazwi et al., 2018), and side marketing that results in loan defaults (Shonhe & Scoones, 2022).

Theoretical Background for Side Marketing

Side marketing is one of the undesirable outcomes of contract farming since it constitutes a breach of contract (Mazwi, 2018). Contract farmers tend to resort to side marketing as a desperate response to unbearable supply chain risks transferred to them by the agro processing merchants (Mutambara & Mujeyi, 2021). There are several theoretical explanations that account for the practice of side marketing such as the market imperfection theory, opportunistic behaviour theory, and the information asymmetry theory.

Market Imperfections Theory

The market imperfection theory state that in most economic activities markets rarely function in a perfect manner due to different factors. The imperfect function of markets leads to inefficiencies of varying nature and severities as well as suboptimal outcomes (Kamoyo et al., 2015). More specifically, market imperfections in the farming industry are a function of price volatility, pests and disease outbreaks, and unpredictable weather conditions (Mutambara & Mujeyi, 2021). Drawing from this theory, side marketing happens as a result of market structure flaws (Dupa, 2022). Contract farming frequently entails the agro-processing merchant's concentration of power (Mutambara & Mujeyi, 2021), resulting in unequal bargaining power between buyers and producers (Mazwi et al., 2020). Farmers may resort to side marketing as a technique to counteract this power imbalance and gain higher prices for their produce (Mazwi et al., 2018). Farmers might perhaps negotiate better conditions with other agro-processing merchants (Mutambara & Mujeyi, 2020), or embrace higher prices in the open market by expanding their marketing channels (Koomson et al., 2022).

Opportunistic Behaviour Theory

The opportunistic behaviour theory states that individuals exploit advantages for personal gains (Granja & Wollni, 2018). Farmers' opportunistic behaviour, according to this view, drives side marketing. Farmers may participate in side marketing if they believe the benefits of violating the contract outweigh the potential costs or penalties (Ikeda & Natawidjaja, 2022). Higher selling prices offered by alternative agro-processing merchants, unhappiness with contract terms (Mutambara & Mujeyi, 2021), or a lack of trust in the contractor's capacity to fulfil its responsibilities are some of the factors influencing the decision by farmers to indulge in side marketing. Farmers may also engage in side marketing if they believe the risk of contract defaulting detection or enforcement is minimal (Ikeda & Natawidjaja, 2022).

Information Asymmetry Theory

Information asymmetry theory describes the knowledge imbalance between the parties involved in a transaction (Akerlof, 1970). Information asymmetry often leads to uneven power distribution (Tang et al., 2022). Generally, in farming business information asymmetry has been found to lead to a bottleneck in the controlling of epidemics such as disease outbreaks (Si, Zhang, Yao, Liu & Lu, 2021). In the context of contract farming, asymmetry in information between merchants and farmers may contribute to side marketing (Tang et al., 2022). Farmers may participate in side marketing if they have better market information or believe the merchant is not completely aware of current market conditions (Ton t al., 2018; Maertens & Velde, 2017; Mwambi et al., 2016; Zhang, 2012). Farmers who take advantage of this information advantage may be able to achieve greater prices or better terms through side marketing (Mazwi et al., 2018).

Hypotheses Development

Age

Age refers to the passage of time that a human being has been in existence since birth (Roebuck, 1979). It is generally measured in chronological progression and is often marked by the stages one has gone through (Rose et al., 2012; Clark-Kazak, 2009). The farmers can be categorised as youth or mature. Youthful farmers are those adults below the age of 40 years, while mature farmers are those that are 40 years and above. Most behaviour of individuals in various economic sectors have been a function of age. One such behaviour is side marketing.

Side marketing is common among the two age groups (Koomson et al., 2022), but drawing from the opportunistic behaviour theory it is likely that the youth practice side marketing more than the mature farmers. The youth tend to be exposed to a lot of information through various platforms such as the internet and social media platforms such as WhatsApp where ubiquitous information about trading opportunities is easily available (Gibson, 2022). It is through this ubiquitous information that the youthful farmers may attain information about better marketing opportunities than the ones they are currently offered in their contract terms. This may be tempting enough to drive them into side marketing. Therefore, it is plausible to suggest that;

H1: *Farmers' age predicts the practice of side marketing.*

Gender

Gender refers to the behaviour and expectations emanating from social and cultural roles along the basis of sex (Lindqvist et al., 2000). Gender plays a very important role in segmenting markets and understanding general consumer behaviour (Kennelly et al., 2001). Likewise gender has also been influential in other sectors such as labour and entrepreneurial practices, as well as business operations (Conroy & Weiler, 2016; Neergaard et al., 2005; Elizabeth & Baines, 1998).

Contract farming has of late been gender inclusive as many government policies to address gender disparities are being implemented across all the sectors of the economy (Mazwi et al., 2020; Mazwi et al., 2019). However, the practices of side marketing within contract farming arrangements are likely to vary on the basis of gender. Drawing from the opportunistic behaviour theory males are likely to practice side marketing than their female counterparts. Males are generally keen to grab opportunities than females especially in male dominated societies like Zimbabwe (Mazwi, 2020). In the cases where there are opportunities for gaining more it is likely that male farmers would side market, while their female counterparts due to general conformance to fidelity issues among general life issues, may stick to the contract terms. It is therefore anticipated that;

H2: *Farmers' gender predicts the practice of side marketing.*

Education

Education refers to skills, and knowledge acquisition through either training, research or teaching in either formal; or informal ways (Maher, 2011). The level of education has a bearing on various behaviors across various disciplines (Günther et al., 2022). There are several ways in the extant literature of categorizing the levels of education, but in Zimbabwe, education is categorized into primary, secondary and tertiary levels.

There is likely to be a differentiation in the practice of side marketing on the basis of the levels of education. Drawing from the information asymmetry theory it is likely that the most educated farmers are privy to the trading information in terms of the spot market prices and the other channels of reaching the market. They are therefore likely to be tempted with the better prices that may be offered by the spot markets and end up practicing side marketing (Mazwi et al., 2018). In contrast, the least educated farmers may have less exposure to the trending practices in the market. Due to that ignorance, they may stick to their original contracts thinking that they may be the best available option in the market. It is therefore likely that side marketing would be high among those with tertiary education and decreases as we approach to the primary level of education. It is therefore projected that;

H3: *Farmers' level of education predicts the practice of side marketing.*

Marital Status

Marital status has been defined from a multiplicity of angles such as religious, traditional, and legal. Traditionally a marriage exists when a man pays a bride price for a woman (Chisango et al., 2022), religiously marriage comes into existence when a man and a woman are joined into a holy matrimony before an ordained priest (Mugweni et al., 2012; Ngundu, 2008), and legally marriage comes into being when a man and a woman who are not related decides to share the same bedroom (Schmidt, 1990) including in the cases of co-habiting (Mujuzi, 2023a, Mujuzi, 2023b).

In contract farming, side marketing is likely to be high among the single farmers than the married farmers. Single farmers are highly opportunistic and as a result they may explore several options to maximize their economic gains. In the case of married people there is a lot of restraint among the couples that may inhibit the wayward behaviour of seeking unorthodox gains from practices (Bartley et al., 2005; Sillars & Kalbflesch, 1989) such as side marketing. It is therefore suggested that;

H4: *Farmers' marital status predicts the practice of side marketing.*

Landholding Structure

Land holding structure refers to the ownership levels that farmers have on the land that they till (Scoones et al., 2022). The most common land holding structures in Zimbabwe are communal, small scale, and small-scale commercial farms (Mukucha et al., 2023). These landholding structures have differential land sizes (Shonhe et al., 2022), access to resources (Scoones et al., 2022), and demand for farming support through contract farming arrangements (Mutambara & Mujeyi, 2021).

It is anticipated that the land holding structure predicts the side marketing practices of contracted farmers. Side marketing practice is likely to be high among the communal farmers since these farmers are mostly former peasant farmers. These farmers are associated with poor agronomic practices and low-quality soil structures that makes their yields inherently very low (Mutambara & Mujeyi, 2021). Due to lower yields that may not enable them to recoup their investments as well as their peasantry background (Mazwi et al., 2020), they are likely to indulge in opportunistic behaviour so as to maximize their gains. It is therefore anticipated that;

H5: *Farmers' landholding structure predicts the practice of side marketing.*

METHODOLOGY

Population and Sampling

The population for this study were contract farmers in the Mashonaland Central province of Zimbabwe. Bindura district is one of the largest tobacco producing region in the country due to its favourable agro-ecological climate that is suitable for commercial tobacco farming (Ruckert et al., 2022). There were around 47000 registered tobacco farmers in Mashonaland Central province of Zimbabwe in 2023. A sampling frame was initially sought from tobacco contractors in the province, but none of them was willing to divulge the personal details of their clients. This was despite the researchers formally approaching them and vividly expressing that the information was required solely for academic purposes.

After failing to get the sampling frames from the agro-processing merchants, the researchers directly approached the farmers using the using the snowballing sampling technique. Snowballing sampling involves getting the initial respondent and then use that respondent to gain access to the other respondents (Saunders et al., 2018). Although this method had a weakness in that it does not give each potential respondent an equal opportunity for inclusion in the study (Bryman, 2016), it was the most viable option for most conservative societies in Africa where participation is surveys is viewed with great scepticism (Kanyanda et al., 2021; Renn, Hoeft, Lee et al., 2019; Zimbalist, 2018).

Data Collection Procedures

Data was collected from the contracted tobacco farmers in the province of Mashonaland central. The selected farmers were contacted at their homesteads. The homesteads were chosen because they are far away from the auction floors or the offices of merchants. Proximity to the merchants increases the risk of non-participation by the contracted farmers or worse still they may not reveal the information that they deem to be unpleasant to their partnering merchants. In collecting data some ethical considerations were observed. For instance, farmers were requested to fill the consent form before they partook in the study. Furthermore, they were assured that the data was being collected specifically for academic purposes. Confidentiality of the data was enhanced by leaving out individual particulars on the data collection instrument.

Data Analysis Procedures

Data analysis was conducted using logistic regression analysis. Logistic regression analysis is a statistical tool for modelling the relationship between a single or multiple independent variables and a binary dependent variable (Hair, Jr et al., 2020; Hair et al., 2019). Logistic regression analysis estimates the probability of a certain event occurring based on the variability of the values of the independent variables (Field, 2018).

RESULTS

Sample Characteristics

The demographic characteristics of the surveyed sample comprised of age, marital status, and level of education as shown in Table 1.

The sample consisted mostly of the farmers in the 40-49 age group (26.3%), followed by those in the 50-59 age group (24.8%), and lastly the 18-29 age group (9.0%). Most of the surveyed farmers were married (52.0%). The majority of the respondents had some primary education qualification (38.0%), followed by those with a secondary education qualification (34.3%), and lastly a tertiary education qualification (27.8%). Lastly, the surveyed farmers from communal lands were 53.0%, and smallholder lands (47.0%).

Table 1. Demographics

Category	Frequency	Percent	Valid Percent
18-29	36	9.0	9.0
30-39	66	16.5	16.5
40-49	105	26.3	26.3
50-59	99	24.8	24.8
60+	94	23.5	23.5
Male	228	57.0	57.0
Female	172	43.0	43.0
Primary	152	38.0	38.0
Secondary	137	34.3	34.3
Tertiary	111	27.8	27.8
Single	192	48.0	48.0
Married	208	52.0	52.0
Communal	212	53.0	53.0
Smallholder	188	47.0	47.0
Non side-marketing	235	58.8	58.8
Side-marketing	165	41.3	41.3
Total	**400**	**100**	**100**

Hypotheses Testing

A logistic regression was performed to ascertain the influence of various contract farmer demographics on the side-marketing practice. More specifically H^1 predicted that incidences of side-marketing increase as the focus moves from the mature farmers to the youthful farmers, H^2 suggested that side-marketing is the function of age, H^3 predicted side-marketing to be influenced by the level of education, H^4 suggested that side-marketing practice is linked to marital status, while H^5 anticipated side-marketing to be a function of the land holding structure. Prior to the analysis of the statistical significance of the hypothesized relationship, the overall model was analysed for model fit using the Omnibus Tests of Model Coefficients. The results from that test are shown in Table 2.

The results shown in Table 2 indicated that the binary logistic model was statistically significant, $\chi2$ (5) = 151.121, $p < .001$. This indicated the validity of the results that would be attained from hypotheses testing. The logistic regression model was further assessed using the Hosmer and Lemeshow goodness of fit statistical test. The results are shown in Table 3.

Table 2. Omnibus tests of model coefficients

Chi-square	df	Sig.
151.121	5	.000
151.121	5	.000
151.121	5	.000

Table 3. Hosmer and Lemeshow test

Chi-square	df	Sig.
3.075	8	.930

The Hosmer and Lemeshow test further provided some evidence that the model was fit. The Hosmer and Lemeshow statistics indicated the absence of differences between the predicted model and the observed model, $\chi 2$ (8) = 3.075, p = .930. The amount of variance explained by the model was then assessed using the Nagelkerke R Square and the results are shown in Table 4.

The results shown in Table 4 as indicated by the Nagelkerke R^2 coefficient revealed that the model explained 42.4% of the variance in side-marketing practices. The correctness of the cases classification was then assessed and displayed in Table 5.

The results shown in Table 5 indicated that the logistic regression analysis model correctly classified 75.3% of the cases. Having ensured that the cases were correctly and accurately classified on a higher frequency, the analysis proceeded to assess the statistical significance of the relationships specified in the study's hypotheses. The results from hypotheses testing are shown in Table 6.

The results from binary the logistic regression analysis shown in Table 6 revealed that side-marketing practices are a function of contract farmers' age, B = -696, p < .001, gender, B= -1.436, p < .001, education, B = -.598, p < .001, marital status, B = -1.443, p < .001, and landholding structure, B = -.717, p < .001. The odds ratios indicated that side marketing was likely to be practised by the youthful farmers more

Table 4. Model summary

-2 Log likelihood	Nagelkerke R Square
391.084[a]	.424

a. Estimation terminated at iteration no. 5 since parameter estimates changed by less than .001.

Table 5. Classification table

Observed		Predicted		
		Side Marketing		Percentage Correct
		Non-Side-Marketing	Side-Marketing	
Side marketing	Non-Side marketing	186	49	79.1
	Side-marketing	50	115	69.7
Overall percentage				75.3

a. The cut value is .500

Table 6. Variables in the equation

Variable	B	S.E.	Wald	df	Sig.	Exp(B)	95% C.I.for EXP(B)	
							Lower	Upper
Age	-.696	.108	41.446	1	.000	.499	.403	.616
Gender	-1.436	.267	29.035	1	.000	.238	.141	.401
Education	-.598	.159	14.083	1	.000	.550	.402	.751
Marital status	-1.443	.258	31.346	1	.000	.236	.143	.392
Landholding structure	-.717	.194	13.582	1	.000	.488	.334	.715
Constant	8.435	.924	83.261	1	.000	4606.022		

a. Variable(s) entered on step 1: Age, Gender, Education, Marital status, Landholding structure.

than the mature farmers, OR = .499, 95%, CI [.403, .616], male farmers more than female farmers, OR = .238, 95%, CI [.141, .401], the less educated more than the educated farmers, OR = .550, 95%, CI [.402, .751], the unmarried farmers more than the unmarried farmers, OR = .236, 95%, CI [.143, .392], and the communal farmers more than the smallholding farmers, OR = .488, 95%, CI [.334, .715].

DISCUSSION

The study was conducted to determine the demographics of the tobacco contract farmers that are likely to practice side marketing. This is against the background of side marketing emerging as one of the threats against the sustainability of contract farming schemes in general, and more specifically the tobacco sector. This is likely due to the fact that side marketing is a breach of contract. Breaches in contract are likely to lead to supply chain disruptions as the flow of goods would be less than the planned rate (Mukucha & Chari, 2022).

The results revealed that the youth are more likely to practice side marketing than the mature farmers. This likely to be attributed to the youth being highly adventurous and risk taking (Brymer & Feletti, 2020; Lee & Tseng, 2015; Schlegelmilch & Ollenburg, 2013). Risk taking in general can be for positive or negative adventures. In the case of side marketing the youth are likely to take the risks that increase their incomes despite the legal and economic consequences that follow. This is in sync with previous research by Koomson et al. (2022) that found out that as the farmers grow in age, they tend to do less of side marketing.

The results also indicated that males in the tobacco farming industry are more likely to practice side marketing than their female counterparts. This may be attributed to several reasons such as higher levels of risk taking among the males (Boniface, 2006) and bearing the traditional burdens of being bread winners (Gonalons-Pons & Gangl, 2021). On the other side women were found to be less prone to loan defaults through side marketing. Perhaps this can be accounted for by the fact that traditionally women were less involved in contract farming activities (Mazwi et al., 2019). The idea that women are still new to farming, they may not have made deliberate decisions to exploit the existing loopholes.

The level of education of contracted farmers was also found to contribute to the retrogressive practice of side marketing. The less educated farmers were found to practice side marketing more than their educated counterparts. This may be attributed to the fact that the less educated may not be appreciative of the importance of honouring contracts since they may rarely understand the contents of most of the contracts they sign. This is particularly so because the contracts are written in English which most farmers particularly those from the communal areas may not appreciate (Mazwi et al., 2018). Moreover, even where the contracted farmers understand English, they may still not comprehend the legal jargons that characterise most of the contracts (Mazwi et al., 2018). Therefore, they may default without appreciating the attendant consequences.

The results also indicated that side marketing practice is low among the married farmers than their unmarried counterparts. This result confirms the validity of the suggestion that single farmers are more opportunistic than their married counterparts. Family persons are generally cautious in their decision making and their behaviours since the resultant consequences of their actions have a bearing on the entire family (Bartley et al., 2005). In the event of a merchant embarking on loan recovery strategies such as asset seizures (Mutambara & Mujeyi, 2021) the whole family is likely to suffer and the defaulting farmer may face the unbearable process of being blamed by his close family members (Kassegn & Endris, 2022).

Finally, side marketing was found to be a function of landholding structure, with communal farmers being highly likely to indulge in side marketing than the farmers with other forms of landholdings like A1 and A2 farmers. Generally, land productivity in communal areas is very low due to poor soil structure and primitive agronomic practices (Mukucha et al., 2023; Mutambara & Mujeyi, 2021). Poor yields may entail that all the proceeds may go towards settling the debt leaving the farmer with little or no income for survival (Mazwi et al., 2018). This scenario is likely to lead communal farmers into side marketing in order to augment their residual income (Kamoyo et al., 2018).

POLICY IMPLICATIONS

The results revealed that side marketing is high among the youth, males, tertiary educated, unmarried, and communal farmers. This implies that tobacco merchants may reduce the risk of side marketing through carefully choosing the right type of farmers to enter into contracts with. This may mean taking additional measures when dealing with high-risk farmers such as the males, youth, and unmarried individuals. Where possible tobacco merchants must contract females. In cases where there are limited numbers of interested females, as has been the case in the past, then males can be contracted on condition that they jointly enter into contracts with their female spouses. This brings in the element of giving preferences to married individuals as supported by the results in this study. Married people tend to give checks and balances to each other especially when one of them and in most cases males, is tempted to practice side marketing. Lastly, despite that the future lies with the youth, extraordinary measures should be put in place when entering into farming contracts with the youth. It would be prudent to include the parents of the youth as guarantors for the contracts signed. These guarantors would always ensure that their children conform to the signed contracts.

The other policy implication would be to avoid contracting communal in favour of smallholder farmers where possible. However, this option is severely curtailed by the political and moral obligations to the communal farmers who are clearly the majority of the citizens in the country. It would be a political suicide for any government to overlook any corporate policy that side-lines the communal farmers. Moreover, the corporate social responsibilities of most business entities include alleviating the economic well-being of the disadvantaged societies such as communal farmers. Therefore, agro-processing merchants would need to implement extensive supplier development initiatives with communal farmers beyond the traditional contract farming.

REFERENCES

Akerlof, G. A. (1970). 4. The market for 'lemons': Quality uncertainty and the market mechanism. *The Quarterly Journal of Economics*, *66*(3), 488. Advance online publication. doi:10.2307/1879431

Alemu, D., Guinan, J., & Hermanson, J. (2021). Contract farming, cooperatives and challenges of side selling: Malt barley value-chain development in Ethiopia. *Development in Practice*, *31*(4), 496–510. doi:10.1080/09614524.2020.1860194

Bartley, S. J., Blanton, P. W., & Gilliard, J. L. (2005). Husbands and Wives in Dual-Earner Marriages: Decision-Making, Gender Role Attitudes, Division of Household Labor, and Equity. *Marriage & Family Review*, *37*(4), 69–94. doi:10.1300/J002v37n04_05

Boniface, M. (2006). The meaning of adventurous activities for 'women in the outdoors. *Journal of Adventure Education and Outdoor Learning*, *6*(1), 9–24. doi:10.1080/14729670685200711

Brymer, F., & Feletti, F. (2020). Beyond risk: The importance of adventure in the everyday life of young people. *Annals of Leisure Research*, *23*(3), 429–446. doi:10.1080/11745398.2019.1659837

Chen, J., & Chen, Y. J. (2021). The impact of contract farming on agricultural product supply in developing economies. *Production and Operations Management*, *30*(8), 2395–2419. doi:10.1111/poms.13382

Chisango, T., Mafa, I., & Maunganidze, L. (2022). Investigating the Sexist Implications of Bride Price (Lobola) in Zimbabwe. *Society*, *59*(5), 564–575. doi:10.1007/s12115-022-00743-7

Clark-Kazak, C. R. (2009). Towards a working definition and application of social age in international development studies. *The Journal of Development Studies*, *45*(8), 1307–1324. doi:10.1080/00220380902862952

Conroy, T., & Weiler, S. (2016). Does gender matter for job creation? Business ownership and employment growth. *Small Business Economics*, *47*(2), 397–419. doi:10.1007/s11187-016-9735-8

Dupa, H. J. P. (2022). Reasons of Engagement and Impact of Contract Farming to Smallhold Farmers: A Literature Review. *International Journal of Advanced Research in Economics and Finance*, *4*(1), 56–60. https://myjms.mohe.gov.my/index.php/ijaref/article/view/17435

Elizabeth, C., & Baines, S. (1998). Does gender affect business 'performance? A study of microbusinesses in business services in the UK. *Entrepreneurship and Regional Development, 10*(2), 117–135. doi:10.1080/08985629800000007

Field, A. (2018). *Discovering Statistics Using IBM SPSS Statistics*. SAGE Publications Ltd.

Gibson, K. (2022). Bridging the digital divide: Reflections on using WhatsApp instant messenger interviews in youth research. *Qualitative Research in Psychology, 19*(3), 611–631. doi:10.1080/14780887.2020.1751902

Gonalons-Pons, P., & Gangl, M. (2021). Marriage and Masculinity: Male-Breadwinner Culture, Unemployment, and Separation Risk in 29 Countries. *American Sociological Review, 86*(3), 465–502. doi:10.1177/00031224211012442 PMID:34149053

Günther, J., Overbeck, A. K., Muster, S., Tempel, B. J., Schaal, S., Schaal, S., Kühner, E., & Otto, S. (2022). Outcome indicator development: Defining education for sustainable development outcomes for the individual level and connecting them to the SDGs. *Global Environmental Change, 74*, 102526. doi:10.1016/j.gloenvcha.2022.102526

Hair, J. F., Black, W. C., Babin, B. J., & Anderson, R. E. (2019). *Multivariate data analysis* (8th ed.). Cengage.

Hair, J. F. Jr, Page, M., & Brunsveld, N. (2020). *Essentials of Business Research Methods (4th ed.)*. Routledge.

Hambloch, C. (2022). Contract farming and everyday acts of resistance: Oil palm contract farmers in the Philippines. *Journal of Agrarian Change, 22*(1), 58–76. doi:10.1111/joac.12462

Ikeda, S., & Natawidjaja, R. S. (2022). The sustainability of contract farming with specialised suppliers to modern retailers: Insights from vegetable marketing in Indonesia. *Agriculture, 12*(3), 380. doi:10.3390/agriculture12030380

Kamoyo, M., Muranda, Z., & Chikuya, T. (2015). Agricultural export crop participation, contract farming and rural livelihood in Zimbabwe: The case of cotton farming in Rushinga District. IOSR. *Journal of Economics and Finance, 6*(6), 110–120. doi:10.9790/5933-0661110120

Kanyanda, S., Markhof, Y., Wollburg, P., & Zezza, A. (2021). Acceptance of COVID-19 vaccines in sub-Saharan Africa: Evidence from six national phone surveys. *BMJ Open, 11*(12), e055159. doi:10.1136/bmjopen-2021-055159 PMID:34911723

Kassegn, A., & Endris, E. (2022). Factors affecting loan repayment rate among smallholder farmers got loans from the Amhara Credit and Saving Institution: In the case of Habru District, Amhara Regional State, Ethiopia. *International Area Studies Review*, *25*(1), 73–96. doi:10.1177/22338659211040993

Kaur, P., Singla, N. & Singh, S. (2021). Role of Contract Farming in Crop Diversification and Employment Generation: Empirical Evidence from Indian Punjab. *Millennial Asia*, *12*(3), 350-366. . doi:10.1177/09763996211051300

Kennelly, I., Merz, S. N., & Lorber, J. (2001). What Is Gender? *American Sociological Review*, *66*(4), 598–605. doi:10.1177/000312240106600407

Koomson, J. E., Donkor, E., & Owusu, V. (2022). Contract scheme for rubber production in Western region of Ghana: Why do farmers side sell? *Forests, Trees and Livelihoods*, *31*(3), 139–152. doi:10.1080/14728028.2022.2079007

Lee, T. H., & Tseng, C. H. (2015). How personality and risk-taking attitude affect the behavior of adventure recreationists. *Tourism Geographies*, *17*(3), 307–331. do i:10.1080/14616688.2014.1000955

Li, J., Qing, P., Hu, W., & Li, M. (2022). Contract farming, community effect, and farmer valuation of biofortified crop varieties in China: The case of high-zinc wheat. *Review of Development Economics*, *26*(2), 1035–1055. doi:10.1111/rode.12847

Lindqvist, A., Sendén, M. G., & Renström, E. A. (2021). What is gender, anyway: A review of the options for operationalising gender. *Psychology and Sexuality*, *12*(4), 332–344. doi:10.1080/19419899.2020.1729844

Lysons, K. & Farrington, B. 2020. *Procurement and supply chain management.* Pearson UK.

Maertens, M., & Velde, K. V. (2017). Contract-farming in Staple Food Chains: The Case of Rice in Benin. *World Development*, *95*, 73–87. doi:10.1016/j.worlddev.2017.02.011

Maher, T. C. (2011). Defining Education. *The Journal of Continuing Higher Education*, *59*(2), 108–109. doi:10.1080/07377363.2011.583184

Mango, L., & Kugedera, A. T. (2022). Pitfalls and prospects of contract farming to smallholder tobacco farmers in Shamva District, Zimbabwe. *International Journal of Agriculture Environment and Food Sciences*, *6*(1), 189–196. doi:10.31015/jaefs.2022.1.24

Mazwi, F. (2020). Sugar production dynamics in Zimbabwe: An analysis of contract farming at Hippo Valley. *Review of African Political Economy*, *47*(166), 568–584. doi:10.1080/03056244.2020.1832022

Mazwi, F., Chambati, W., & Mudimu, G. T. (2020). Tobacco Contract Farming in Zimbabwe: Power Dynamics, Accumulation Trajectories, Land Use Patterns and Livelihoods. *Journal of Contemporary African Studies*, *38*(1), 55–71. doi:10.1080/02589001.2020.1746752

Mazwi, F., Chambati, W., & Mutodi, K. (2018). *Contract farming arrangement and poor resourced farmers in Zimbabwe. Policy brief. Sam Moyo African Institute for agrarian studies*. SMAIAS.

Mazwi, F., Chemura, A., Mudimu, G. T., & Chambati, W. (2019). Political Economy of Command Agriculture in Zimbabwe: A State-led Contract Farming Model. *Agrarian South - the Journal of Political Economy*, *8*(1–2), 232–257. doi:10.1177/2277976019856742

Meemken, E. M., & Bellemare, M. F. (2020). Smallholder farmers and contract farming in developing countries. *Proceedings of the National Academy of Sciences of the United States of America*, *117*(1), 259–264. doi:10.1073/pnas.1909501116 PMID:31836695

Mugweni, E., Pearson, S., & Omar, M. (2012). Traditional gender roles, forced sex and HIV in Zimbabwean marriages. *Culture, Health & Sexuality*, *14*(5), 577–590. doi:10.1080/13691058.2012.671962 PMID:22472019

Mujuzi, J. D. (2023). Distribution of property at the termination of *de facto* unions (marriages by cohabitation/repute) in some African countries. *International Journal of Law, Policy and the Family*, *37*(1), ebad015. Advance online publication. doi:10.1093/lawfam/ebad015

Mujuzi, J. D. (2023). Marriage by Cohabitation (Common Law Marriage) in Seychelles: Emerging Issues. In Cohabitation and the Evolving Nature of Intimate and Family Relationships. Emerald Publishing Limited. doi:10.1108/S1530-35352023009

Mukucha, P. & Chari, F. (2021). The influence of supplier development, in the form of contract farming, on performance in Zimbabwean tobacco industry. *Cogent Business & Management*, *8*(1). . doi:10.1080/23311975.2021.1968732

Mukucha, P., & Chari, F. (2022). Supply chain resilience: The role of supplier development in the form of contract farming in fast-food outlets in Zimbabwe. *Continuity & Resilience Review*, *4*(3), 280–299. doi:10.1108/CRR-03-2022-0006

Mukucha, P., Chari, F., & Manzini, S. (2023). An assessment of differential supplier category performance in the agro processing industry: A case for supplier development in the Zimbabwean tobacco industry. *Journal of Future Sustainability, 3*(4), 223–232. doi:10.5267/j.jfs.2023.2.003

Mukucha, P., Tsekea, S., Jaravaza, D. C. & Jaravaza, N. (2024). Contract farming: a comparison of production and marketing contracts as sourcing strategies in the fast food restaurant industry. *Journal of Foodservice Business Research*, 1-18. https://doi.org/. doi:10.1080/15378020.2024.2303540

Mutambara, J., & Mujeyi, K. (2021). Enhancing competitiveness of Zimbabwe's cotton production under contract farming. *African Journal of Science, Technology, Innovation and Development, 13*(3), 283–291. doi:10.1080/20421338.2020.1744270

Mwambi, M. M., Oduol, J., Mshenga, P., & Saidi, M. (2016). Does contract farming improve smallholder income? The case of avocado farmers in Kenya. *Journal of Agribusiness in Developing and Emerging Economies, 6*(1), 2–20. doi:10.1108/JADEE-05-2013-0019

Neergaard, H., Shaw, E., & Carter, S. (2005). The impact of gender, social capital and networks on business ownership: A research agenda. *International Journal of Entrepreneurial Behaviour & Research, 11*(5), 338–357. doi:10.1108/13552550510614999

Ngarava, S. (2020). Impact of the fast-track land reform programme (FTLRP) on agricultural production: A tobacco success story in Zimbabwe? *Land Use Policy, 99*, 105000. doi:10.1016/j.landusepol.2020.105000

Ngundu, O. A. (2008). *Mission churches and African marriage in Southern Rhodesia 1890-1970: contemporary African Christian dilemmas over marriage in historical and theological perspective* [Doctoral dissertation]. University of Cambridge. doi:10.17863/CAM.100699

Obi-Egbedi, O., Osanyinlusi, O. I., & Adekunle, O. B. (2022). Does contract farming really matter in cassava farms productivity in Iseyin Local Government Area, Oyo State, Nigeria? *Agricultura Tropica et Subtropica, 55*(1), 65–73. doi:10.2478/ats-2022-0008

Otsuka, K., Nakano, Y., & Takahashi, K. (2016). Contract farming in developed and developing countries. *Annual Review of Resource Economics, 8*(1), 353–376. doi:10.1146/annurev-resource-100815-095459

Pangapanga-Phiri, I., Mungatana, E., & Mhondoro, G. (2024). Does contract farming arrangement improve smallholder tobacco productivity? Evidence from Zimbabwe. *Heliyon*, *10*(1), e23862. Advance online publication. doi:10.1016/j.heliyon.2023.e23862 PMID:38205304

Pratama, I. P., Winarso, H., Hudalah, D., & Syabri, I. (2021). Extended urbanization through capital centralization: Contract farming in palm oil-based agroindustrialization. *Sustainability (Basel)*, *13*(18), 10044. doi:10.3390/su131810044

Pryzgoda, J., & Chrisler, J. C. (2000). Definitions of gender and sex: The subtleties of meaning. *Sex Roles*, *43*(7/8), 553–569. doi:10.1023/A:1007123617636

Renn, B. N., Hoeft, T. J., Lee, H. S., Bauer, A. M., & Areán, P. A. (2019). Preference for in-person psychotherapy versus digital psychotherapy options for depression: Survey of adults in the US. *NPJ Digital Medicine*, *2*(1), 6. doi:10.1038/s41746-019-0077-1 PMID:31304356

Roebuck, J. (1979). When does "old age begin? The evolution of the English definition. *Journal of Social History*, *12*(3), 416–428. doi:10.1353/jsh/12.3.416

Rose, M. R., Flatt, T., Graves, J. L., Greer, L. F., Martínez, D. E., Matos, M., Mueller, L. D., Shmookler Reis, R. J., & Shahrestani, P. (2012). What is aging? *Frontiers in Genetics*, *3*, 134. doi:10.3389/fgene.2012.00134 PMID:22833755

Ruckert, A., Ciurlia, D., Labonte, R., Lencucha, R., Drope, J., Nhamo, N., Kadungure, A., & Mlambo, Z. (2022). *The Political Economy of Tobacco Production and Control in Zimbabwe: A Document Analysis*. https://tobacconomics.org/files/research/763/political-economy-of-tobacco-production-and-control-in-zimbabwe-draft-6-final.pdf

Ruml, A., & Qaim, M. (2020). Effects of marketing contracts and resource providing contracts in the African small farm sector: Insights from oil palm production in Ghana. *World Development*, *136*, 105110. doi:10.1016/j.worlddev.2020.105110

Ruml, A., & Qaim, M. (2021). New evidence regarding the effects of contract farming on agricultural labor use. *Agricultural Economics*, *52*(1), 51–66. doi:10.1111/agec.12606

Saunders, M., Lewis, P., & Thornhill, A. (2018). *Research Methods for Business Students* (8th ed.). Pearson Education Limited.

Schlegelmilch, F., & Ollenburg, C. (2013). Marketing the adventure: Utilizing the aspects of risk/fear/thrill to target the youth traveller segment. *Tourism Review*, *68*(3), 44–54. doi:10.1108/TR-03-2013-0010

Schmidt, E. (1990). Negotiated spaces and contested terrain: Men, women, and the law in colonial Zimbabwe, 1890–1939. *Journal of Southern African Studies, 16*(4), 622–648. doi:10.1080/03057079008708254

Scoones, I., Mavedzenge, B., Murimbarimba, F., & Sukume, C. (2018). Tobacco, contract farming, and agrarian change in Zimbabwe. *Journal of Agrarian Change, 18*(1), 22–42. doi:10.1111/joac.12210

Shiringo, L., Chundu, M., & Sithole, P. (2022). Contributing of smallholder tobacco contract farming to economic and human development of rural communities in Zimbabwe: A case of Igava community, Marondera District. *Open Journal of Social Sciences, 10*(9), 36–57. doi:10.4236/jss.2022.109004

Shonhe, T., & Scoones, I. (2021). Private and state-led contract farming in Zimbabwe: Accumulation, social differentiation and rural politics. *Journal of Agrarian Change, 22*(1), 118–138. doi:10.1111/joac.12473

Si, R., Zhang, X., Yao, Y., Liu, L., & Lu, Q. (2021). Influence of contract commitment system in reducing information asymmetry, and prevention and control of livestock epidemics: Evidence from pig farmers in China. *One Health, 13*, 100302. doi:10.1016/j.onehlt.2021.100302 PMID:34430696

Sillars, A. L., & Kalbflesch, P. J. (1989). Implicit and Explicit Decision-Making Styles in Couples. In D. Brinberg & J. Jaccard (Eds.), *Dyadic Decision Making*. Springer. doi:10.1007/978-1-4612-3516-3_8

Tang, C., Zhang, H., & Xie, J. (2022). Optimal contract design in contract farming under asymmetric effort information. *Sustainability (Basel), 14*(22), 15000. doi:10.3390/su142215000

Ton, G., Vellema, W., Desiere, S., Weituschat, S., & D'Haese, M. (2018). Contract farming for improving smallholder incomes: What can we learn from effectiveness studies? *World Development, 104*, 46–64. doi:10.1016/j.worlddev.2017.11.015

Vamuloh, V. V., Kozak, R. A., & Panwar, R. (2020). Voices unheard: Barriers to and opportunities for small farmers' participation in oil palm contract farming. *Journal of Cleaner Production, 275*, 121955. doi:10.1016/j.jclepro.2020.121955

Vicol, M. (2018). Potatoes, petty commodity producers and livelihoods: Contract farming and agrarian change in Maharashtra, India. *Journal of Agrarian Change, 19*(1), 135–161. doi:10.1111/joac.12273

White, B., & Wijaya, H. (2022). What kind of labour regime is contract farming? Contracting and sharecropping in Java compared. *Journal of Agrarian Change*, *22*(1), 19–35. doi:10.1111/joac.12459

Zhang, Q. F. (2012). The Political Economy of Contract Farming in China's Agrarian Transition. *Journal of Agrarian Change*, *12*(4), 460–483. doi:10.1111/j.1471-0366.2012.00352.x

Zimbalist, Z. (2018). "Fear-of-the-State Bias" in Survey Data. *International Journal of Public Opinion Research*, *30*(4), 631–651. doi:10.1093/ijpor/edx020

APPENDIX: DATA COLLECTION ITEMS

Section A: Demographics

Please indicate your demographics by encircling the correct box

Variable	Category	
Age	18-29	
	30-39	
	40-49	
	50-59	
	60+	
Gender	Male	
	Female	
Education	Primary	
	Secondary	
	Tertiary	
	Communal	
	Smallholder	

Section B: Side Marketing

Please indicate whether you had been involved in side marketing during the previous agricultural season

Not side marketing	
Side marketing	

Chapter 9
Contemporary Agriculture Marketing Strategies for Smallholder Farmers in a Developing Context:
Echoes From Zimbabwe

Samuel Musungwini
https://orcid.org/0000-0002-9315-6252
Midlands State University, Zimbabwe

Yvonne Madongonda
Midlands State University, Zimbabwe

Hope Hogo
Midlands State University, Zimbabwe

ABSTRACT

For millions of people in developing nations, the agriculture industry is essential to their livelihoods, economic growth, and food security. It does, however, confront several difficulties, such as constrained resource availability, inefficient markets, and susceptibility to environmental fluctuations. Within this framework, integrating emerging technology offers a strong chance to transform marketing for agriculture and promote sustainable growth. This chapter explores the revolutionary prospective of using cutting-edge technologies for sustainable agriculture marketing in underdeveloped nations using a case study in Zimbabwe. Through an analysis of the convergence of technology, agriculture, and marketing in the distinctive circumstances of developing nations, the goal is to furnish stakeholders with significant perspectives and useful recommendations for harnessing innovation for the productive transformation of smallholder agricultural marketing practices.

DOI: 10.4018/979-8-3693-2011-2.ch009

1. INTRODUCTION

Agriculture is one of the primary drivers propelling global economic development in the world its importance as the backbone of economic growth and advancement is monumental because it provides food and nutrition to humanity (Toor, 2021). Agricultural activities are contributing to the creation of jobs, food security, reduction of poverty, and general economic prosperity. Hence the complex effects of agricultural activities on country economies are colossal. Research has shown that in the developing world like sub-Saharan Africa (SSA), the agricultural sector serves as a cornerstone of economic development, livelihoods, and food security for millions of people (Modi, 2019). It [agriculture] influences a country's socioeconomic environment as well as encourages sustainable development for future generations. The largest chunk of agricultural activities in SSA is done by smallholder farmers, hence it faces numerous challenges, including limited access to resources, market inefficiencies, and vulnerability to environmental changes (Misaki et al., 2018). In this context, the integration of emerging technologies presents a compelling opportunity to revolutionize agricultural marketing and drive sustainable development. The goal of sustainable agricultural marketing is to establish a productive and eco-friendly agricultural supply chain that connects farmers with consumer markets (Kassem et al., 2021). It entails actions designed to cut down on wastage, mitigate the adverse effects on the environment, and guarantee equitable trade and revenue distribution.

That is why emerging technologies are being touted as essential ingredients to the transformation of agricultural marketing because they offer creative ways to solve problems and boost productivity (Norton & Alwang, 2020). Among these technologies are Internet of Things (IoT) devices, which gather and send data from sensors positioned in fields, greenhouses, and storage facilities. This allows for continuous surveillance of environmental parameters, crop and livestock health and the state of the soil. Blockchain technology lowers the risk of corruption and encourages moral behavior by ensuring transparency, traceability and accountability across the supply chain (Friha et al., 2021). Agricultural data can be analyzed by AI and ML algorithms to enhance crop productivity, forecast market trends, and facilitate better decision-making.

This book chapter elucidates how these emerging technologies can be tailored and applied to address the specific needs and constraints of agricultural marketing in developing regions like SSA. It analyses the key technologies such as precision agriculture, IoT, drones, blockchain, AI, e-commerce platforms, and renewable energy. Furthermore, it delves into the potential benefits, challenges, and best practices associated with the adoption of these technologies in developing contexts. It sheds light on the opportunities and contemplations for integrating emerging technologies into sustainable agricultural marketing in the developing world. The chapter endeavors to empower policymakers, practitioners, and stakeholders with the knowledge and strategies necessary to drive positive transformation and inclusive growth within the agricultural sector.

2. BACKGROUND AND CONTEXT

The agricultural sector plays a crucial role in the economic development of developing countries especially sub-Saharan Africa (SSA), by directly providing livelihood to more than 50% of the populace and contributing to food security and the gross domestic product (GDP) (Dethier & Effenberger, 2012). However, smallholder farmers, who constitute the bulk of farmers often face challenges in effectively marketing their products and some of the perishable products end up getting bad before they can get to the market (Musungwini, 2018). In recent years, emerging technologies such as mobile phone applications and social media have revolutionized the way businesses market their products (Musungwini et al., 2022). The book chapter explored how smallholder farmers in Zimbabwe can leverage emerging technologies to successfully market their agricultural products for sustainable development.

Zimbabwe, like many other developing countries especially SSA, has a large population of smallholder farmers who rely on agriculture for their livelihoods (Dunjana et al., 2018; Quandt et al., 2020). However, due to limited resources and access to markets, these farmers often struggle to sell their products at fair [lucrative] prices (Kwapong et al., 2020). This does not only affect their income but also hinders the growth of the agricultural sector in the country. With the rapid growth of technology, there is a need to explore how smallholder farmers can utilize these tools to overcome marketing challenges and improve their agricultural practices. This is because research has consistently shown that while these technologies benefit those who have adopted the technologies, however, empirical data have decried the lack of adoption of these technologies among smallholder farmers not only in Zimbabwe but across the SSA region (Khan et al., 2019; Mugwisi et al., 2015; Musungwini, 2016). The chapter offers an engaging exploration of creative marketing approaches designed specifically for smallholder farmers in a developing context. This is an enlightening chapter that examines the possible routes taken by agricultural produce from the fields to the tables of consumers, illuminating the difficulties and achievements encountered by small-scale farmers in developing nations. This chapter provides insightful guidance and insightful anecdotes for anybody navigating the ever-changing field of agriculture marketing in a developing context specifically Zimbabwe and possibly other sub-Saharan countries.

3. OBJECTIVES

Our book chapter will attempt to address the following objectives:

a) To understand the current marketing strategies used by smallholder farmers in Zimbabwe.
b) To identify the potential of emerging technologies such as mobile phone applications and social media in agricultural marketing.
c) To explore successful case studies of smallholder farmers in Zimbabwe who have utilized these technologies for marketing.
d) To analyze the impact of these technologies on the sustainable development of agricultural practices in Zimbabwe.
e) To provide recommendations for smallholder farmers on how to effectively use these technologies for marketing their products.

4. METHODOLOGY

This book chapter utilised a qualitative research approach (Ponelis, 2015), mainly a systematic literature review (Xiao & Watson, 2019) of research articles on smallholder agriculture technology use, challenges and opportunities for growth and literature on case studies of those that have successfully adopted and used emerging technologies in their agricultural products marketing activities. The literature review provides a theoretical framework for understanding the role of emerging technologies in agricultural marketing. Case studies of successful smallholder farmers in Zimbabwe who have utilized mobile phone applications and social media for marketing will be analyzed to gain practical insights. Morse (1995) posits that fewer respondents are required to ensure that useful data are gathered from each individual. Most qualitative researchers have found that after interviewing about 20 members of a single group for an interview-based enquiry with a fairly defined research question, not much additional knowledge is produced.

Marshall et al. (2015) suggested that single case studies should usually encompass a minimum of 15 and a maximum of 30 interview participants to achieve data saturation. That is why this research had targeted to interview 20 participants, however, due to challenges experienced in the field we ended up with 19 participants meeting our criteria for interviewing. This is in line with research that has shown that smaller samples are typically used in qualitative research to enable the in-depth case-oriented analysis that is central to this kind of investigation (Richards & Morse, 2006). Furthermore, qualitative samples are purposefully chosen, meaning they are chosen based on their ability to yield detailed information pertinent to the subject matter being studied (Robinson, 2014). Consequently, "information-rich" instances are chosen using purposive selection rather than probability sampling, which is used in quantitative research (Palinkas et al., 2015). New studies corroborate long-held claims made by qualitative methodologists regarding the superior effectiveness of purposive sampling over random sampling in qualitative investigations (Vasileiou et al., 2018).

Researchers acknowledge that the subject of sample size in qualitative research has long been a contested issue and authorities in qualitative studies contend with the claim that there is no simple response to the subject of how many participants are adequate in a study arguing that the number of participants depends on a wide range of elements including methodological, logistical, and epistemological concerns (Baker et al., 2012). For this reason, and to facilitate the development of a novel and deeply textured comprehension of the phenomenon being studied. Sandelowski (1995), suggests that qualitative sample sizes should be both sufficiently large and sufficiently small enough to allow for thorough, case-oriented scrutiny of the data.

5. LITERATURE REVIEW

The literature review focuses on contextualising smallholder farmers' marketing activities and usage of emerging technologies in a developing context and primacy was given to Sub-Saharan Africa. The literature is divided into three subsections and the first section looks at the current technologies that are being used by smallholder farmers for marketing purposes. The second section identifies, describes and explores the emerging technologies in the Agro sector in Zimbabwe. Lastly, the third section delves into the impact of mobile phones and social media technology on the sustainable development of agricultural practices in Zimbabwe.

5.1 Current Technologies in Use

The potential of technologies in unlocking marketing potential and market access can never be underestimated. According to SNV Zimbabwe (2012), the rise of the usage of mobile technologies and the internet has enhanced efficiency and improved access to information about supplier and output markets. Market information is very critical in agro-value chains as it assists smallholder farmers fuse the voice of the customer (VOC) and the Voice of the Organisation (VOG) for agro-products thereby harnessing the potential benefits of digital agriculture in the farming sector. Market information is pertinent to smallholder farmers to help them contain costs and curtail losses by establishing what customers want and when. According to Heeks (2015) and Heeks et al. (2014), the use of mobile phones has been identified as a key driver towards the attainment of the Sustainable Development Goals (SDGs). Odunze (2015) asserts that farmers in once-stated remote villages also have access to information with the aid of mobile telephony in rural Zimbabwe.

The current technologies being used by smallholder farmers in Zimbabwe comprise mobile technologies and social media platforms. Whatsapp and Facebook are being used for e-commerce business correspondence and market linkages (Irungu et al., 2015). Advisory services are the most common digital agriculture technologies present in Zimbabwe (USAID, 2023). The incorporation of advisory services in the agribusiness value chain has a bearing on the quality of the output which impacts customer satisfaction. The SMS service has been largely incorporated in extension services correspondence in the farming community and appeals to the farming populace with low digital literacy skills.

The telecommunications mobile giant Econet Wireless Zimbabwe with more than 10 million subscribers and 70% market share has revolutionised farming technologies by introducing the Ecofarmer- a farm insurance platform that insures crops against adverse weather conditions like excessive rainfall; Eco-weather for early warning systems for adverse weather conditions (Odunze & Hove, 2022). Netone also introduced Hurudza- Umlimi Omkhulu to support farmer information services. All network service providers, corporations and banking institutions have introduced Digital Financial Services (DFS) technologies among others which facilitate e-business transactions.

The transfer facilities comprise Mukuru, Mama Money, City Hopper by Pick n Pay, InnBucks-Innscor Africa, One-Wallet-Netone, Ecocash-Econet, O'mari-Old Mutual, Texta cash under CABS and Telecash-Telecel (Chitungo & Munongo, 2013; Ndlovu & Ndlovu, 2013; Nyagadza & Chuchu, 2023). These ICT–based technologies have tremendously improved the livelihoods of farming communities. Esoko is yet another agriculture digital technology that was established in Zimbabwe in 2012 it was set up to assist 170,000 smallholder farmers and facilitates trading between farmers and buyers covering 33 commodities and also services 17 produce markets (Odunze & Hove, 2022).

The risk attitude of farmers restricts farmers from harnessing the full potential gains of mobile technologies. The digital divide stifles marketing activity amongst small-scale farmers in rural settlements (He et al., 2021). Digital literacy continues to pose a major inhibiting factor in the adoption of ICT technologies (Emeana et al., 2020). Furthermore, the high data costs and limited access to finance also exacerbate the digital divide between rural and urban populations as well as the rich and poor divide. Despite these challenges grappling the agriculture ecosystem, Zimbabwe is a fertile ground for the growth of digital agriculture given the high mobile money account penetration rates and a growing agile generation-Z which is more techno-savvy and highly literate.

5.2. Emerging Technologies In the Agro Sector in Zimbabwe

The new disruptive technologies trending in the world today have ushered in new ways of conducting marketing business correspondence. The agricultural sector has also not been spared from this radical transformation. Musungwini et al. (2023) affirm that it is now business unusual across all facets of the industry. Technology inclusion for the farming sector is of relevance since agriculture is the backbone of Zimbabwe's economy. A significant two-thirds of Zimbabwe's population is employed in the agricultural sector (O'Donnell, 2013). More than 80% of the rural populace relies on agriculture for survival with the majority 60% constituting the female gender (U.S. Agency for International Development's Support, 2022). The rapid population growth calls for farmers to harness technologies and take measures towards adopting digital agriculture technologies which can help enhance the agribusiness value chains and enable the farmers to also meet the growing food demands in the agribusiness supply chains (Kubzansky, 2012).

The dilemma that continues to plague the smallholder farmers is that of being under-resourced as reported by Marambi et al. (2018). Consequently, they cannot match market demand requirements for the output markets as their archaic tooling and high technology costs hinder them from maximizing their productivity. Furthermore, Zimbabwe is ranked the third most expensive country in the world with data costs averaging USD$4.32 per gigabyte (Agency et al., 2023). The limited internet connectivity is a barrier to the growth of e-commerce. The Government of Zimbabwe has also made efforts to improve technology literacy by increasing access to skills training, computers and the internet in rural and marginalised areas by setting up Containerised Village Information Centres (CVICs) (Agency et al., 2023). More digital initiatives have also been witnessed in the involvement of youth in agriculture and encouraging innovations where POTRAZ is working with the government and using ICT to support agricultural entrepreneurship.

Real-time accurate market data on commodity prices, trade volumes, price trends, bids and offers value can be accessed through the Zimbabwe Mercantile Exchange (ZMX) and mobile phones can be leveraged to provide an avenue through which such information can be channeled to smallholder farmers (Musungwini, 2018). Marketers on the input side grapple with high technology set-up costs and the costs of continuing to generate interest in advisory services amongst farmers. The continued viability of these advisory services is threatened by the existence of low-cost information sources that the SMF can get hold of (ibid). Econet Wireless continues to be a trendsetter in terms of eCommerce technology development; it set up OWNAI- an online marketplace. Ecocash Holdings set up an Ecofarmer Club Trading platform for its subscribers who can then trade crops and livestock on the platform (Odunze, 2015).

Despite the stringent barriers to the development of e-commerce in Zimbabwe, some companies have braved the unfavorable conditions and introduced e-commerce facilities that support the marketing of products including agro-processed. In Zimbabwe, e-commerce is still at its embryonic stage and has seen the budding of e-commerce companies comprising Fresh in a Box, Much Zimbabwe, OK Zimbabwe Limited, Zimall and Mukuru Groceries. E-commerce transactions are conducted mainly on WhatsApp platforms owing to the market reach that the social media platform has.

The use of e-commerce business models has paid off for OK Zimbabwe where it registered a significant increase in online sales by 300% during the COVID-19 era (Ferreira et al., 2020). However, these models only benefit the urban dwellers since internet connectivity is a challenge in rural settlements. The high costs of capital to acquire state-of-the-art farming equipment like drone technologies, EWS, Computer Aided Manufacturing (CAM) for agro-processing, value addition and beneficiation are far from reach for smallholder farmers (Musungwini et al., 2023; Reinecke & Prinsloo, 2017). Drone deliveries in the developed world have gained popularity across different industries. It may take time for e-commerce companies in developing countries like Zimbabwe to adopt drone technologies in their distribution strategy owing to POTRAZ regulations and security threats from potential hijackers. State-of-the-art packaging which complies with the requirements of the new emerging consumer values inclined to sustainable production and consumption cannot be ignored. The contemporary consumer demands organic food products that are produced under conditions that do not harm the environment and the use of biodegradable packaging. This may require farmers to also adopt technologies that support these consumer demands and values.

5.3 The Impact of Mobile Phones and Social Media Technology on the Sustainable Development of Agricultural Practices in Zimbabwe

To fully utilize technology for sustainable agricultural development in Zimbabwe, several factors related to agricultural development must be taken into consideration. These factors include information access, market integration, resource management, and climate-smart practices, which improve farmers' productivity, market accessibility, and climate change resilience (Misaki et al., 2018).

Access to Information

In Zimbabwe, access to agricultural information has been transformed by mobile phones as reported by Masimba et al., (2019). Quandt et al. (2020) posit that mobile applications and SMS services have made it possible for farmers to obtain weather forecasts, market prices, and professional guidance. Farmers are better equipped to choose crops, schedule planting and harvesting, and employ pest control techniques when they have access to this information.

Market Integration

Farmers, buyers, and other players in the agricultural value chain can now communicate and interact directly with one another thanks to social media platforms like Facebook and WhatsApp (Kristjanson et al., 2017; Musungwini et al., 2022). Farmers may advertise their produce, haggle over costs, and plan transportation using these platforms, cutting out middlemen and enhancing market access.

Resource Management

Through the use of their mobile devices, farmers may obtain agricultural extension assistance and guidance on how best to use resources specifically pesticides, fertilizers, and water (Chikozho et al., 2017). Correspondingly, real-time data on pest infestations, nutrient deficits, and soil moisture levels are provided by mobile applications, which let farmers apply inputs sparingly and lessen resource waste and environmental effects.

Climate-Smart Practices

In Zimbabwe, the use of mobile phones and social media technology facilitates the adoption of climate-smart agricultural practices. Farmers have access to weather alerts, early warning systems, and strategies for adapting to climate change through mobile applications and SMS services (CTA, 2019). With this information, they can modify their farming practices, such as planting dates, selecting drought-resistant crop varieties, and applying conservation agriculture techniques.

Challenges

Despite the positive effects that social media and mobile phones have exhibited, there are still obstacles and restrictions. Widespread acceptance and access to these technologies are hampered by low network coverage, illiteracy, power outages, inadequate training, and the high cost of smartphones (Chirume, 2019). Subsequently, thorough verification is necessary to ensure the accuracy of material provided on social networking platforms.

6. EMPIRICAL DATA FROM KEY INFORMANTS

Our research intended to understand the current marketing strategies being employed by smallholder farmers to ensure that their agricultural produce reaches the market. To gather empirical data in the Zimbabwean context we identified as indicated in the methodological section 22 smallholder farmers were interviewed and their demographic profiles are provided for in the table below.

6.1 Demographic Data

The research explored the need for understanding how agricultural produce moves from the farm to the feeding table which is why we explore contemporary agriculture marketing techniques available for exploitation by smallholder farmers. The goal was to draw attention to how crucial it is for smallholder farmers in developing countries to have efficient marketing methods. The chapter highlights the necessity for modern strategies that allow small-scale farmers to successfully connect their agricultural products with consumers, guaranteeing them equitable pricing and long-term means of sustenance. Food production greatly depends on smallholder farmers, particularly in emerging nations. However, farmers frequently have challenges getting their produce into marketplaces and getting paid fairly for it. Modern agricultural marketing techniques seek to overcome these obstacles by giving smallholder farmers creative and effective means of promoting their products.

The researchers had targeted to collect empirical data by interviewing 20 smallholder farmers from four villages in Shurugwi resettlements. The villages were anonymized as Village W, Village X, Village Y and Village Z, and each of these provided five research participants except for Village X which eventually provided four participants. That is why 19 interviewees as shown in Table 1. The interviewees were anonymised as Smallholder Farmer [SHF], then village [W, X, Y and Z] and participant number [1, 2, 3 and so on] to make up SHFW1, SHFW2, SHFX6, SHFY10 and so on up to SHFZ19. The participants were made up of nine males and ten females and were all mature and experienced with all of them belonging to the 40-plus years age category. Table 1 below presents the details of the research participants' data.

The bulk of these farmers had attained the ordinary level which is a standard secondary level except for two who had only attained the Junior certificate level which is lower than the ordinary level. However, it is equally important to highlight that these farmers had attained practical qualifications organised by the agricultural extension services like the Master Farmer certificate and the Advanced Farmer certificate. The interviewed farmers also had farming experience, having been practicing agriculture for a period ranging from at least 13 years up to 33 years.

Table 1. Research participant's data

Village	Smallholder Farmers	Age	Gender	Highest Academic Qualifications	Date of Interview	Types of Produce	Duration of Interview	Farming Experience
Village W	SHFW1	56	Male	O' Level	21 September 2023	Cash crops, food crops and livestock	48 Minutes	27 Years
	SHFW2	61	Female	O' Level	29 September 2023	Market gardening, cash crops and livestock	53 Minutes	32 Years
	SHFW3	52	Male	Certificate in Agriculture	30 September 2023	Poultry, market gardening and livestock	56 Minutes	20 Years
	SHFW4	47	Male	Master Farmer	9 October 2023	Cash crops, food crops and poultry	50 Minutes	25 Years
	SHFW5	55	Female	Master Farmer	10 October 2023	Food crops, poultry and market gardening	59 Minutes	26 Years
Village X	SHFX6	42	Female	Master Farmer	11 October 2023	Food crops, cash crops and livestock	55 Minutes	18 Years
	SHFX7	52	Female	Master Farmer	13 October 2023	Market gardening, cash crops and livestock	58 Minutes	29 Years
	SHFY8	54	Female	Advance Farmer	15 October 2023	Poultry, market gardening and livestock	59 Minutes	33 Years
	SHFY9	57	Male	O' level	17 October 2023	Cash crops, food crops and poultry	55 Minutes	27 Years
Village Y	SHFZ10	53	Male	Advanced Farmer	21 October 2023	Food crops, poultry and market gardening	49 Minutes	26 Years
	SHFZ11	49	Male	Advanced Farmer	23 October 2023	Cash crops, livestock and food crops	53 Minutes	22 Years
	SHFZ12	63	Female	Junior Certificate	27 October 2023	Market gardening, cash crops and livestock	52 Minutes	28 Years
	SHFX13	49	Female	Master Farmer	29 October 2023	Poultry, market gardening and livestock	47 Minutes	18 Years
	SHFX14	44	Female	Master Farmer	31 October 2023	Cash crops, food crops and poultry	48 Minutes	19 Years

continued on following page

Table 1. Continued

Village	Smallholder Farmers	Age	Gender	Highest Academic Qualifications	Date of Interview	Types of Produce	Duration of Interview	Farming Experience
Village Z	SHFY15	47	Female	Advance Farmer	3 November 2023	Poultry, food crops and market gardening	49 Minutes	13 Years
	SHFY16	45	Male	O' level	5 November 2023	Cash crops, food crops and livestock	45 Minutes	17 Years
	SHFZ17	52	Male	Advanced Farmer	8 November 2023	Market gardening, cash crops and livestock	49 Minutes	27 Years
	SHFZ18	49	Male	Advanced Farmer	10 November 2023	Poultry, market gardening and livestock	47 Minutes	22 Years
	SHFZ19	51	Female	Junior Certificate	13 November 2023	Cash crops, food crops and poultry	52 Minutes	18 Years

Because the thrust of the research was on marketing strategies when we collected empirical data, we were more interested in agricultural produce meant for selling hence data on cash crops, market gardening, poultry and livestock meant specifically for selling was collected. The findings show that these farmers were engaged in multiple agricultural production which has resulted in each farmer being involved in more variants and 12 of the 19 smallholder farmers were engaged in cash crops and specifically producing crops like cotton, sunflower, tobacco and other farming activities like food crops like maize, groundnuts, beans and other crops where 11 of the farmers identified food crops as some of the agricultural produce earning them revenue. Other categories such as livestock production had 12 farmers identifying with it as a source of revenue, poultry had 11 formers and market gardening also had 11. The farmers were asked to list at most three of the types of products that they earn more revenue from in order of priority and that is how these categories came to be established.

The data on the agricultural products produced by the farmers for selling is slightly different because the farmers were asked to list the top three products they produce and market in order of priority. There were 19 respondents and since each participant was required to provide three agricultural categories, a total of 57 responses were expected and eventually obtained with 25 responses from males constituting 43.86% and females accounting for 32 responses making up 56.14%. On the categorisation of these responses, we established that 12 farmers identified cash crops as a category of products they produce for selling and that constituted 63.16%

of the participants, on food crops 11 responses were obtained accounting for 57.9%. Other categories were as follows livestock obtained 12, making 63.16%, market gardening and poultry got 11 responses each, accounting for a 57.9% response rate each. The composition of 63.16% of the cash crop category concerning gender and 58.33% of farmers engaged in cash crops were males, while 41.67% were females. On food crops, females accounted for 72.73% of the responses while males constituted 27.27%. On livestock, it was evenly balanced with each gender accounting for 50% of the 12 responses. The females also pipped their male counterparts on market gardening and poultry respectfully. The next section presents the empirical data on the current marketing strategies which were found to be utilised by smallholder farmers in Shurugwi resettlements in the purposefully selected four villages.

Table 2 below provides the summary of these demographic variable data about the research participants.

6.2 The Current Marketing Strategies Used by Smallholder Farmers in Zimbabwe

The research discovered that there is a broad spectrum of marketing techniques frequently used by smallholder farmers to sell their agricultural goods. Among the current marketing techniques employed by smallholder farmers in Zimbabwe include the use of local marketplaces. The smallholder farmers in this research indicated that they have established agricultural farmers' marketing places in their areas, especially at established shopping centres. At such places, agricultural produce is sold directly to customers and small companies throughout the neighborhood. Participants also pointed out that they have also managed to set up cooperatives whereby they join forces with other farmers and group together. By joining forces with other smallholder farmers and even merging with other agricultural cooperatives, these farmers were jointly promoting and selling their goods, opening up new markets and negotiating better pricing.

The research established that the smallholder farmers in Shurugwi resettlements were using established farmers' markets in urban areas like Kudzanai and Mutapa markets to market and sell their agricultural products. The farmers alleged that attending farmers' markets allows them to sell their goods and engage in direct conversation with customers. These markets are characterised by pure market forces where farmers [suppliers] intersect with customers [demanders], hence there is a bidding element and the quality products attract more buyers engaged in a bidding contest and the highest bidder takes the product/s. The participants also talked about selling their agricultural produce through dealers and intermediaries. These dealers engage in purchasing produce from farmers and selling it to larger marketplaces or processing facilities like the Agricultural Marketing Authorities (AMA). The research

Table 2. Categorisation of respondents' data

	Males Number	%	Females Number	%	Total	%
Sex	9	47.37%	10	52.63%	19	100%
Age category: 40-45	2		1		3	
46-50	3		2		5	
51-55	2		5		7	
56-60	1		1		2	
60+	1		1		2	
Total	**9**	**47.37%**	**10**	**52.63%**	**19**	**100%**
Educational level: Junior Cert			2		2	
Ordinary level	3		1		4	
Certificate in Agriculture	1		0		1	
Master farmer	1		5		6	
Advanced farmer	5		1		6	
Total	**9**	**47.37%**		**52.63%**	**19**	**100%**
Farming experience: Below 16 Years	1		0		1	
16 – 20 Years	3		2		5	
21 - 25 years	2		2		4	
26 – 30 Years	2		5		7	
Above 30 Years	1		1		2	
Total	**9**		**10**		**19**	**100%**
Agricultural products: Cash crops	*7*	*58.33%*	*5*	*41.67%*	*12*	*63.16%*
Food crops	*3*	*27.27%*	*8*	*72.73%*	*11*	*57.9%*
Livestock	*6*	*50%*	*6*	*50%*	*12*	*63.16%*
Market gardening	*4*	*36.36%*	*7*	*63.64%*	*11*	*57.9%*
Poultry	*5*	*45.45%*	*6*	*54.55%*	*11*	*57.9%*
	25	**43.86%**	**32**	**56.14%**	**57**	

also established that there were a few farmers who were engaging in direct sales to different businesses. These farmers were offering their agricultural products directly to supermarkets, restaurants, butcheries, hotels, or other food-related enterprises.

These farmers also indicated that in the last ten years, some farmers had engaged in different forms of contract farming with different agricultural business operators. Contract farming is the practice of establishing agreements to supply specific quantities of particular agricultural products at predetermined pricing with agribusinesses or food firms. These agribusinesses usually [but not always] supply the farmers with capital for agricultural inputs and other implements. In this case, farmers are assured of a ready market for their agricultural produce hence they only focus on their productivity. The research participants also indicated they were getting used to engaging with customers using online platforms like WhatsApp, and Facebook and making phone calls but this was said to be a bit limited. The next section focuses on identifying the potential of emerging technologies such as mobile phone applications, social media applications and platforms to facilitate agricultural marketing for smallholder farmers in Zimbabwe.

6.3 To Identify the Potential of Emerging Technologies Such as Mobile Phone Applications and Social Media in Agricultural Marketing

This section presents data on how social media platforms along with other emerging technologies, such as smartphone applications may have a big impact on how smallholder farmers may market their agricultural products. The following salient features underscore the potential of such technologies to access real-time agricultural market data. Respondents suggested that smallholder farmers can obtain real-time agricultural produce market data, such as prices, demand patterns, and customer preferences, through mobile phone applications. Because of this, farmers are better equipped to decide what to grow and when and where to sell their goods. These technologies can also facilitate direct marketing of products where farmers can directly market and sell their agricultural goods to customers, restaurants, and food enterprises by using various social networking platforms. Farmers may improve sales, cultivate client relationships, and define their brand identity with the aid of this direct engagement.

Farmers suggested that they can receive agricultural extension services via mobile applications, which give them access to useful data on pest control, weather forecasts, standard procedures, and crop wellness surveillance. These emerging technologies can facilitate the establishment of agriculture produce supply distribution value chain transparency. By enabling consumers to track the provenance of products, mobile applications and social media can improve supply chain transparency in the agriculture industry. This may help consumers develop trust and confidence in the high-quality and moral production methods. The respondents suggested that farmers may reach a larger client base outside their local markets by using social media platforms and mobile applications to support online sales and delivery of agricultural products. Farmers can use social media platforms as a medium to interact with customers, get their opinions, and customize their products to suit their tastes. By harnessing the potential of mobile phone applications and social media, the smallholder farmers in Shurugwi resettlements can expand their market reach, improve market access, increase sales, and build stronger connections with consumers and other stakeholders in the agricultural value chain.

The research also established that there were a few select smallholder farmers in Shurugwi resettlement areas who were successfully using emerging platforms and technologies in their agricultural marketing strategies. The usage of mobile phone gadgets to access marketplaces is one instance. Smallholder farmers in Shurugwi resettlements have started using mobile phones to connect with potential customers to obtain market information like prices and the supply and demand of agricultural commodities thereby enabling them to plan the sale of their produce. They have been able to access more markets and get better rates for their agricultural products as a consequence of using these technologies. Some of these smallholder farmers have started engendering social media platforms for promotional purposes of their agricultural products. Some of the farmers have made use of social media sites like Twitter, Facebook, and WhatsApp to advertise their goods, establish connections with prospective customers, and grow their clientele. This has made it possible for them to directly take orders from clients and present their produce to a wider market base.

6.4 The Impact of These Technologies on the Sustainable Development of Agricultural Practices in Zimbabwe

These advances in technology have had a major impact on Zimbabwe's smallholder farmers' agricultural practices' ability to flourish sustainably. They have facilitated access to agricultural product market information. Some respondents suggested that the use of social media platforms and mobile devices has made it easier for smallholder farmers to obtain market information thereby giving them the power to decide when and where to sell their goods. As a result of using these emerging technologies, agricultural marketing strategies have become more effective and lucrative, supporting the long-term viability of agriculture. Better market accessibility has ensued as smallholder farmers may now connect with buyers outside of their local communities thanks to the usage of chatbots, social media, and mobile phones. Increased selling prices for agricultural products and higher revenue for farmers have resulted from this greater market access and decreased reliance on middlemen.

According to the respondents, the farmers are now more prepared to anticipate variations in the weather, modify their agriculture methods, and reduce the negative effects of the changing climate on the productivity of agriculture because they have adapted to the implementation of these technological advances. In general, the application of developing technology in agriculture has promoted more effective and ecologically friendly farming techniques, improved market access, and helped Zimbabwe's smallholder farmers' agricultural practices evolve sustainably. These technological advancements have been crucial in empowering smallholder farmers and advancing the nation's productive and sustainable agricultural growth.

6.5 How to Effectively Use These Technologies for Marketing Their Products

According to the respondents, the farmers are now more prepared to anticipate variations in the weather, modify their agriculture methods, and reduce the negative effects of the changing climate on the productivity of agriculture because they have adapted to the implementation of these technological advances. In general, the application of developing technology in agriculture has promoted more effective and ecologically friendly farming techniques, improved market access, and helped Zimbabwe's smallholder farmers' agricultural practices evolve sustainably. These technological advancements have been crucial in empowering smallholder farmers and advancing the nation's productive and sustainable agricultural growth.

The selected participants pointed out that there were isolated cases of very few smallholder farmers who engaged in product promotion and branding using emerging technologies. Those farmers were showcasing their agricultural products on social media platforms, sharing their success stories, and interacting with prospective buyers. As a result of using digital platforms, these farmers were managing to pique consumers' attention. They were using the technologies to share interesting content such as pictures and/or videos of agricultural products as well as product descriptions.

7. DISCUSSION

The research results suggest that some farmers are using their mobile phones to obtain market intelligence information. This enables farmers to increase market reach and boost sales opportunities. Another dimension brought about by emerging technologies is the increase of financial inclusion for smallholder farmers as mobile phones are enabling them to leverage mobile payment technologies like InBucks, Telecash, Ecocash and others to streamline transactions. Our research discovered that some of the smallholder farmers from Shurugwi resettlements were using social media platforms to conduct product promotion. This results in an increase in awareness of their agricultural products and recognition. The use of social media provides avenues for farmers to highlight their agricultural products, share success stories and interact with prospective customers. The use of social media facilitates direct sales and customer involvement by enabling direct communication with buyers using social media channels.

The use of chatbots, artificial intelligence, and other technologies gives farmers access to the latest farming techniques, weather forecasts, and real-time market data so they can make well-informed choices. Emerging technologies also offer automated customer query facilities and we believe farmers can leverage the capabilities of these technologies and offer product details and streamline sales transactions with the use of chatbots. The adoption and usage of precision agriculture can boost productivity and resource efficiency by using drones for the surveillance of crops, controlling pests, and regulating irrigation. This can improve the marketing of products as we understand that quality products can market themselves through viral marketing and those customers who buy the products can go on to talk positively about them. The research participants indicated they were getting used to engaging with customers using online platforms like WhatsApp, and Facebook and making phone calls but this was said to be a bit limited. Making use of e-commerce websites and online platforms to expand their customer base and offer their goods directly to customers. With the use of these strategies, smallholder farmers can expand their marketing horizons, build rapport with potential consumers, and reach a wider audience to increase sales.

While the research did not find smallholder farmers in Shurugwi resettlements using some of the technologies in place, however, it is possible smallholder farmers in Zimbabwe can use emerging technologies like mobile phones, social media platforms, chatbots, artificial intelligence, mixed reality, augmented reality, and virtual reality in several ways to market their products sustainably. For instance, blockchain technology can assist with product traceability can be accomplished, as using blockchain technology enables customers to keep track of the origins of agricultural products how they are produced and the path they travel to reach the destination. This may foster a culture of trust and transparency. The use of blockchain technology may also facilitate smart contracts which enables safe and transparent transactions to occur between smallholder farmers and their clientele.

While the findings show that emerging technologies such as Virtual Reality (VR), Augmented Reality (AR) and Mixed Reality (MR) are a far cry for smallholder farmers in Shurugwi resettlements, it is clear that there is a need for capacity building and training of these farmers to be able to embrace these technologies as they have the potential to broaden their horizon and open new lucrative markets for the farmers. Training of these farmers in digital literacy and the use of technology for agricultural marketing by introducing mixed reality, augmented reality and virtual reality technologies will ensure increased success of their agricultural outcomes.

Adoption and use of these emerging technologies enable customers to explore agricultural items in a virtual setting through augmented reality which may improve their buying experience and comprehension of the product. Implementation of these cutting-edge technologies has the potential to expand smallholder farmers' accessibility to markets, boost their productivity, encourage sustainable farming methods, and open up new business possibilities to venture into according to a case study of smallholder farmers in Zimbabwe. Smallholder farmers can attain sustainable agricultural marketing practices and transcend conventional obstacles to market access by skillfully utilizing these tools.

8. RECOMMENDATIONS

Farmers can also benefit from the usage of chatbots to learn about market prices, weather predictions, and the latest developments in farming. Farmers may make well-informed judgments regarding the best times and locations to sell their goods as well as how to maximize their farming techniques by communicating with chatbots. The practice of precision agriculture has made use of drones to monitor crop health, locate pest infestation regions, and improve watering techniques. Farmers may be able to increase crop yields and make better decisions about their farming operations thanks to cutting-edge disruptive technology. There are just a few instances of how some of Zimbabwe's smallholder farmers have effectively marketed their agricultural products using cutting-edge technologies. If all smallholder farmers can leverage emerging technologies, they will be able to raise their standard of living and get over long-standing obstacles to market access by riding on the evolving wave of disruptive technologies.

As some farmers have pointed out there is a need for better crop management techniques that have resulted from the usage of emerging technologies. Through the use of precision agriculture, farmers may be able to monitor and manage the conditions of their crops and livestock, thereby enabling them to spot and isolate problem plants, field sections, sick livestock and improve insect and irrigation methods as well as animal disease treatment. As a result of the more focused use of resources, production will increase and the environmental effect will decrease. This is in line with research such as Cruz et al. (2017) and Loures et al. (2020), which established that emerging technologies such as precision farming and resource management technologies and drones have been used to monitor crop and livestock health for swift identification of the genesis of bacteria and diseases and make a swift response with treatment and improve the overall agricultural practices including irrigation techniques for those practicing it.

The adoption and use of emerging technologies have the potential to improve knowledge exchange and smallholder farmers' capabilities. Therefore, leveraging the use of Chatbots and other digital platforms can make it easier for smallholder farmers to exchange their knowledge amongst themselves as well as with experts thereby developing their [smallholder farmers] capabilities. They can further use chatbots to handle customer support inquiries offer product details and subsequently expedite purchases. By providing smallholder farmers access to data on market trends, weather predictions, and best farming techniques, they [emerging technologies] enable them [smallholder farmers] to make better and informed decisions and implement productive and sustainable agricultural practices.

Research has also shown that other technologies such as blockchain technology can be used to manage agricultural product supply chain logistics to guarantee transaction transparency and cut down on distribution process inefficiencies. Hence, the book chapter also recommends that there is a need to examine how blockchain technology might be used to improve confidence among customers and traceability of agricultural commodities by transparently revealing the provenance and production means of agricultural commodities. The farmers whose products have been grown under sustainable, environment and climate-friendly conditions may be more marketable and their market safeguarded using blockchain technology. Hence, there is also a need to explore how blockchain technology can be used to create smart contracts with purchasers that guarantee safe and open transactions. The research also recommends that the use of social media can help to establish connections with other smallholder farmers in other different areas in Zimbabwe or other countries in the region, customers as well and agricultural industry players to forge alliances and take advantage of new business prospects. Smallholder farmers in developing countries can use emerging technology to promote their produce more successfully, reach a wider audience, and achieve better business outcomes by putting these suggestions into practice. It would be prudent for all smallholder farmers to harness emerging technologies to conduct direct marketing and sales. These technologies can enable smallholder farmers to engage in direct communication and sales negotiations with prospective customers by using social media and mobile devices.

However, for the success of the usage of emerging technologies amongst smallholder farmers in Shurugwi resettlements of Zimbabwe, there is a need for building the knowledge capacity of the farmers as well as a cross-pollination of agricultural knowledge amongst farmers in different regions as well as with different experts. Therefore, it is vital to rope in those organisations that provide such technological services and others that provide agricultural extension services together and organise training workshops with the farmers. Thus, smallholder farmers should then take part in workshops and training sessions that emphasize understanding digital technologies and the application of modern technologies to agricultural marketing.

9. CONCLUSION

Our findings suggest that the farmers who are making use of digital platforms and technologies like mobile phones and social media platforms are removing the need for middlemen in the agricultural value chain as these emerging technologies bring smallholder farmers and their customers closer together directly. The farmers alleged that the use of digital technologies and applications for mobile devices can boost their revenue margins. The use of emerging technologies can also boost the creation and coordination of farmers' cooperatives which enable those farmers to use their co-operatives as leverage to bargain for better prices for their agricultural commodities. This can be attained by aggregating their produce and negotiating higher prices with buyers as cooperatives provide smallholder farmers more negotiating strength. Encouraging smallholder farmers to process and add value to their agricultural produce by indicating that they use organic manure like cow dung as the fertilizer for their crop produce and mainly use indigenous traditional knowledge for their livestock treatment which is all climate user friendly and a sustainable means of agricultural production. This can raise the market value of their agricultural produce and open up new revenue streams. This is known as value-added product promotion. Forming alliances with larger supermarkets or agricultural enterprises can open up new markets for smallholder farmers and guarantee steady demand for their goods. Smallholder farmers in Shurugwi resettlements may make more informed choices about production and marketing if they have access to current and trustworthy data about agricultural market prices, consumer demand and trends.

Mobile Technologies such as mobile applications can provide smallholder farmers in Shurugwi resettlements with access to weather forecasts, market data, and extension services through mobile applications thereby enabling them to make well-informed decisions. As a consequence, smallholder farmers may enhance their ability to sustain themselves, reduce expenses, explore new markets and increase output by utilising these cutting-edge technologies. Through the application of modern agriculture marketing techniques, smallholder farmers in Shurugwi resettlements', Zimbabwe at large and other developing countries in sub-Saharan Africa can raise their standard of living, increase their incomes, and support the growth of both the agricultural industry and their communities as a whole.

REFERENCES

Baker, S. E., Edwards, R., Adler, P., Becker, H. S., & Doucet, A. (2012). *How many qualitative interviews is enough? Expert voices and early career reflections on sampling Expert voices*. National Centre for Research Methods Review.

Chirume, S. (2019). The use of Zimbabwe Open University's MyVista platform in e-communication, accessing and uploading learning materials, and assessing students' work. *American Educational Research Journal*, *7*(1), 1–13. doi:10.30918/AERJ.64.18.079

Chitungo, S., & Munongo, S. (2013). Extending the Technology Acceptance Model to Mobile Banking Adoption in Rural Zimbabwe. *Journal of Business Administration and Education*, *3*(1), 51–79.

Cruz, J. R. D., Baldovino, R. G., Bandala, A. A., & Dadios, E. P. (2017, October 18). Water usage optimization of Smart Farm Automated Irrigation System using artificial neural network. *2017 5th International Conference on Information and Communication Technology, ICoIC7 2017*. 10.1109/ICoICT.2017.8074668

CTA. (2019, June-August). Digitalising extension: Smart Advisory Services For Farmers. *Spore*, (194), 1–48.

Dethier, J.-J., & Effenberger, A. (2012). Agriculture and development: A brief review of the literature. *Economic Systems*, *36*(2), 175–205. doi:10.1016/j.ecosys.2011.09.003

Dunjana, N., Zengeni, R., Muchaonyerwa, P., & Wuta, M. (2018). Typological characterisation of farms in a smallholder food-cash crop production system in zimbabwe – opportunities for livelihood sustainability. *Journal of Agriculture and Rural Development in the Tropics and Subtropics*, *119*(2), 11–22.

Emeana, E. M., Trenchard, L., & Dehnen-Schmutz, K. (2020). The revolution of mobile phone-enabled services for agricultural development (m-Agri services) in Africa: The challenges for sustainability. *Sustainability (Basel)*, *12*(2), 485. doi:10.3390/su12020485

Ferreira, M. J., Moreira, F., Pereira, C. S., & Durão, N. (2020). The Digital Transformation at Organizations – The Case of Retail Sector. *Advances in Intelligent Systems and Computing, 1159 AISC*, 560–567. https://doi.org// doi:10.1007/978-3-030-45688-7_56/COVER

Friha, O., Ferrag, M. A., Shu, L., Maglaras, L., & Wang, X. (2021). Internet of Things for the Future of Smart Agriculture: A Comprehensive Survey of Emerging Technologies. *IEEE/CAA Journal of Automatica Sinica*, *8*(4), 718–752. doi:10.1109/JAS.2021.1003925

He, W., Zhang, Z., & Li, W. (2020, June). (Justin), & Li, W. (2021). Information technology solutions, challenges, and suggestions for tackling the COVID-19 pandemic. *International Journal of Information Management*, *57*, 102287. Advance online publication. doi:10.1016/j.ijinfomgt.2020.102287 PMID:33318721

HeeksR. (2015). *ICT4D 2016: New priorities for ICT4D policy, practice and WSIS in a Post-2015 World*. doi:10.2139/ssrn.3438433

Heeks, R., Morgan, S., & Nicholson, B. (2014). *IT sourcing and development : new drivers, models and impacts*. Issue October.

Irungu, K. R. G., Mbugua, D., & Muia, J. (2015). Information and Communication Technologies (ICTs) Attract Youth into Profitable Agriculture in Kenya. *East African Agricultural and Forestry Journal*, *81*(1), 24–33. doi:10.1080/00128325.2015.1040645

Kassem, H. S., Alotaibi, B. A., Ghoneim, Y. A., & Diab, A. M. (2021). Mobile-based advisory services for sustainable agriculture: Assessing farmers' information behavior. *Information Development*, *37*(3), 483–495. doi:10.1177/0266666920967979

Khan, N. A., Qijie, G., Ali, S., Shahbaz, B., & Shah, A. A. (2019). Farmers' use of mobile phone for accessing agricultural information in Pakistan: A case of Punjab province. *Ciência Rural*, *49*(10), 1–12. doi:10.1590/0103-8478cr20181016

Kristjanson, P., Bryan, E., Bernier, Q., Twyman, J., Meinzen-Dick, R., Kieran, C., Ringler, C., Jost, C., & Doss, C. (2017). Addressing gender in agricultural research for development in the face of a changing climate: Where are we and where should we be going? *International Journal of Agricultural Sustainability*, *15*(5), 482–500. doi:10.1080/14735903.2017.1336411

Kubzansky, M. (2012). *Old Problems, New Solutions : Harnessing Technology and Innovation in the Fight Against Global Poverty*. Global Economy and Development.

Kwapong, N. A., Ankrah, D. A., Boateng-Gyambiby, D., Asenso-Agyemang, J., & Fening, L. O. (2020). Assessment of agricultural advisory messages from farmer-To-farmer in making a case for scaling up production: A qualitative study. *The Qualitative Report*, *25*(8), 2011–2025. doi:10.46743/2160-3715/2020.4241

Loures, L., Chamizo, A., Ferreira, P., Loures, A., Castanho, R., & Panagopoulos, T. (2020). Assessing the effectiveness of precision agriculture management systems in mediterranean small farms. *Sustainability (Basel)*, *12*(9), 1–15. doi:10.3390/su12093765

Marambi, S., Musungwini, S., & Mzikamwi, T. (2018). Strategy for the adoption and effective utilization of mobile phone technologies in smallholder agriculture in Zimbabwe. In *Proceedings of 6th International Conference on M4D Mobile Communication for Development* (pp. 15–16). Universitetstryckeriet, Karlstad. www.researchjournal.co.in

Marshall, B., Cardon, P., Poddar, A., Fontenot, R., Marshall, B., Cardon, P., Poddar, A., & Does, R. F. (2015). *Does Sample Size Matter in Qualitative Research ? A Review of Qualitative Interviews in is.* doi:10.1080/08874417.2013.11645667

Masimba, F., Appiah, M., & Zuva, T. (2019). Influencing factors of mobile technology adoption by small holder farmers in Zimbabwe. *Advances in Intelligent Systems and Computing, 1046*, 125–134. doi:10.1007/978-3-030-30329-7_12

Misaki, E., Apiola, M., Gaiani, S., & Tedre, M. (2018). Challenges facing sub-Saharan small-scale farmers in accessing farming information through mobile phones: A systematic literature review. *The Electronic Journal on Information Systems in Developing Countries, 84*(4), e12034. doi:10.1002/isd2.12034

Modi, R. (2019). *The Role of Agriculture for Food Security and Poverty Reduction in Sub-Saharan Africa.* The Palgrave Handbook of Contemporary International Political Economy. doi:10.1057/978-1-137-45443-0_25

Morse, J. M. (1995). *The significance of saturation.pdf.*

Mugwisi, T., Mostert, J., & Ocholla, D. N. (2015). Access to and Utilization of Information and Communication Technologies by Agricultural Researchers and Extension Workers in Zimbabwe. *Information Technology for Development, 21*(1), 67–84. Advance online publication. doi:10.1080/02681102.2013.874317

Musungwini, S. (2016). A model for harnessing the power of the mobile phone technology to improve smallholder agriculture in Zimbabwe. *Proceedings of the 5th International Conference on M4D Mobile Communication Technology for Development: M4D 2016, General Tracks, December*, 237–252.

Musungwini, S. (2018). Mobile phone use by Zimbabwean smallholder farmers : A baseline study. *The African Journal of Information and Communication, 22*(22), 29–52. doi:10.23962/10539/26171

Musungwini, S., Gavai, P. V., Munyoro, B., & Chare, A. (2023). Emerging ICT technologies for agriculture, training, and capacity building for farmers in developing countries: A case study in Zimbabwe. *Advances in Environmental Engineering and Green Technologies*, (January), 12–30. doi:10.4018/978-1-6684-6413-7.ch002

Musungwini, S., van Zyl, I., & Kroeze, J. H. (2022). The perceptions of smallholder farmers on the use of mobile technology: A naturalistic inquiry in Zimbabwe. *Lecture Notes in Networks and Systems, 439 LNNS*, 530–544. doi:10.1007/978-3-030-98015-3_37

Ndlovu, I., & Ndlovu, M. (2013). Mobile banking the future to rural financial inclusion: Case study of Zimbabwe. *IOSR Journal of Humanities And Social ....* http://iosrjournals.org/iosr-jhss/papers/Vol9-issue4/K0947075.pdf

New Business Models in the Accommodation Industry - Benchmarking of Rules and Regulations in the Short-term Rental Market. (2019). In New Business Models in the Accommodation Industry - Benchmarking of Rules and Regulations in the Short-term Rental Market. World Tourism Organization (UNWTO). doi:10.18111/9789284421084

Norton, G. W., & Alwang, J. (2020). Changes in Agricultural Extension and Implications for Farmer Adoption of New Practices. *Applied Economic Perspectives and Policy*, *42*(1), 8–20. doi:10.1002/aepp.13008

Nyagadza, B., & Chuchu, T. (2023). *Mobile Fin-Tech Ecosystem Shaping Financial Inclusion in Zimbabwean Banking and Financial Services Markets Metadata of the chapter that will be visualized online.* doi:10.1007/978-3-031-23863-5

O'Donnell, M. (2013). Using ICT to enhance marketing for small agricultural producers. *USAID Briefing Paper,* 1–6.

Odunze, D. (2015). *An Analysis of the Impact of the Use of Mobile Communica tion Technologies by Farmers in Zimbabwe. A Case Study of Esoko and EcoFarmer Platforms*. https://www.researchgate.net/publication/330347260

Odunze, D., & Hove, M. T. (2022). *An Analysis of the Impact of the Use of Mobile Communica tion Technologies by Farmers in Zimbabwe. A Case Study of Esoko and EcoFarmer Platforms*. Academic Press.

Palinkas, L. A., Horwitz, S. M., Green, C. A., Wisdom, J. P., Duan, N., & Hoagwood, K. (2015). Purposeful Sampling for Qualitative Data Collection and Analysis in Mixed Method Implementation Research. *Administration and Policy in Mental Health*, *42*(5), 2–4. doi:10.1007/s10488-013-0528-y PMID:24193818

Ponelis, S. R. (2015). Using interpretive qualitative case studies for exploratory research in Doctoral studies : A case of information systems research in small and medium enterprises. *International Journal of Doctoral Studies*, *10*, 535–550. doi:10.28945/2339

Quandt, A., Salerno, J. D., Neff, J. C., Baird, T. D., Herrick, J. E., Terrence McCabe, J., Xu, E., & Hartter, J. (2020). Mobile phone use is associated with higher smallholder agricultural productivity in Tanzania, East Africa. *PLoS One, 15*(8), e0237337. doi:10.1371/journal.pone.0237337 PMID:32760125

Reinecke, M., & Prinsloo, T. (2017). The influence of drone monitoring on crop health and harvest size. *2017 1st International Conference on Next Generation Computing Applications, NextComp 2017*, 5–10. 10.1109/NEXTCOMP.2017.8016168

Richards, L., & Morse, J. M. (2006). Qualitative research design. Read Me First for a User's Guide to Qualitative Methods, 73–104. doi:10.1186/1472-6963-11-23

Robinson, O. C. (2014). Sampling in Interview-Based Qualitative Research: A Theoretical and Practical Guide. *Qualitative Research in Psychology, 11*(1), 25–41. doi:10.1080/14780887.2013.801543

Sandelowski, M. (1995). Sample size in qualitative research. *Research in Nursing & Health, 18*(2), 179–183. doi:10.1002/nur.4770180211 PMID:7899572

Toor, M. D. (2021). Nutrients and Their Importance in Agriculture Crop Production: A Review. *Indian Journal of Pure & Applied Biosciences, 9*(1), 1–6. doi:10.18782/2582-2845.8527

U.S. Agency for International Development's Support. (2022). *Adapting food systems in Africa*. Author.

Vasileiou, K., Barnett, J., Thorpe, S., & Young, T. (2018). *Characterising and justifying sample size sufficiency in interview-based studies: Systematic analysis of qualitative health research over a 15-year period*. Academic Press.

Xiao, Y., & Watson, M. (2019). Guidance on Conducting a Systematic Literature Review. *Journal of Planning Education and Research, 39*(1), 93–112. doi:10.1177/0739456X17723971

Chapter 10
Alternative Dispute Resolution Techniques for Intertwined Artisanal Mining-Communal Farming Conflict

Annabel Banda
https://orcid.org/0000-0002-0144-6026
Gwanda State University, Zimbabwe

Mthuthukisi M. Ncube
National University of Science and Technology, Zimbabwe

ABSTRACT

The chapter discusses challenges of artisanal mining (AM), a nature-based livelihood strategy, in promoting sustainable agricultural practices. It highlights how AM competes for biodiversity and ecosystem services and causes environmental damage and a shift from traditional sustainable agricultural practices land-use tenure to uncontrolled itinerant AM. Methodologically, the chapter reviewed secondary systematic-literature review (SLR) of articles articulating the use of alternative dispute resolution (ADR) techniques. These include facilitated mediation, negotiation, and arbitration that yield land-use zoning agreements (conciliation) to attain synergy and market convergencies. It recommends transforming negative-synergy and implementing zoning strategies that prevent land-use conflicts, conserve biodiversity, and extricate competing AM from communal farming to attain sustainability. The resultant, unbridled development would protect ecosystem-service producing biodiversity, converge AM-agricultural markets and sustainably, leading to sustainable agricultural growth and conservation.

DOI: 10.4018/979-8-3693-2011-2.ch010

1. INTRODUCTION

Artisanal mining and communal farming are two nature-based livelihood strategies inextricably embedded on biodiversity's provisioning, regulating, supporting and cultural ecosystem services. They have however not been entirely socially, economically and environmentally compatible, raising sustainability of biodiversity at the centre of the seemingly polar opposite livelihood strategies whose interaction often sparks conflicts large and small among rural communities (Chari, Novukela & Ngcamu, 2022; Funoh, 2014). Tracking the drivers of land-use conflicts is necessary if lasting resolutions are to be achieved among actors involved in the dichotomised rural economy. Prime among the often-hidden factors driving conflicts among artisanal mining and communal farming communities is the pursuit of, and abuse of structures of power whose occupation results in classifications. These predispose such power structures to access to resources for self-aggrandisement. Kamncube's SPCR 4Δs Conflict Model depicted in Figure 1 below indicates various interdependencies among factors driving conflicts, including among artisanal miners and farming communities' land-use conflicts.

2. BACKGROUND OF THE STUDY

Individual and community access and use of natural resources in rural communities is under an ever-demanding global trends more focused on extraction and economic gains among competing livelihoods (Shackleton, 2020). People derive livelihood benefits from agriculture, livestock, and natural resources in ecosystems governed by a number of ecological (e.g., drought, floods, exhausted soils) and social factors (e.g., legislation, privatization, over-exploitation) factors (Moyo, Ncube & Mamhute, 2021; Shackleton, 2020; Agrawal and Ostrom, 2001; Chimhowu and Woodhouse, 2006). Losing access can render such communities vulnerable and have a negative impact on their resilience and general well-being (Shackleton, 2020).

Mining has negative impacts on rural farming communities and their means of subsistence worldwide, contributing to social-ecological change and conflict (Harlow, Hurley, Fox, Vargas-Guerra & Gibson, 2019; Bebbington, Bebbington, Bury, Lingan, Muñoz & Scurrah, 2018; Mtero, 2017; Andrews, 2018; Issah and Umejesi, 2018; Mnwana and Bowman, 2018; Kitula, 2006; Hilson, 2002). This is one factor influencing land-access and people's capacity to use natural resources. Across the world, rural farming communities depend on access to land and natural resources it offers for their subsistence. Livelihoods of millions of people are dependent on crop and cattle production, and the exploitation of non-timber forest products (NTFPs) on communal lands (de Sherbinin, VanWey, McSweeney, Aggarwal, Barbieri, Henry

& Walker, 2020; Mishra & Mishra, 2017; Shackleton & Luckert, 2015; Scoones, 1998). Thus, destroying these puts their livelihoods at risk, compelling those affected to defend their turf, even violently

In addition to providing food security, financial savings, and occasionally supplementary income, livestock and crops are significant cultural assets in many rural poor households (Aliber and Hart, 2009; Baiphethi and Jacobs, 2009). Additionally, unadulterated ecosystems offer essential services vital to human well-being, such as fruit, fuelwood, medicinal plants, and other natural resources collectively referred to as provisioning services or NTFPs (Shackleton, 2020; Shackleton & Luckert, 2015; Shackleton et al., 2007; de Groot, Wilson & Boumans, 2002). Artisanal mining has tended to be deleterious on farming land and other land-uses, decimating pastures and fostering environmental changes that are detrimental for agro-based livelihoods. Despite artisanal mining's beneficial effects to households and the economy, it has negative effects, particularly at the communal level where one clear consequence of mining has been the influx of itinerant mining communities into gold-rush locations in pursuit of jobs, economic opportunities, attracting people away from agriculture (Moyo, Ndlovu, Francis & Ncube, 2018; Tenkorang & Osei-Kufuor, 2013). Rising competition, not only for land use and dependence on the environment with negative externalities thereto, but also for housing, food, and employment drives up costs, rise in social vices like transactional sex, prostitution and crime, which increase demand for quick income that agriculture does not provide (Moyo et al., 2018; Akabzaa & Darimani, 2010).

The above result in outcomes that negate on social, economic and environmental sustainability elements of communities wherever the two livelihood strategies, regarded as detrimental to lives and the environment co-exist (Moyo et al., 2022; Tenkorang & Osei-Kufuor, 2013; Akabzaa & Darimani, 2001). There is need to harmonise and reconcile artisanal mining and communal farming livelihood strategies through deliberate rather than reactive conflict transformative models through the deployment of Alternative Dispute Resolution (ADR) mechanisms. This approach would be useful in dismantling the notions of superiority between artisanal mining and various strands of agriculture based on social class, power, and socio-economic structures to which individuals belong. The ADR would be helpful in ensuring mining and farming structures facilitate mediation, negotiation that results in conciliation among the actors in artisanal mining and communal farming.

3. RESEARCH OBJECTIVES

1. To highlight the role of traditional land-use tenure in conserving biodiversity and ecosystem services supporting agricultural production

2. To describe the increasing shift from agro-based livelihoods towards quick-wins offered by artisanal mining
3. To expose the paradox of artisanal mining and agricultural practices as unsustainable
4. To propose the alternative dispute resolution (ADR) as zoning strategy for competing land-uses

4. KAMNCUBE'S SPCR 4ΔS CONFLICT MODEL

The author-proposed model, the SPCR depicts the interaction of **structures**, and the use of **power** derived therefrom to increase **class**-based identities that help society gain access to **resources** they desire. This happens in a manner skewed in favour of one or another societal group or class of people, resulting in conflicts large and small. The model explains the occurrence of conflict as an outcome of the pursuit of **structures**, the (ab)use of **class power**, to gain access to biodiversity **resources**. Such land-use leaves a trail of destruction that pits one group or class in society against another, all in pursuit of resources.

The matrix above explains how in pursuit of a **central element**, **a target factor** is sought with a view of achieving an **intended means** to acquiring an **envisaged outcome**, resulting in conflicts. For example, *in pursuit of occupation of esteemed*

Figure 1. Kamncube's SPCR 4Δs conflict model (Mthuthukisi Ncube's conceptualisation of conflict driver)

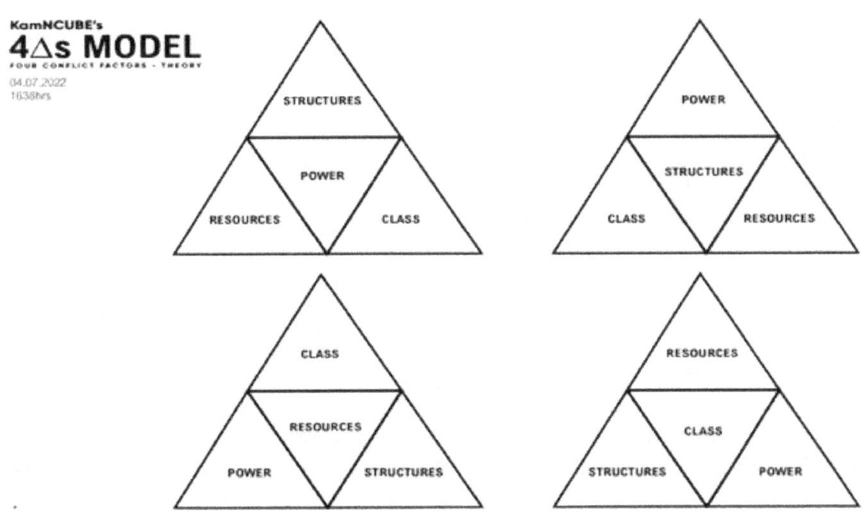

Table 1. Tabular explanation of Kamncube's SPCR 4Δs conflict model

Central Pursued Element	Target Factor	Intended Means	Envisaged Outcome
Structure	Power	Class	Resources
Power	Structures	Resources	Class
Class	Resources	Power	Structures
Resources	Class	Structures	Power

structures, power has been used, legitimately and illegitimately to create or belong to a given class that has access to certain privileges and resources. Secondly, *in pursuit of power (social, political, and economic), certain structures are (ab)used for their resources with a view of accessing certain classes for prestige as an outcome.* Thirdly, *in pursuit of a given class, resources have been deployed to attain power (rightly or wrongly so), which power is then (ab)used as a means to an end, which entails the occupation of structures desired for beneficial power purposes.* Fourthly, *in pursuit of resources, class has been the target factor. Such class is then (ab)used to yield access to structures of privilege and ultimately power as an envisaged outcome.*

Communal farmers tussle to occupy biodiversity structures of influence like water bodies and pastures while artisanal miners ride on the power and dominance of legal frameworks governing mining activities. This turn creates social classes where these groups have reduced or increased access to envisaged economic resources. It is important to understand how provisioning, regulating, supporting and cultural ecosystems services, communal farming land-use tenure, and artisanal mining interface in biodiversity conflicts. Kamncube's SPCR 4Δs Conflict Model helps identify interactions among the SPCR conflict factors, which channel alternative dispute resolution (ADR) efforts to separate artisanal mining from communal farming using zoning for each activity.

5. ARTISANAL-MINING, LAND-USE TENURE AND ECOSYSTEM SERVICES

Citing the World Bank Group (2001), Funoh (2014) posits small-scale mining is largely a poverty driven activity, typically practiced in the poorest and most remote rural areas of a country by a largely itinerant, poorly educated populace with few employment alternatives. Artisanal mining can be an unregistered or unlicensed mining and is carried out by lone individuals or small groups using a minimal amount of technology, equipment, inadequate health care and safety measures, and low environmental protection (Hentshel *et al.*, 2001). Not only in developing countries do individuals only survive on Artisanal mining but do also venture in farming activities. Agriculture is the engine of economic growth in many African countries, with a quarter of the global arable land in its possession. Africa has an estimated 30% of gross domestic product, and 60% of labour force employment attributed to the continent's agricultural sector (Chari, Novukela & Ngcamu, 2022).

In many developing countries, artisanal and small-scale mining (ASM) is largely a poverty-driven activity that plays an important socio-economic role and has thus experienced exponential growth in recent years, with millions of families in rural areas dependant on it for their livelihood (Chari et al., 2022). An estimated 13 million people from about 30 countries are directly engaged in small-scale mining, a significant proportion of whom are women and children. A further 80 to 100 million people across the developing world could depend on small-scale mining for some aspects of their livelihoods. Many have seen it as a panacea to poverty, unemployment, and unsustainable livelihoods (Adonten-Kissi and Adonteng-Kissi 2017). Only sixteen studies on the interactions between ASM and agriculture were found in a review by Ofosu *et al.* (2020), which emphasizes both synergistic and antagonistic relationships between the two major rural livelihood strategies (Moyo, Ncube & Ndlovu, 2022; Mkodzongi & Spiegel, 2019). However, Malone (2021) argues it is a common misconception that mining competes for scarce resources and land or has other detrimental effects on other livelihood strategies (Pijpers, 2011).

However, the influx of artisanal miners in communities has led to direct competition for resources with small-scale farmers, leading to community protests and in some cases, violent conflict (Crawford and Botchwey, 2017). Small holder farming and artisanal mining are both significant contributors to Zimbabwe's economy; with smallholder farming employing nearly 70%, while mining employs 7.1% of the country's economically active population (World Bank, 2019). However, there has been a major rise in conflict between miners and farmers primarily brought on by land use and tenure issues of entitlement. The link between artisanal mining activities and economic sustainability of agricultural food supply chain is accounted for by the conflict theory. Propagated by Karl Marx in 1867, the conflict theory claims society is in a state of perpetual conflict because of competition for limited resources (Cross-man, 2019). According to McCafferty (2006), the conflict theory focuses on competition for social, political, or economic supremacy, which finds expression in mining and agricultural conflict.

The conflict theory opines that individuals and groups within society have differing amounts of material and non-material resources. The powerful group dictates the terms and conditions to the oppressed majority because they have control over resources (McCafferty, 2006). Inspired by Karl Marx's Structural Conflict Theory, Ncube suggested that **structures** (S) have been manipulated to gain **power** (P), which is used in the **classification** (C) of society with a view of limiting or regulating access to **resources** (R) resulting in conflicts large and small as explained by Kamncube's subsequent SPCR 4Δs Conflict Model (Ncube, unpublished). The model depicts the root causes of conflict as pursuit or abuse of **structures**, pursuit of or abuse of **power**, pursuit of or abuse of **class**, with a view of gaining access to desired **resources**.

That artisanal mining and agriculture are the main livelihood strategies pursued by the country towards attaining the Middle Upper Income economy under Vision 2030 has sparked fierce conflict between them. What has exacerbated the conflict is the high unemployment and climate change that has drawn village farmers into ASM. Although local organizations have stepped in to arbitrate on these disputes, it is not clear whether the land can be divided based on zones. Although mining has helped grow the country's economy, it has also ruined the environment for farmers.

This book chapter presents an overview of the conflict between artisanal mining and the farming community, and suggests the adoption of transformative leadership and Alternative Dispute Resolution (ADR) techniques to resolve these conflicts peacefully. The Fast Track Land Redistribution Programme in the 2000s generated significant new modes of "belonging" in rural lands, which prompted modifications to farming along with new artisanal mining activities as articulated by the late agricultural expert Sam Moyo (Mkodzongi and Spiegel, 2019). Years later, mining-farming interface still influence rural labour trajectory and demand close examination (Moyo, Ncube, and Ndlovu, 2022). Artisanal gold mining, and small-scale communal agriculture remain two of Zimbabwe's economic foundations (Moyo *et al.*, 2022). Despite being prohibited by Zimbabwe's colonial-era mining laws, which continue to influence modern mining policies (Spiegel, 2015; Dube *et al.*, 2016), ASM has grown to be a common economic activity carried out by socially diverse groups with a variety of educational backgrounds and socioeconomic statuses (Mabhena, 2012; Mpofu and Mpofu, 2017). Over a million people's livelihoods directly depend on ASM across the nation (PACT, 2015; Mkodzongi and Spiegel, 2019).

Individuals may engage in a variety of ASM activities, including diamond or gold rushes, which are characterized by unsteady communities that are conflict-prone and are often transitory and driven by recession with examples such as gold-mining nations of Zimbabwe, Peru, Venezuela, and Bolivia (Hentschel *et al.*, 2003). Communities with substantial population swings and initial instability may either go away after a few years or develop into permanent communities alone (Hentschel *et al.*, 2003).

Mining and farming have frequently coexisted in Zimbabwe with many other informal revenue generating tactics (Sachikonye, 2011; Mkodzongi and Spiegel, 2019). In 1990s and early 2000s, participation in the ASM industry in Zimbabwe evolved into a seasonal, part-time, and permanent source of income (Mondlane and Shoko, 2003; Maponga and Ngorima, 2003). With the adverse effects of Economic Structural Adjustment Programmes (ESAPs) and drought-restricted employment prospects in agriculture, national increases in ASM activity began to appear in the 1990s (Mabhena, 2012; Mkodzongi and Spiegel, 2019).

Since the early 2000s, the resource boom has brought up significant issues regarding how projects affect social dynamics and conflict as well as livelihoods (Bebbington *et al.*, 2008; Brain, 2017; Malone, Smith and Zeballos, 2021). As opponents rally around environmental concerns, requests for benefits and compensation, and demands for territorial sovereignty and local authority, large-scale extractive activities have historically been accompanied by conflict (Helwege, 2015; Paredes, 2016; Malone *et al.*, 2021). However, very little empirical studies have been done to gain insight into how Zimbabwe's Fast Track Land Reforms changed livelihoods of newly resettled peasant farmers, legally designated as "A1 farmers," despite the rise in ASM activities that followed (Mabhena, 2012; Moyo, Ncube, and Ndlovu, 2022). Discussions about artisanal mining in Zimbabwe have frequently centred on allegations of political favouritism and elite takeover of natural resources (Mawowa, 2013). In contrast, narratives about the connections between mine and farming in Africa more generally have a tendency to portray artisanal mining as an off-farm endeavour that is replacing or destroying peasant farming rather than enhancing it.

Such perspectives have occasionally given rise to calls for fundamentally rethinking the nature of mining-farming linkages in African contexts, with much critical research focusing on West Africa and in some cases pointing to significant ASM-generated funds complementing and supporting investment into smallholder farming (Bryceson and Jnsson 2009; Banchrigah and Hilson 2010; Maconachie, 2011; Maconachie and Hilson 2011; Hilson, 2016; Moyo *et al.*, 2022). The two activities are increasingly intertwined in many rural areas of Zimbabwe although some studies have highlighted recent disputes between farmers and miners with lengthy histories of miner-farmer contestations (Musemwa, 2009). This is despite well-documented environmental changes, destruction and land degradation associated with ASM (Ncube-Phiri, Mucherera, and Ncube, 2015; Mkodzongi and Spiegel, 2019; Moyo *et al.*, 2022).

A few exceptions do look at disagreements involving how ASM interacts with other sources of income. Occasionally, illegal miners seized land and displaced crops (Gilbert and Albert 2016). In other instances, ASM activity affected traditional subsistence livelihoods, forcing local communities to adopt ASM to make a living, setting off a vicious cycle that further weakened traditional livelihoods (Hennessy, 2015; Ofosu *et al*, 2020). Other instances include the harm caused by ASM-driven deforestation to livelihoods reliant on the forest resources (Theije and Salman 2018) and ASM lead poisoning pollution to fishing (Hook 2019).

Communal farmers livestock have perished in abandoned pits of unregistered artisanal miners. Unemployment surge in the country is causing most people especially males to engage in artisanal gold mining (Moyo & Chinembiri, 2019). Artisanal miners have created danger for communal farmers livestock leaving behind remnants of chemicals like mercury being eaten by livestock, the chemicals getting into water sources hence being hazardous to both humans and animals some burning the veld so as to use gold sensors thus reducing the grazing area and biodiversity. In Matabeleland South some Non-Governmental Organization (ZimPro) are trying to engage the two parties (miners and farmers) to minimize conflict (Moyo & Chinembiri, 2019). The conflict therefore would require a holistic approach with different stakeholders (miners, farmers, environmental management agency, police, ministry of health) engaging to minimize the conflict.

From the review of literature, it has emerged that the impacts of artisanal mining are mainly felt in five areas. These are pollution, land degradation, loss of land, labour migration, and social ills (Jnr *et al.* 2014; Poku 2016). The production process in artisanal mining has the potential to affect the environment in several ways. The pollution index shows that mine sites are "heavily polluted" with abnormally high concentrations of chemicals (Eludoyin *et al.*, 2017). Illegal miners pollute water bodies destroy farmlands and they do not reclaim the land (Poku, 2016). The opened land portions are rarely filled up, causing them to be dangerous causes of accidents to victims that fall in them, or serve as habitats for reptiles and dangerous animals (Eludoyin *et al.*, 2017). For agricultural food supply chains, generally, these are linked directly by one or more of the upstream or down stream flows of products, services, and information (Sodhi and Tang, 2012).

Mining operations further affect water quality through infiltration of waste water into ground water thereby polluting it with chemicals. There is existing biological evidence linking this pollution to reduction in crop yields (Aragon and Rud, 2012; Poku, 2016). Mining activities consume large tracts of land, there by eating into and reducing, agricultural land for farming community. The negative impacts from the mining activities can be permanent and render previously fertile agricultural land barren. This sets up a direct competition with small-scale farmers for control and use of land (Slack, 2013). Large numbers of people are displaced by mining operations resulting in loss of farmlands (Slack, 2013). Additionally, Jnr *et al.* (2014) found out that farmers in Ghana who were displaced by mining activities were not compensated in the form of alternative land. The areas lost represented a substantial portion of the affected communities' farmlands. A significant reduction in food security is inevitable in areas where artisanal mining is practiced. All these issues and problems caused by artisanal mining has frustrated the farmers to a greater extent and caused them to retaliate in order to preserve their agricultural land.

6. ARTISANAL MINING-COMMUNAL FARMING IMPACTS ON THE ENVIRONMENT

Communal livestock farming is a type of agriculture where livestock farmers share a common area and manage their holdings as cooperative businesses. A portion of these units is situated on privately held, leased, or state-owned property (Molieleng, Fourie and Nwafor, 2021). This farming strategy, which is still in use in many southern African rural areas, is the oldest and most reliable way to increase household farming security. Communal livestock husbandry then becomes a means of ensuring food security, since over 75% of the world's poorest population live in rural areas and 2.5 billion people live on tiny farms and are solely dependent on agriculture for their living (Molieleng et al., 2021).

For instance, over 10,000 people rely on Ghana's forest reserves for their food and livelihood, and their degradation is largely linked to mining (Adjei, 2017). Only 1.2 million of Ghana's 8.3 million hectares of forest estate remain today, primarily due to the country's decision to let mining operations within the forests after it gained independence in 1957 (Adjei, 2017). The eviction of indigenous people from their communal and ancestral lands where they had been engaged in economic activities like farming to sustain their families is underscored as an impact of artisanal mining on communal farming (Adjei, 2017). The rangeland management policies in southern Africa are shaped by a three-tiered conceptual model of development, a model that applies the modernization theory of development to two compelling environmental narratives: the tragedy of the commons, and land degradation (Rohde, Moleele, Mphale, Allsopp, Chanda, Hoffman, Magole and Young, 2006).

Raising of livestock poses a risk to the ecosystem (Dopelt, Radon and Davidovitch, 2019), with main effects evident through the high rates of greenhouse gas (GHG) emissions, water and land pollution, and deforestation. van Noordwijk, Minang, Dewi, Hall and Rantala (2009) assert that 96% of the world's deforestation is caused by livestock farming that produces feed, fodder, and grazing areas. Additionally, the management of animal manure and the use of pesticides and herbicides are significant causes of water and land contamination (Grossi, Goglio, Vitali and Williams, 2019; Neeliah, Rajkomar, Dookun-Saumtully and Ramkissoon, 2006). The two main greenhouse gasses that come from animal agriculture are methane and nitrous oxide. Grossi et al. (2019) argue that nitrous oxide, which is produced when manure and fertilizers are stored, contributes to global warming 265 times more than carbon dioxide, whereas methane influences global warming 28 times more than carbon dioxide. Consequently, the production of livestock is also negatively impacted by climate change (Grossi et al., 2019).

The primary driver of forest clearing is the rising demand for animal products, which inevitably leads to the need for new pastures to accommodate the growing animal population (Dopelt, Radon, Davidovitch, 2019). Expanding agriculture is the primary cause of deforestation worldwide, accounting for 96% of it. Indirect causes of deforestation at the local level include political, cultural, and socioeconomic variables such as poor governance and policies, corruption, landlessness and ambiguous allocation rights, migration, rural poverty, and a lack of capital and funding (Dopelt, Radon, Davidovitch, 2019; van Noordwijk, Minang, Dewi, Hall and Rantala, 2009). In Africa, agriculture was found to be the primary cause of deforestation where poor farmers and communities depend on forest lands for agricultural and fuelwood collection, often adopt slash-and-burn/fallow techniques were found to be the primary cause of deforestation in the Congo Basin rainforest (Dopelt et al., 2019; van Noordwijk et al., 2009).

Most of the developing countries especially in the southern Africa have adopted the communal land tenure system (Gottlieb & Grobovšek, 2019). The principle here is "use it or lose it", if not in use, communal owners lose their land and are not allowed to transfer it, while rentals may be allowed, there is risk of losing ones' land. It is governed by customary law and varies among countries. Land property rights are not complete as the land either belongs to the state or to the community. When people acquire full property rights, it was noted that they would be a large shift of unskilled workers from agriculture into non-agricultural productivity (Gottlieb & Grobovšek, 2019). Additionally, the restrictive nature of land-rich farmers who limit the amount of land they rent out to highly-skilled land-poor individuals adds to the negative effects of communal land ownership due to risk of expropriation. Thus, skilled land-poor people end up paying exorbitant rentals and being allocated little land. If this misallocation is removed, they will be an estimated one-fifth increase in agricultural productivity. The Communal Land Act of Zimbabwe section number nine (9e), states that, 'a rural district council may, with the approval of the Minister, issue permit authorizing any person or class of persons to occupy and use, subjects to Regional, Town and Country Planning Act [Chapter 29:12] and any other order issued in terms thereof, any portion of communal land within the area of such rural district council, where such occupation or use is for any of the following purposes – (e) any purpose whatsoever which, in the opinion of the rural district council, is in the interests of inhabitants of the area concerned (Legislation, 2023). This could be the reason why in Zimbabwe gold panning is rampant and some rural houses have been demolished by such acts.

After the attainment of independence in 1980 wildlife habitats decreased however due to the introduction of Communal Areas Management Program for Indigenous Resources (CAMPFIRE) the loss slowed down (Taylor, 2009). Recently though there has been pressure on habitats and other natural resources due to deteriorating socio-economic conditions in the country. Food security among smallholder farmers is threatened in Zimbabwe, especially due to drought hence there are not able to generate extra income. Low yields therefore are compensated by extensification rather than intensification (Marongwe et al., 2011). This threatens the distribution of already thin resources of labour and inputs and of land degradation as farmers move to marginal and fragile environments.

A great deal of these communal farmers practice unsustainable soil and crop management activities. Sustainability of food production in southern Africa is mostly threatened by water scarcity and degradation of soil through salinization, water logging, soil erosion and nutrient depletion. Yield reduction has also been aggravated by conventional tillage which results in fertile top soil erosion and the expansion of cropping into unsuitable areas like steep slopes and river banks (Marongwe et al., 2011). In order to increase yields, reduce effects of low rainfall, and thus, conserve moisture through reduced soil disturbance, Zimbabwe communal farmers have been introduced to conservation agriculture (conservation farming), natively called "Ga-Tshompo" under the Pfumvudza Zimbabwean government program. The main principles with this conservation agriculture are (i) minimum mechanical soil disturbance (ii) maintenance of ground cover with organic matter and (iii) diversification of crop species grown in rotation, sequence or association (Marongwe et al., 2011).

This was implemented by non-governmental organisations in Zimbabwe in 2003-2004. This was introduced in order to curb low productivity, and hence improve food security and to address the draught-power shortages. There is increased adoption of conservation farming by communal farmers in Zimbabwe, mostly inspired by inputs-supply, and ready markets for outputs (Marongwe et al., 2011). A number of advantages of practicing conservation farming abound in Zimbabwe, namely; climate change mitigation through increased seed germination and reduced moisture stress; mitigation of land degradation by reduction in soil loss and increased soil carbon content. Farmers have lamented the increased need for weeding and base preparation labour, hence the need to come up with a locally adapted machinery to move from totally manual system (Marongwe et al., 2011).

7. MATERIALS AND METHODS

The study was informed by a documentary review of existing literature based on the preferred reporting items for systematic analyses focusing on a combination of 30 PRISMA-based documents including peer-reviewed journal articles, books, edited book chapters, relevant policy, laws, programmes and implementation strategies related to resource conservation, protected area management, threats and challenges to biodiversity conservation. The Zimbabwe Mines and Mineral Act of 1961 gives more power to mineral prospects than farming and other land-

Table 2. Comparison of improved communal land farming and artisanal gold mining impact on conservation and biodiversity

Variable	Communal Land Farming (Conservation Farming)	Artisanal Mining
Climate change factors	Addressed through increased rainfall use efficiency by employing techniques that increase water infiltration and reduce moisture evaporation from the soil	No climate change factors employed
Mitigation land degradation	Soil loss due to erosion reduced, (5t/ha/year) in mulch ripped conservation agriculture systems.	Increased soil loss
Soil carbon content	Increased soil carbon content	Reduced soil carbon content
Biodiversity	Increased due to intensification rather than extensification	Reduced due to burning of bush and land degradation

uses (Moyo & Chinembiri, 2019). The Act has been on hold on 61 proposed amendments in Parliament since 2015 although there is a headway in the Bill awaiting Presidential Assent that mandates miners to seek the consent of the land owner before prospecting, which is a move towards balancing the two-nature based livelihoods.

8. RESULTS AND DISCUSSION: POSSIBLE RESOLUTIONS TO ARTISANAL MINERS-FARMERS CONFLICT

The conflict between artisanal miners and farmers indicated above can be successfully resolved by using the transformative leadership approach. Involving those who are fuelling the dispute, educating them to better understand both sides, and helping them to identify common ground that will allow both sides to coexist in the same area. It is possible to urge artisanal miners to abide with environmental laws such as controlling chemically contaminated process water so that it does not contaminate natural water sources like wells and rivers used by farmers and their livestock. Additionally, miners can fence and enclose only the active sites if their pegged blocks leaving the idle portions open for usage by local farmers as grazing areas and farming operations. It is possible to appropriate dispute resolution due to its contextual nature, offering routes for resolving disputes through the use of alternative dispute resolution methods. In most cases, state courts or traditional court systems are used to resolve land disputes motivate miners to plant trees after closing the pits they would have left open while mining.

According to the Mines and Minerals Act, miners must obtain the landowner's consent before prospecting on any parcel of land that is smaller than 200 hectares. Such cooperation and understanding will result in a win-win situation that allow both parties to exist without interfering with each other's business. Alternative Dispute Resolution, often abbreviated as ADR, entails a range of approaches aimed at resolving disputes devoid of confrontation (Shamir, 2016; Afolabi *et al.*, 2019; Nwachukwu, 2020). The ADR as an 'alternative' dispute settlement mechanism takes place outside the statutory court system, typically not akin to statutory court procedures (Kumari, 2020). Particularly in light of the complexity of the customary land realm, the state courts are sometimes characterized by exacting and bureaucratic judicial norms and processes that cause delays in the justice delivery process (Crook *et al.*, 2011; Oppong-Kusi, 2019). The conventional customary courts, on the other hand, are frequently criticized for their inability to handle the scope and complexity of contemporary property conflicts in the context of waning respect for traditional authorities (Agheyisi, 2019; Kalabamu, 2021).

Alternative Dispute Resolution (ADR) procedures are anticipated to offer a compromise between these two points of view in order to address the shortcomings of both approaches to the settlement of land disputes. Scholars (Kuusaana *et al.*, 2013; Aiyedun and Ordor, 2016) argue that ADR mechanisms can help in the decongestion of the statutory courts overwhelmed with land cases, offer a cheaper, faster, and peaceful form of justice for the ordinary citizen, particularly the rural poor. Rural poor often do not have access to the state justice system either because of lack of resources or because of long physical distance to formal courts.

The conflicts between miners and farmers can be resolved through facilitation, a different approach to conflict resolution in which a third party is specially prepared, neutral, impartial, objective, and answers questions from group members to help the group achieve what it has proposed while taking into account the opinions of participants. The alternate approach of conflict resolution known as negotiation involves the parties directly engaging in a discourse to attempt and seek a resolution to their differences. In order to discuss and examine a wide range of issues, negotiation offers a framework that may be used for a variety of goals. By negotiating effectively, you can accomplish your objectives, win the support and participation of others, and create or enhance productive working relationships. In this instance, it has been noticed that conflicts generally result from artisanal mining's negative effects, such as pollution, land degradation, and societal issues, all of which may be discussed and accommodated by both parties. Zoning agreements involving the traditional leadership land use-tenure system could be used to resolve encroachment between these two livelihood strategies to prevent competitive clashes (Corbett, O'Faircheallaigh & Regan, 2017).

9. REDUCING BIODIVERSITY LOSSES

Zimbabwe's southern Matabeleland has historically been a cattle-producing region, with grain farming serving as a major source of income (Moyo, Ncube and Ndlovu, 2022). In response to the increasingly severe impacts of climate change on ecosystem-based livelihoods, diminishing grazing pastures, and rain-fed agriculture, agricultural communities in the same region have turned to artisanal small-scale gold mining (ASM) (Moyo et al., 2022). In what has been referred to as the tragedy of ecological services that unites incongruent livelihood options, ecosystem-based rural livelihoods such as irrigation-based communal farming, rain-fed agriculture, and artisanal mining now share common ground among vulnerable rural populations in drought-prone areas (Ncube, Moyo & Mamhute, 2021; Shackleton, 2020). However, inadequately controlled, ASM frequently encroaches and straddles beyond farming boundaries with detrimental effects on other land-use activities, which is detrimental to food security, sustainable livelihoods and the environment (Moyo et al., 2022; Shackleton, 2020).

Ecosystems known as agro-ecosystems are those in which the makeup of living things has been purposefully chosen by humans for agricultural purposes (MA, 2005). Agro-ecosystems are different from unmanaged ecosystems because they are deliberately changed, and frequently extensively managed, to provide food, fibre, and other products (Kremen and Miles, 2012). Resultantly, they have economic, environmental, and human community components by nature, which makes the maintenance of biodiversity within an agro-ecosystem necessary to ensure the continued supply of goods and services.

Sustainable agriculture has been more popular since the 1987 release of the Brundtland Report. However, the definition of sustainable agriculture is somewhat nebulous, making its application and utilization exceedingly challenging. For human survival and, by extension, for any human activity, an agriculture that can consistently supply food and other resources to an increasing global population is essential (Velten, Leventon, Jager & Newig, 2015; Rivera-Ferre, Ortega-Cerdà, & Baumgärtner, 20131). However, a number of issues, such as climate change, a high rate of biodiversity loss, land degradation due to compaction, erosion, pollution, and salinization, pollution of water resources, rising production costs, a declining number of farms and, consequently, poverty and a decline in the rural population, threaten agriculture's ability to meet human needs both now and in the future (Velten, Leventon, Jager & Newig, 2015; Rivera-Ferre, Ortega-Cerdà, & Baumgärtner, 20131). In addition to facing these challenges, agriculture's current practices during the past few decades have been a major contributor to all of these difficulties (Velten et al., 2015; Rivera-Ferre et al., 2013).

Climate change, energy, water, and land shortages are threatening agriculture production, leading to increased food insecurity and hunger in affected regions, particularly in Africa. Sustainable agriculture, an eco-system approach, is recommended to balance soil, water, plants, environment, and living organisms. Innovative technologies and research should be developed to ensure sustainable agriculture and productivity, while resource conservation and best management practices are used. Sustainable Agricultural Development presents successful experiences in sustainable farming, water and land resource management, and innovative livestock production, promoting resource conservation and tackling food security challenges.

Livestock contributes to nearly a fifth of global greenhouse gas emissions and may suffer from climate change effects. To address this, improved soil quality and resource conservation technologies are needed. However, selective breeding for higher productivity and yield can cause health and welfare stresses. A contraction and convergence model could help develop countries increase sustainable consumption (Behnassi, Shahid & D'silva, 2011).

These views are underscored by Muhie (2022) who suggests agrarian scientists and legislators around the world are irritated by the unsettling rate at which the world's population is growing and the amount of food required to feed this rapidly expanding human population. Scholars contend that today's population is stealing natural resources from the next generation, potentially jeopardizing their access to wholesome food and clean air (Umesha, Manukumar & Chandrasekhar, 2018Müller-Lindenlauf, 2009). The reasons behind this problem are numerous and can be stated, but one must stand out: the inadequate adoption of innovative techniques and strategies for sustainable agriculture. Climate smart agriculture (CSA), organic farming, biodynamic farming, sustainable intensification, and regenerative agriculture are few innovative techniques for adoption. Other cutting-edge sustainable agriculture methods include precision agriculture, integrated nutrient management (INM), integrated farming systems (IFS), and integrated pest management (IPM) (Muhie, 2022; Umesha et al., 2018). Sustainabke agricultural practices include the following:

Agroforestry

Agroforestry, which Mbow, Van Noordwijk, Luedeling, Neufeldt, Minang & Kowero (2014); Lasco, Delfino, Catacutan, Simelton & Wilson (2014) unpack as the practice of growing trees alongside crops to enhance ecological and economic systems and offers benefits like increased soil organic matter, agricultural yields, carbon sequestration, water storage, agrobiodiversity, and farmer income is a sustainable agricultural practice that is recommended. Variables like agroecosystem type, tree phylum, and management practices impact carbon sequestration (Mbow et al., 2014;

Lasco et al., 2014). Agroforestry is cost-effective and climate-smart, especially in drought conditions, as trees search for nutrients and water (Paul, Weber & Knoke, 2017; Zomer, Neufeldt, Xu, Ahrends, Bossio, Trabucco & Wang, 2016).

Another sustainable agricultural practice is the *Mixed Cropping* is an indigenous farming method that diversifies food supply, reduces risk of decline, pests, and illness, and promotes biodiversity preservation, thereby reducing global warming and promoting sustainable food production (Singh, 2023; Fraser, Frausin & Jarvis, 2015).

Crop Rotation

The other sustainable agricultural practice is *Crop Rotation* is a traditional method of growing plant species in succession on the same land, aiming to address agroecological issues like soil degradation and global warming caused by short rotation and mono-cropping. It is effective in sequestering carbon and reducing greenhouse gas emissions. Crop rotation alters soil structure, organic carbon concentration, nutrient cycling, and disease presence, improving crop health and production.

Cover Cropping

Additional to the above, there is *Cover Cropping,* which is a sustainable method for improving soil health, agroecosystem, and microbial biomass (Singh, 2023; Cooper, Hama-Aziz, Hiscock, Lovett, Dugdale, Sünnenberg & Hovesen, 2017). It involves growing non-cash crops to prevent soil erosion and nutrient loss while cover crops can be non-leguminous or leguminous, and can be grown alongside primary crops to provide living mulch (Cooper et al., 2017). Popular cover crop species include sun hemp, sorghum, velvet bean, rye, oat, pea, vetch, and clover (Cooper et al., 2017).

Intercropping

Sustainable agriculture also includes *Intercropping*, or growing multiple crop species simultaneously on the same area, is a useful application of fundamental ecological concepts like diversity, rivalry, and simplification. It is one of the farming methods that is very productive. Since various crops have varying degrees of climatic adaptability, intercropping decreases climate-driven crop failure. Intercrops improve the biodiversity, productivity, resilience, and stability of an agroecosystem by effectively using natural resources like soil, light, water, and nutrients (Ning, Qu, He, Yang, Chen, Luo & Cai, 2017). Intercropping grains and legumes improves the efficiency of cultivation in nutrient-restricted environments. Legumes collaborate symbiotically with rhizobium microbes to fix nitrogen in the soil. Leguminous crops decrease N2O emissions from agricultural regions while also increasing the release and recycling of mineralizable nitrogen-containing compounds in soil (Singh, 2023; Scalise, Pappa, Gelsomino & Rees, 2017).

Integrated Crop-Animal Farming

Integrated Crop-Animal Farming is one of the many sustainable agricultural practices smallholder producers in Asia use, which benefits agrobiodiversity, food diversity, and land resource management. Rice-fish culture is a crucial farming technique that enhances food production's sustainability, intensification, output, and profitability. Fish and ducks consume weeds, plant hoppers, and insects, reducing the need for chemical fertilizers in conventional rice cultivation.

Soil Carbon Sequestration

Soil carbon sequestration, also known as carbon farming or regenerative agriculture, involves caring for land to increase soil carbon storage and absorption. This climatologically renewable resource is crucial for soil functionality and ecosystem services. Improving this resource involves positive eco-system and soil carbon expenditure.

10. LIMITATIONS OF THE STUDY

The study was limited only to secondary data reviewed from the preferred reporting items for systematic review and meta-analysis conducted. It did not involve primary data, which could have generated new insights and location specific context, hence the findings represent findings and sentiments obtained from secondary empirical studies reviewed.

11. FURTHER STUDY

Deriving from the limitation of the chapter, future studies may need to focus on a given geographic location to conduct a similar study that is informed by primary data collection to test the application of artisanal mining-farming conflict theory explained by Kamncube's SPCR **4Δs Conflict Model** and how the alternative dispute resolution could best address the biodiversity and land-use conflicts to attain sustainability of the livelihood strategies.

12. CONCLUSION

The intertwined nature-based livelihood strategies of artisanal mining and communal farming stand to be more effective operating as harmonised and synergised than stand-alone strategies due to the rampaging effects of climate change on environmental change. While biodiversity presents with good opportunities for both agricultural and mining activities, the ecosystem services driving from the same biodiversity has often generated conflicting outcomes pitting artisanal miners and local communal farming villagers. Current land-use conflicts can and should be resolved by deploying alternative dispute resolution mechanism (ADR) that would effectively conduct zoning of each land-use. This is possible through the involvement of the Traditional Leaders in collaboration with local districts offices in each area such that there is avoidance of itinerancy and maintenance of orderly activities, farming and mining alike by known and traceable locals or the use of a local permit system that allows one to conduct mining responsibly through traditional and legal policies as a collaborative framework.

REFERENCES

Adjei, E. (2007). *Impact of mining on livelihoods of rural households. A case study of farmers in the Wassa Mining Region, Ghana* [Master's thesis]. Geografisk institutt.

Adonteng-Kissi, O., & Adonteng-Kissi, B. (2017). Living with conflicts in Ghana's Prestea mining area: Is community engagement the answer? *Journal of Sustainable Mining*, *16*(4), 196–206. doi:10.1016/j.jsm.2017.12.005

Agrawal, A., & Ostrom, E. (2001). Collective action, property rights, and decentralization in resource use in India and Nepal. *Politics & Society*, *29*(4), 485–514. doi:10.1177/0032329201029004002

Aliber, M., & Hart, T. G. (2009). Should subsistence agriculture be supported as a strategy to address rural food insecurity? *Agrekon*, *48*(4), 434–458. doi:10.1080/03031853.2009.9523835

Andrews, N. (2018). Land versus livelihoods: Community perspectives on dispossession and marginalization in Ghana's mining sector. *Resources Policy*, *58*, 240–249. doi:10.1016/j.resourpol.2018.05.011

Bebbington, A., Bebbington, D. H., Bury, J., Lingan, J., Muñoz, J. P., & Scurrah, M. (2018). Mining and social movements: Struggles over livelihood and rural territorial development in the Andes. *World Development*, *36*(12), 2888–2905. doi:10.1016/j.worlddev.2007.11.016

Bebbington, A., & Humphreys Bebbington, D. (2018). Mining, movements and sustainable development: Concepts for a framework. *Sustainable Development (Bradford)*, *26*(5), 441–449. doi:10.1002/sd.1888

Behnassi, M., Shahid, S. A., & D'silva, J. (2011). *Sustainable agricultural development*. Springer. doi:10.1007/978-94-007-0519-7

Cappelli, S. L., Domeignoz-Horta, L. A., Loaiza, V., & Laine, A. L. (2022). Plant biodiversity promotes sustainable agriculture directly and via belowground effects. *Trends in Plant Science*, *27*(7), 674–687. doi:10.1016/j.tplants.2022.02.003 PMID:35279365

Chari, F., Novukela, C., & Ngcamu, B. S. (2022). Artisanal mining versus sustainability of agricultural food supply chains: Effects of the conflicts in Southern Zimbabwe. *Development in Practice*, *32*(3), 349–360. doi:10.1080/09614524.2021.1937543

Chimhowu, A., & Woodhouse, P. (2006). Customary vs private property rights? Dynamics and trajectories of vernacular land markets in Sub-Saharan Africa. *Journal of Agrarian Change*, *6*(3), 346–371. doi:10.1111/j.1471-0366.2006.00125.x

Corbett, T., O'Faircheallaigh, C., & Regan, A. (2017). 'Designated areas' and the regulation of artisanal and small-scale mining. *Land Use Policy*, *68*, 393–401. doi:10.1016/j.landusepol.2017.08.004

Crawford, G., & Botchwey, G. (2017). Conflict, collusion and corruption in small-scale gold mining: Chinese miners and the state in Ghana. *Commonwealth and Comparative Politics*, *55*(4), 444–470. doi:10.1080/14662043.2017.1283479

de Groot, R. S., Wilson, M. A., & Boumans, R. M. (2002). A typology for the classification, description and valuation of ecosystem functions, goods and services. *Ecological Economics*, *41*(3), 393–408. doi:10.1016/S0921-8009(02)00089-7

de Sherbinin, A., VanWey, L. K., McSweeney, K., Aggarwal, R., Barbieri, A., Henry, S., Hunter, L. M., Twine, W., & Walker, R. (2020). Rural household demographics, livelihoods and the environment. *Global Environmental Change*, *18*(1), 38–53. doi:10.1016/j.gloenvcha.2007.05.005 PMID:19190718

de Theije, M., & Salman, T. (2018). Conflicts in marginal locations: Small-scale gold-mining in the Amazon. *The Pick*, 261.

Diringer, S. E., Feingold, B. J., Ortiz, E. J., Gallis, J. A., Araújo-Flores, J. M., Berky, A., Pan, W. K. Y., & Hsu-Kim, H. (2015). River transport of mercury from artisanal and small-scale gold mining and risks for dietary mercury exposure in Madre de Dios, Peru. *Environmental Science. Processes & Impacts*, *17*(2), 478–487. doi:10.1039/C4EM00567H PMID:25573610

Dopelt, K., Radon, P., & Davidovitch, N. (2019). Environmental effects of the livestock industry: The relationship between knowledge, attitudes, and behavior among students in Israel. *International Journal of Environmental Research and Public Health*, *16*(8), 1359. doi:10.3390/ijerph16081359 PMID:31014019

Eludoyin, A. O., Ojo, A. T., Ojo, T. O., & Awotoye, O. O. (2017). Effects of artisanal gold mining activities on soil properties in a part of south-western Nigeria. *Cogent Environmental Science*, *3*(1), 1305650. doi:10.1080/23311843.2017.1305650

Fraser, J. A., Frausin, V., & Jarvis, A. (2015). An intergenerational transmission of sustainability? Ancestral habitus and food production in a traditional agro-ecosystem of the Upper Guinea Forest, West Africa. *Global Environmental Change*, *31*, 226–238. doi:10.1016/j.gloenvcha.2015.01.013

Funoh, K. N. (2014). *The impacts of artisanal gold mining on local livelihoods and the environment in the forested areas of Cameroon* (Vol. 150). CIFOR.

Gilbert, D., & Albert, O. B. (2016). Illegal small-scale gold mining in Ghana: A threat to food security. *Journal of Food Security*, *4*(5), 112–119.

Grossi, G., Goglio, P., Vitali, A., & Williams, A. G. (2019). Livestock and climate change: Impact of livestock on climate and mitigation strategies. *Animal Frontiers*, *9*(1), 69–76. doi:10.1093/af/vfy034 PMID:32071797

Harlow, D. E., Hurley, K., Fox, A., Vargas-Guerra, A., & Gibson, J. (2019). *Small-scale & artisanal mining: Impacts on biodiversity in Latin America*. The Cadmus Group and USAID.

Hennessy, L. (2015). Where there is no company: Indigenous peoples, sustainability, and the challenges of mid-stream mining reforms in Guyana's small-scale gold sector. *New Political Economy*, *20*(1), 126–153. doi:10.1080/13563467.2014.914158

Hentschel, T. (2003). *Artisanal and small-scale mining: challenges and opportunities*. Academic Press.

Hentschel, T., Hruschka, F., & Priester, M. (2002). *Global report on artisanal and small-scale mining*. Report commissioned by the Mining, Minerals and Sustainable Development of the International Institute for Environment and Development. http://www.iied. org/mmsd/mmsd_pdfs/asm_global_report_draft_jan02.pdf

Hilson, G. (2002). An overview of land use conflicts in mining communities. *Land Use Policy*, *19*(1), 65–73. doi:10.1016/S0264-8377(01)00043-6

Issah, M., & Umejesi, I. (2018). Risks and vulnerability in uranium mining: A synthesis of local perspectives in the Great Karoo region of South Africa. *The Extractive Industries and Society*, *5*(3), 284–293. doi:10.1016/j.exis.2018.04.002

Jnr, S. D., Cieem, G., Ayensu-Ntim, A., Twumasi-Ankrah, B., & Barimah, P. T. (2016). Effects of loss of agricultural land due to large-scale gold mining on agriculture in Ghana: The case of the Western Region. *Br. J. Res*, *2*(6).

Kitula, A. G. N. (2006). The environmental and socio-economic impacts of mining on local livelihoods in Tanzania: A case study of Geita District. *Journal of Cleaner Production*, *14*(3-4), 405–414. doi:10.1016/j.jclepro.2004.01.012

Kremen, C., & Miles, A. (2012). Ecosystem services in biologically diversified versus conventional farming systems: Benefits, externalities, and trade-offs. *Ecology and Society*, *17*(4), art40. doi:10.5751/ES-05035-170440

Lasco, R. D., Delfino, R. J. P., Catacutan, D. C., Simelton, E. S., & Wilson, D. M. (2014). Climate risk adaptation by smallholder farmers: The roles of trees and agroforestry. *Current Opinion in Environmental Sustainability*, *6*, 83–88. doi:10.1016/j.cosust.2013.11.013

Ma, M. (2005). Chapter 11: Biodiversity Regulation of Ecosystem Services. Ecosystems and human well-being: Current State and Trends.

Malone, A., Smith, N. M., & Zeballos, E. Z. (2021). Coexistence and conflict between artisanal mining, fishing, and farming in a Peruvian boomtown. *Geoforum*, *120*, 142–154. doi:10.1016/j.geoforum.2021.01.012

Mbow, C., Van Noordwijk, M., Luedeling, E., Neufeldt, H., Minang, P. A., & Kowero, G. (2014). Agroforestry solutions to address food security and climate change challenges in Africa. *Current Opinion in Environmental Sustainability*, *6*, 61–67. doi:10.1016/j.cosust.2013.10.014

Mishra, S. K., & Mishra, P. (2017). Determinants of households' resistance against land acquisition for mining: Experiences at Talcher coalfields in India. *Land Use Policy*, *66*, 10–17. doi:10.1016/j.landusepol.2017.04.024

Mkodzongi, G. (2013). New people, new land and new livelihoods: A micro-study of Zimbabwe's fast-track land reform. *Agrarian South - the Journal of Political Economy*, *2*(3), 345–366. doi:10.1177/2277976013517320

Mkodzongi, G., & Spiegel, S. (2019). Artisanal gold mining and farming: Livelihood linkages and labour dynamics after land reforms in Zimbabwe. *The Journal of Development Studies*, *55*(10), 2145–2161. doi:10.1080/00220388.2018.1516867

Mnwana, S., & Bowman, A. (2018). Mine mechanisation and distributional conflict in rural South Africa. *Resources Policy*, *59*, 227–237. doi:10.1016/j.resourpol.2018.07.008

Molieleng, L., Fourie, P., & Nwafor, I. (2021). Adoption of climate smart agriculture by communal livestock farmers in South Africa. *Sustainability (Basel)*, *13*(18), 10468. doi:10.3390/su131810468

Moyo, F., Ncube, M., & Ndlovu, T. (2022). *The Competing Nature-Based Livelihood-Strategies: Artisanal Small-Scale Mining (ASM).* Perspectives in Agricultural-Communities in Umzingwane District.

Moyo, F., Ndlovu, T., Francis, B., & Ncube, M. (2018). The effects of artisanal mining on irrigation farming-the case of Umzinyathini Irrigation Scheme in Umzingwane District, Southern Matabeleland, Zimbabwe. *African Journal of Public Affairs*, *10*(2), 139–162.

Mtero, F. (2017). Rural livelihoods, large-scale mining and agrarian change in Mapela, Limpopo, South Africa. *Resources Policy*, *53*, 190–200. doi:10.1016/j.resourpol.2017.06.015

Müller-Lindenlauf, M. (2009). *Organic agriculture and carbon sequestration. Possibilities and constrains for the consideration of organic agriculture within carbon accounting systems*. Natural Resources Management and Environment Department, Food and Agriculture Organization of the United Nations.

Ncube, M., Moyo, F., & Mamhute, S. T. (2021). Unpacking Negative Externalities of Social Capital in the Sustainability of Smallholder Rural Irrigation Farming: The Case of Rozva Irrigation Scheme in Bikita District, Zimbabwe. Academic Press.

Neeliah, H., Rajkomar, B., Dookun-Saumtully, A., & Ramkissoon, J. (2006). *Food Agriculture and Natural Resources Policy Analysis Network*. FANRPAN.

Ning, C., Qu, J., He, L., Yang, R., Chen, Q., Luo, S., & Cai, K. (2017). Improvement of yield, pest control and Si nutrition of rice by rice-water spinach intercropping. *Field Crops Research*, *208*, 34–43. doi:10.1016/j.fcr.2017.04.005

Paredes, M. (2016). The glocalization of mining conflict: Cases from Peru. *The Extractive Industries and Society*, *3*(4), 1046–1057. doi:10.1016/j.exis.2016.08.007

Paul, C., Weber, M., & Knoke, T. (2017). Agroforestry versus farm mosaic systems–Comparing land-use efficiency, economic returns and risks under climate change effects. *The Science of the Total Environment*, *587*, 22–35. doi:10.1016/j.scitotenv.2017.02.037 PMID:28189309

Peters, K. A. (2010). Creating a sustainable urban agriculture revolution. *J. Envtl. L. & Litig.*, *25*, 203.

Poku, K. O. (2016). *Effects of Illegal Mining on the Agriculture Sector in Ghana.* Policy Hub. https://blogs.ubc.ca/kop7865/2016/11/30/effects-of-illegal-mining-on-the-agriculture-sector-in-ghana

Rivera-Ferre, M. G., Ortega-Cerdà, M., & Baumgärtner, J. (2013). Rethinking study and management of agricultural systems for policy design. *Sustainability (Basel)*, *5*(9), 3858–3875. doi:10.3390/su5093858

Rohde, R. F., Moleele, N. M., Mphale, M., Allsopp, N., Chanda, R., Hoffman, M. T., Magole, L., & Young, E. (2006). Dynamics of grazing policy and practice: Environmental and social impacts in three communal areas of southern Africa. *Environmental Science & Policy*, *9*(3), 302–316. doi:10.1016/j.envsci.2005.11.009

Scalise, A., Pappa, V. A., Gelsomino, A., & Rees, R. M. (2017). Pea cultivar and wheat residues affect carbon/nitrogen dynamics in pea-triticale intercropping: A microcosms approach. *The Science of the Total Environment*, *592*, 436–450. doi:10.1016/j.scitotenv.2017.03.012 PMID:28340454

Scoones, I. (1998). *Sustainable rural livelihoods: a framework for analysis*. Academic Press.

Shackleton, R. T. (2020). Loss of land and livelihoods from mining operations: A case in the Limpopo Province, South Africa. *Land Use Policy*, *99*, 104825. doi:10.1016/j.landusepol.2020.104825

Shackleton, S., & Luckert, M. (2015). Changing livelihoods and landscapes in the rural Eastern Cape, South Africa: Past influences and future trajectories. *Land (Basel)*, *4*(4), 1060–1089. doi:10.3390/land4041060

Singh, A. (2023). *Shweta Sharma*. Academic Press.

Singh, R., & Singh, G. S. (2017). Traditional agriculture: A climate-smart approach for sustainable food production. *Energy, Ecology & Environment*, *2*(5), 296–316. doi:10.1007/s40974-017-0074-7

Slack, K. (2013). *The growing battle between mining and agriculture*. Oxfam Politics of Poverty.

Tenkorang, E. Y., & Osei-Kufuor, P. (2013). The impact of gold mining on local Farming Communities in Ghana. Journal of Global Initiatives: Policy, Pedagogy. *Perspective*, *8*(1), 3.

Umesha, S., Manukumar, H. M., & Chandrasekhar, B. (2018). Sustainable agriculture and food security. In *Biotechnology for sustainable agriculture* (pp. 67–92). Woodhead Publishing. doi:10.1016/B978-0-12-812160-3.00003-9

van Noordwijk, M., Minang, P. A., Dewi, S., Hall, J., & Rantala, S. (2009). *Reducing Emissions from All Land Uses: The case for a whole landscape approach*. Academic Press.

Velten, S., Leventon, J., Jager, N., & Newig, J. (2015). What is sustainable agriculture? A systematic review. *Sustainability (Basel)*, *7*(6), 7833–7865. doi:10.3390/su7067833

Zomer, R. J., Neufeldt, H., Xu, J., Ahrends, A., Bossio, D., Trabucco, A., van Noordwijk, M., & Wang, M. (2016). Global Tree Cover and Biomass Carbon on Agricultural Land: The contribution of agroforestry to global and national carbon budgets. *Scientific Reports*, *6*(1), 29987. doi:10.1038/srep29987 PMID:27435095

Chapter 11
Enhancing Food Security Through Sustainable Agriculture:
A Case Study of the Pfumvudza/Intwasa Program in Zimbabwe

Tendai Kaponda
https://orcid.org/0009-0003-3287-4335
Central China Normal University, Wuhan, China

Option T. Chiwaridzo
https://orcid.org/0000-0001-9755-9123
University of Science and Technology Beijing, China

ABSTRACT

The Pfumvudza programme is an example of how sustainable agriculture can empower smallholder farmers and contribute to global food security. Usage of zero-tillage and mulching has improved soil health and yields, reduced water usage, and boosted the livelihoods of farmers. Pfumvudza offers valuable lessons for other nations grappling with food insecurity. Its principles can be adapted to diverse agroecological contexts, while its theoretical implications can inform research on long-term impact assessment, and the influence of policy on such programs. The study recommends continuous monitoring and evaluation, tailoring practises to different environments, integrating technology and capacity building among farmers, encouraging community participation and securing policy support, and prioritising climate-resilient and gender-inclusive practises. Future research directions include investigations into the program's adaptability, explorations of its social and economic implications scrutinising its climate change resilience, and fostering global collaboration for knowledge exchange.

DOI: 10.4018/979-8-3693-2011-2.ch011

INTRODUCTION

Food security sits at the heart of the Sustainable Development Goals and government policies in sub-Saharan Africa (SSA) (Bjornlund et al., 2022). Emerging after devastating famines in the 1970s (Gerlach, 2015), the concept has become crucial for a region where smallholder agriculture remains a cornerstone of both household food security and rural development (Gollin, 2014 cited in Abegunde et al., 2020). However, climate change increasingly threatens the vital role played by smallholder farmers in ensuring food security across SSA (Kapari et al., 2023). Declining crop yields, coupled with environmental degradation and resource depletion, paint a worrying picture (Mujere, 2022). Rapid population growth, land shortages, and poor soil quality further contribute to the region's vulnerability (Kondwakwenda et al., 2022). The World Bank estimates that approximately 9% of the global population faces food insecurity (2021). Governments and partners are turning to Conservation Agriculture as a potential solution (Mavesere & Dzawanda, 2022). However, rural communities remain highly vulnerable due to their reliance on climate-sensitive livelihoods and rain-fed agriculture (Kapari et al., 2023; Denison et al., 2016). The situation is further compounded by a poorly functioning rural economy with inadequate infrastructure, limited market access, and insufficient support services, leaving smallholders isolated and unable to fully integrate into agricultural value chains (ILO, 2008).

To effectively contribute to national food security according to HLPE (2013), smallholder farmers must:

1. Increase food production: Enhancing availability is vital to address food insecurity.
2. Provide income and livelihood: Sustainable livelihoods are essential for long-term food security.
3. Contribute to diverse diets: Ensuring access to a variety of nutritious food is crucial.
4. Serve as a buffer against market shocks: Smallholder agriculture can provide a safety net during periods of market instability.

Addressing these challenges and empowering smallholder farmers is key to achieving food security in sub-Saharan Africa. By implementing effective policies and investing in infrastructure, market access, and support services, governments and partners can create an enabling environment for smallholder success and ultimately, a food-secure future for the region. The Food and Agriculture Organization (FAO) has identified three fundamental principles of conservation agriculture that farmers can follow as they engage in the conservation agriculture process. These three guidelines summarise the knowledge producers and conservationists have about managing our natural resources sustainably. Some of these guidelines include minimal soil disturbance, topsoil management to establish a long-lasting organic soil cover and diversified crop rotation.[1] The practice of conservation agriculture is known by a variety of terms, including no-till, zero till, and conservation tillage, and is even referred to by local names like Pfumvudza in Zimbabwe (FAO, 2011; Government of Zimbabwe, 2020; Kassam et al., 2011; Mujere, 2021 as cited in Mavesere & Dzawanda, 2022)

This paper explores the legacy of the *Pfumvudza* Programme in Zimbabwe, delving into its sustainable agricultural ideas and its advantageous effect on food safety in the nation. Moreover, it examines the capacity for the worldwide dissemination of *Pfumvudza's* sustainable practices, providing valuable lessons for countries striving to deal with food security in an increasingly uncertain world. As we embark on this exploration, it becomes obvious that *Pfumvudza/Intwasa* stands as a beacon of hope, illuminating the path towards a more sustainable and food-secure future for Zimbabwe and potentially for other countries dealing with comparable hardships.

BACKGROUND OF THE STUDY

Global Food Security Challenge

Food insecurity is an essential global challenge, with almost one in ten human beings worldwide suffering from continual hunger (FAO, 2021). Climate change is one of the devastating results of how human activity has shown to be revering years of progress and is now impacting food security across the globe (EAA, 2023). The impact of climate change on food security varies based on where the population lives (ibid). The rising threat of climate change amplifies existing dangers to food security and livelihoods. From the increasing frequency and intensity of extreme weather events to declining agricultural yields and reduced production in vulnerable regions, the impact is multifaceted. Rising health and sanitation risks, growing water scarcity, and intensifying conflicts over scarce resources further compound the threat, leading to new humanitarian crises and increasing displacement across the globe (IPCC, 2007).

The agroecosystems, which impact cultivated crops, animals, woodlands, and aquatic species, are under pressure from all of these sources. Lack of regular access to adequate safe and nourishing food for normal growth and a healthy life is known as food insecurity. A person who experiences food insecurity has probably gone without food for a day or longer, running out of food and experiencing hunger. The intensifying droughts in Zimbabwe, fuelled by climate change, pose a significant obstacle to achieving the Sustainable Development Goals. Specifically, these droughts impact SDG 1 (poverty eradication), SDG 2 (zero hunger), and SDG 3 (good health and well-being), hindering progress towards a more sustainable and equitable future (Frischen et al., 2020). In many countries, particularly in Sub-Saharan Africa, food security remains a precarious goal due to a combination of factors: climate change, economic instability, population growth, and insufficient agricultural practices. While smallholder agriculture has the potential to drive development in sub-Saharan Africa, its success is hampered by the vulnerability of these farmers to climate change and its fluctuations. Climate change and variability exacerbate pressure on land and water resources, leading to disruptions in farming activities and negatively impacting the livelihoods of smallholder farmers (Serdeczny et al., 2017; Pillay, 2016).

Zimbabwe's Agricultural Landscape

Zimbabwe, a predominantly agricultural nation in southern Africa, relies heavily on farming for its rural population's livelihood. With over 69% of the population residing in rural areas, agriculture serves as the backbone of their economic survival. The nation's annual maize requirement stands at 2.2 million tonnes, with 1.8 million allocated for human consumption and 350,000 tonnes dedicated to livestock feed (Mucheri, 2021). This underscores the crucial role of agriculture in ensuring food security and economic stability for Zimbabwe's rural population. Zimbabwe has a turbulent history marked by periods of economic decline, political instability, and recurring droughts, all of which have contributed to food insecurity. The country's agricultural sector, which has long been the backbone of its economy, has faced numerous setbacks. Historically, conventional farming techniques struggled to address unpredictable weather patterns and limited resources, leading to inconsistent crop yields and food shortages. However, amidst these challenges, an indigenous and time-examined agricultural practice known as '*Pfumvudza*' has performed a pivotal role in Zimbabwe's quest to achieve food security. This historical practice, which has been handed down through generations, holds the promise of a more sustainable and food-secure future for the country.

Over time, Zimbabwean farmers have harnessed the know-how of *Pfumvudza* to mitigate the adverse outcomes of climate change and recurrent droughts, yielding improved food security outcomes. As the global community grapples with similar agricultural challenges, the *Pfumvudza* Programme has garnered attention as a potential model for sustainable agriculture that would be adapted and applied in different countries. This study delves into the legacy and ability of the *Pfumvudza* Programme in Zimbabwe, analysing the sustainable agricultural concepts it embodies and the impact it has had on the nation's food security. Moreover, it explores the possibilities for disseminating *Pfumvudza's* practices on an international scale, offering precious lessons for countries striving to deal with food security concerns in a more and more unpredictable world.

In the 2019/2020 farming season, conventional farming methods yielded 1.16 tonnes of maize per hectare. The introduction of the Pfumvudza concept began to gain recognition in 2020, leading to a significant increase in maize yields in the 2020/2021 season. Pfumvudza farming practices produced 5.20 tonnes per hectare, a notable contrast to the 0.82 tonnes per hectare from conventional methods during the same season. However, in the subsequent 2021/2022 season, there was a decrease in maize yield under Pfumvudza, with 1.39 tonnes per hectare, although still higher compared to the 0.82 tonnes per hectare from conventional farming methods.

The Pfumvudza initiative in Zimbabwe was implemented to address low productivity and production, enhance farmer and household resilience to climate shocks, and ultimately ensure food security (FAO, 2020). This Programme aligns with National Vision 2030, which aims to uplift the poor through agriculture and integrate them into the mainstream economy (National Development Strategy, 2020). However, despite its implementation, limited research has investigated Pfumvudza's effectiveness as a resilient strategy for promoting conservation agriculture and improving food security, particularly in the context of drought resistance (Mavesere & Dzawanda, 2023). This research seeks to address this gap by exploring the effectiveness of Pfumvudza as a strategy for achieving these objectives, specifically focusing on its drought-resistant capabilities. This research seeks to shed light on how indigenous knowledge and sustainable agricultural practices may be harnessed to enhance food security, not only in Zimbabwe but potentially in different countries dealing with comparable challenges. It is however worthwhile to note that in the Zimbabwean context, the term Pfumvudza implies two things; one being the CA practice that dates back to several generations and the second one is a government Programme in which the Government of Zimbabwe is distributing agricultural inputs to citizens, especially the vulnerable communities in rural and farming areas. It is

Table 1. Maize yield comparison: Conventional farming vs. Pfumvudza implementation (2019/2020 - 2021/2022 seasons)

Season	Farming Method	Maize Yield (Tonnes per Hectare
2019/2020	Conventional	1.16
2020/2021	Pfumvudza/Intwasa	5.2
	Conventional	1.16
2021/2022	Pfumvudza/Intwasa	1.39
	Conventional	0.82

Source: Vhera (2022)

against this notion that the study focuses on implementing Pfumvudza/Intwasa a CA practice in the fight against food insecurity.

RESEARCH METHODS

This study examines the Pfumvudza Programme, a climate-smart agricultural initiative in Zimbabwe, through a qualitative analysis of secondary data. Existing literature, including academic articles, policy documents, Programme reports, and media coverage, will be analysed to build a comprehensive understanding of the Programme's objectives, implementation strategies, and perceived impact on food security. By leveraging publicly available documents and existing case studies, this research aims to gain insights into Pfumvudza's implementation in Zimbabwe. Triangulation across various data sources will further enhance the credibility and comprehensiveness of the findings, ensuring a well-rounded understanding of the Programme's impact (Braun & Clarke, 2006). By interpreting the identified themes within the context of existing research and theoretical frameworks, this study aims to draw critical insights into Pfumvudza's contribution to food security in Zimbabwe. Based on the findings, recommendations will be formulated for potential Programme improvements, policy considerations, and future research directions. By employing a rigorous secondary data approach, this research seeks to contribute valuable knowledge towards enhancing food security and promoting sustainable agricultural practices across the world.

SUSTAINABLE AGRICULTURE

Sustainability rests on the principle that we must meet the needs of the present without compromising the ability of future generations to meet their own needs (UC Agriculture & Natural Resources, 2021). "Sustainable agriculture" denotes farming methodologies that integrate and prioritize sustainability, emphasizing the capacity to uphold a process over an extended period (Pepin, 2022). This term characterizes an agricultural framework capable of perpetual sustenance, ensuring the supply of food and fibre to cater to both current and future generations (Ibid). Sustainable agriculture emphasizes environmentally friendly methods, reducing reliance on non-renewable inputs, prioritizing local communities' needs, and aiming to meet nutritional requirements while providing long-term employment and income, minimizing vulnerability to external factors, and fostering inclusive rural institutions (GIZ, 2015).

Sustainable agriculture practices are crucial in addressing the growing demand for food while protecting the environment and ensuring long-term resource security (Arora, 2018). These techniques aim to conserve natural resources and prevent depletion, ensuring sufficient food production for future generations. Sustainable agriculture plays a vital role in tackling global food security and nutrition challenges (ibid). By increasing crop production, sustainable agriculture contributes to meeting the needs of a growing population. This is a key component of food security, ensuring adequate food supplies for all (HLPE Report, 2016). Furthermore, sustainable practices promote efficient resource use, (ibid). In the face of climate change and its devastating consequences, many countries and organizations are actively seeking sustainable agricultural methods that enhance the resilience and adaptability of smallholder farmers (Jakachira et al., 2022). Strengthening resilience helps ensure a stable food supply and mitigate the negative impacts of climate change on agriculture and food security.

CLIMATE-SMART AGRICULTURE

The responsibility lies with Sub-Saharan nations to implement climate-smart agriculture (CSA) practices in their agricultural sectors to reduce greenhouse gas emissions and adapt to a changing climate (Kapari et al., 2023). CSA, a subset of sustainable agriculture, specifically targets adaptation and mitigation of climate change impacts. It emerged as a response to climate change and is a key strategy for achieving food security and sustainably increasing agricultural output, according to the World Bank (2021), Jakachira et al. (2022), and the FAO (2014).

This integrative approach tackles the interconnected challenges of food security and climate change, pursuing three objectives:

1. Sustainable growth in agricultural productivity and equitable increases in farm income, food security, and development.
2. Adaptation and resilience-building of agricultural and food security systems to climate change at various levels.
3. Reduction of greenhouse gas emissions from agriculture.

In essence, CSA enhances food production while ensuring environmentally responsible and long-term sustainable farming practices. It goes beyond merely increasing food output but emphasizes resource preservation. Agriculture is a significant contributor to greenhouse gas emissions, and CSA aims to minimize these emissions through more sustainable and efficient practices. It addresses climate change challenges by striving to increase agricultural productivity, build resilience in farming systems, and reduce greenhouse gas emissions, all while guaranteeing food security and equitable increases in farm incomes. This approach promotes a balance between agricultural development and environmental stewardship in a changing climate.

Climate-smart agriculture (CSA) is a key strategy employed by the Government of Zimbabwe (GOZ) to build resilience against climate change among farmers nationwide, ensuring their ability to withstand extreme weather events and contribute to national food security for all citizens (Manguwo, 2023). This coordinated effort, initiated in 2015 with the national climate-smart agriculture response strategy, involves collaboration between the GOZ and various partners to mitigate the detrimental impacts of climate change on agricultural production, the lifeblood of 70% of Zimbabwe's population (Climate Technology Center & Network, 2017).

To ensure farmer buy-in, CSA content has been integrated into the nation's post-secondary agriculture education system (ibid). CSA encompasses agricultural practices that sustainably increase productivity, promote climate adaptation and resilience, and reduce greenhouse gas emissions (Jakachira et al., 2022). Zimbabwe utilizes a diverse range of CSA techniques, including conservation agriculture, regenerative agriculture, and organic farming. Pfumvudza/Intwasa, a specific conservation agriculture practice (Machingura, 2020), is currently the most popular method among smallholder farmers and is known by its respective Shona and Ndebele names (Jakachira et al., 2022). The Pfumvudza Programme, launched in the 2020-21 cropping season, has received significant government support and high-level political backing. This national drive, further bolstered by international agencies such as donors and the UN's FAO, positions Pfumvudza as a potential solution for achieving the elusive goal of national food security, after years of reliance on food imports due to successive droughts (Scoones, 2021).

PFUMVUDZA/INTWASA CONCEPT

Pfumvudza, a term often used for the conservation agriculture concept, was devised by the FfF, a local NGO, to address the annual cereal needs of a household of six members using a small plot of land (Mujere, 2021). The essence of *Pfumvudza* lies in its adherence to sustainable agricultural concepts that emphasize efficient land use, conservation of resources, and resilience in the face of climate challenges. Small-scale farmers across Zimbabwe have embraced this practice, spotting its capability to ensure food security while protecting the environment. *Pfumvudza* entails techniques like mulching, minimum tillage, and the cultivation of drought-resistant, excessive-yielding crops, all of which contribute to increased crop productivity. *Pfumvudza*, as described by Mujere (2021), is a Shona expression denoting the emergence of new foliage in spring, symbolizing the commencement of a new agricultural planting season. It isn't simply the latest innovation but a time-venerated way of life in Zimbabwean agriculture. This technique has been practised for generations, surpassed down through communities as a sustainable method of farming. It embodies the standards of conservation, resilience, and environmental stewardship which are vital to sustainable agriculture. In the face of recurrent droughts, financial instability, and other challenges, Zimbabweans have trusted the understanding of *Pfumvudza* to ensure food safety.

The *Pfumvudza* approach includes preparing a small piece of land for planting, through a system that consists of mulching, minimum tillage, and the usage of organic matter. Pfumvudza's uniqueness lies in its remarkably small plot size, measuring just 16 meters by 39 meters (0.06 hectares). This compact area allows easy manual preparation, weeding, mulching, and watering, even with harvested rainwater during mid-season dry spells or droughts (Mujere, 2021; Scoones, 2021; Machingura, 2020). This key feature of Pfumvudza holds significant potential for enhancing food security through several key benefits:

- Intensified land use: By focusing farming activities on a small plot, Pfumvudza encourages efficient utilization of limited land resources, maximizing productivity per unit area.
- Resource optimization: The compact size facilitates efficient use of crucial resources like water, fertilizers, and labour, leading to more sustainable and cost-effective farming.
- Climate resilience: The manageable plot size enables effective preparation, management, and maintenance, even under challenging weather conditions like mid-season dry spells or drought. This makes Pfumvudza a potentially valuable tool for building resilience against climate change.
- Increased food production: Careful planning and management within the small plot allows farmers to cultivate diverse crops, maximizing food production from a limited space.

Overall, Pfumvudza's unique size offers a promising approach to enhancing food security by promoting intensive land use, resource optimization, and climate resilience. Farmers focus on planting drought-resistant and high-yielding crops, which include maize, millet, and sorghum. This sustainable technique promotes water conservation, reduces soil erosion, and improves soil fertility, all of which are critical elements in ensuring long-term agricultural sustainability. One of the extraordinary factors of *Pfumvudza* is its adaptability to the challenges of weather trade. In a world where erratic weather patterns and changing rainfall distribution threaten agricultural productivity, *Pfumvudza* serves as a beacon of hope. By utilising sustainable agricultural practices honed over generations, Zimbabwean farmers can deal with unfavourable climate conditions and reduce the chance of crop failure.

CONSERVATION AGRICULTURE PRINCIPLES IN THE CONTEXT OF PFUMVUDZA

According to Zimbabwe's Department of Agricultural Technical and Extension Service (AGRITEX), *Pfumvudza* embodies six principles of which three of them are stemmed from CA principles. According to AGRITEXS' *Pfumvudza* brochure, the following are six principles: [1] minimum soil disturbance, [2] mulching, [3] crop rotation, [4] timely implementation of all operations, [5] precise operations, and [6] efficient use of inputs.

Minimum Soil Disturbance

Minimum soil disturbance involves digging planting stations or making rip lines instead of ploughing. Digging plant stations, a zero-tillage practice synonymous with precise planting, presents a significant opportunity to improve food security, as advocated by the FAO (2023). This approach minimizes soil disturbance through reduced or eliminated tillage and allows for the planting of crops in specific locations. As research by Guibodoni et al. (2023) indicates, this technique effectively reduces soil erosion, a crucial factor in preserving soil health and preventing nutrient loss. Furthermore, enriching the soil with organic matter from cover crops and organic fertilization enhances fertility, water retention, and nutrient availability. This improvement leads to increased crop productivity and resilience to adverse environmental conditions. The combined benefits of enhanced soil health and reduced erosion contribute to agricultural sustainability, ensuring consistent crop yields and improved food production. In essence, implementing these techniques aligns with the FAO's mission to achieve global food security by making agriculture more sustainable, resilient, and productive.

Mulching

Preparing land for planting under the Pfumvudza concept involves slashing or rolling weeds, incorporating previous crop residue, and directly seeding through mulch (ibid). In an interview, Professor Jiri highlighted the role of mulching in enhancing biodiversity and natural biological processes above and below the ground, leading to improved water use efficiency, nutrient utilization, and ultimately, increased crop production (Manguwo, 2023). Indeed, the practice of mulching plays a critical role in bolstering food security. By suppressing weed growth, conserving soil moisture, and regulating temperature, the mulched layer fosters a favourable environment for crop growth. This results in improved soil quality, reduced water loss (particularly crucial in drought-prone regions), and enhanced resilience against pests and diseases. Professor Jiri's emphasis on the promotion of biodiversity and natural biological processes further underscores the potential of mulching to create a healthier and more self-sustaining agricultural ecosystem. This translates to more efficient nutrient use, higher crop yields, and ultimately, a more stable and sustainable food supply, directly contributing to food security objectives (Manguwo, 2023).

Efficient Use of Inputs

Pfumvudza's key principle is "more with less," emphasizing efficient resource utilization to achieve high yields from limited inputs (Manguwo, 2023; Munjere, 2022). This approach promotes lower production costs, making it accessible to a wider range of farmers and preserving traditional agricultural practices. Intensive crop production on smaller plots, a hallmark of Pfumvudza, allows for better land management and resource conservation (Munjere, 2022). This leads to increased crop yields and reduced environmental impact, contributing to long-term food security and sustainable agriculture. By maximizing output with minimal inputs, Pfumvudza aligns with the United Nations' Sustainable Development Goal of zero hunger (Manguwo, 2023; Munjere, 2022). It promotes food availability and accessibility, particularly in regions where traditional agriculture is dominant.

Timely Implementation

Timely and adequate preparatory activities are essential for maximizing the benefits of the Pfumvudza Programme, according to Bhasera J., the former Permanent Secretary in the Ministry of Lands, Water and Rural Resettlement (Machingura, 2020). This statement underscores the crucial role of early land preparation, including digging planting basins and acquiring inputs before the season begins, in the context of enhancing food security in agricultural-dependent economies.

Planting with the first effective rains, facilitated by early preparation, offers several significant advantages:

- Maximized crop yields: Utilizing the entire growing season allows for optimal crop development and production.
- Reduced risk of crop failure: Planting at the opportune time minimizes the impact of adverse weather conditions and pests.

- Sustainable farming practices: Efficient resource utilization through timely preparation contributes to long-term food security.

Therefore, Bhasera's recommendation emphasizes the importance of proper planning and timing in agricultural activities as a key strategy for achieving food security (Machingura, 2020).

Crop Rotation

Through the implementation of crop rotation with species diversification, a multifaceted approach to agricultural sustainability is fostered, ultimately leading to enhanced food security on a global scale. This practice, as championed by Volsi et al. (2022) and Suding (2015), offers a multitude of benefits that contribute to a robust and resilient agricultural system.

Firstly, the diversification of crops serves as a potent weapon against the persistent threat of pests, diseases, and weeds that plague monoculture systems. This diversified approach disrupts the life cycles of harmful organisms, effectively reducing their populations and mitigating the need for chemical interventions. Consequently, crop yields are safeguarded, ensuring a bountiful harvest and contributing to the overall stability of the food supply chain. Furthermore, by employing crop rotation with species diversification, a remarkable improvement in soil fertility and conservation is achieved. This practice nurtures a healthy soil ecosystem through the introduction of various crops with diverse nutrient requirements and root structures. This diversification fosters a vibrant microbial community, leading to enhanced soil fertility and organic matter content. As a result, the long-term sustainability of crop production is assured, minimizing the risk of soil degradation and crop failure.

The economic benefits of this approach are equally noteworthy, with Volsi et al. (2022) highlighting the significant increase in profitability experienced by farmers who adopt crop rotation with species diversification. This financial incentive serves as a powerful motivator, encouraging wider adoption of these sustainable practices. As more farmers embrace this approach, a ripple effect of positive environmental and economic outcomes is generated, further solidifying the foundation for a resilient and sustainable food system. Beyond the individual farmer, the large-scale implementation of crop rotation with species diversification addresses critical food security concerns on a global scale. By promoting stable and increased grain production through sustainable means, this approach helps to ensure a reliable food supply, mitigating the risk of food scarcity and hunger. Moreover, the adoption of these practices safeguards the environment for future generations, ensuring that the land remains fertile and productive for years to come. The multifaceted benefits of crop rotation with species diversification, as championed by Volsi et al. (2022) and Suding (2015), offer a compelling solution for achieving agricultural sustainability and enhancing food security on a global scale. By embracing this sustainable approach, we can ensure a bountiful harvest for generations to come, while simultaneously safeguarding the delicate balance of our planet's ecosystems.

This technique specializes in optimizing the usage of available resources, including limited rainfall and arable land, whilst minimizing environmental degradation. The essence of *Pfumvudza* is to utilize small plots of land efficiently (Maseya, 2020), ensuring that each inch of arable soil contributes to food production. The *Pfumvudza* technique, as described by Maseya (2020), demonstrates a proactive approach to enhancing food security by optimizing resource usage and minimizing environmental degradation. By focusing on the efficient utilization of small plots of land, *Pfumvudza* ensures that every inch of arable soil contributes to food production. This strategy is particularly crucial in regions with limited rainfall and arable land, as it maximizes the yield from available resources. By adopting such sustainable agricultural practices, not only can communities produce more food on their limited land, but they also reduce the risk of soil degradation, erosion, and loss of biodiversity. Consequently, this approach promotes long-term food security by increasing agricultural productivity while preserving the environment, making it a valuable tool in the fight against food scarcity and hunger (ibid).

Successful application of the '*Pfumvudza*' concept also depends on in-depth training and precise application of the four principles popularised by FfF (Mucheri, 2021).

In the "on-time" stage, the sequence of agricultural activities listed, from land preparation to harvesting, is crucial in enhancing food security. Each step plays a vital role in optimizing crop production and ensuring a stable food supply. Land preparation, including soil testing, sets the foundation by determining soil quality and nutrient content, allowing for better crop selection and tailored fertilization practices (FAO, 2017). Holing and liming improve soil structure and pH balance, contributing to healthy plant growth (Holland et al., 2019). Mulching favourably influences the soil moisture regime by controlling the surface evaporation rate; in summer, mulching conserves soil moisture by reducing the evaporation rate, which is especially important in arid regions (Kader et al., 2017). The procurement of quality inputs and effective fertility management ensures that crops receive essential nutrients (FAO, 2017). Weed and pest control minimize yield losses (Oerke, 2006).

Precision in making holes, fertilizer application, and planting is essential for optimizing agricultural practices and enhancing food security. When farmers use advanced precision techniques, such as precision agriculture, they can minimize the wastage of resources like water, seeds, and fertilizers. Precision hole-making ensures that each seed is planted at the ideal depth and spacing, optimizing crop growth.

Figure 1. Adopted from the foundation for farming (FfF)

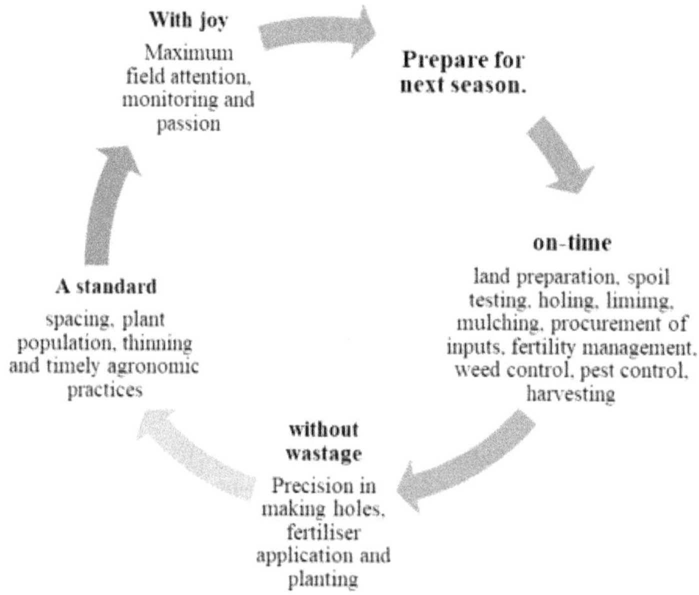

Additionally, precise fertilizer application targets the specific needs of plants, reducing excess usage and minimizing environmental impact. These practices ultimately lead to increased crop yields and improved resource efficiency, thus contributing to food security (Bramley & Trengove, 2013). By adopting precision farming methods, we can produce more food with the same or even fewer resources, making a significant contribution to addressing the challenges of feeding a growing global population.

Implementing standard spacing, appropriate plant population, timely thinning, and sound agronomic practices in agricultural production plays a critical role in enhancing food security. By maintaining proper spacing between crops, farmers can optimize the efficient use of resources such as land, water, and nutrients. This helps increase crop yields and minimizes competition between plants for these vital resources. Moreover, the use of appropriate plant populations ensures that the available land is utilized to its full potential. Timely thinning, which involves removing excess plants, enables the remaining crops to grow vigorously, reducing competition for nutrients and sunlight. Furthermore, the application of sound agronomic practices, including proper fertilization and pest management, ensures the health and productivity of crops. These strategies collectively result in higher agricultural productivity, greater food availability, and improved access to nutritious food, thus contributing to enhanced food security

"Maximum field attention, monitoring, and passion" are key elements in enhancing food security. By promoting these factors, we can significantly improve the efficiency and productivity of agriculture. The passionate engagement of farmers and agricultural professionals, along with rigorous monitoring and attention to fields, leads to better crop management and reduced losses due to pests, diseases, or adverse weather conditions. As a study by the Food and Agriculture Organization (FAO) suggests, close monitoring and timely interventions can increase crop yields by preventing crop damage and loss, ultimately contributing to food security (FAO, 2020). This approach not only ensures a more reliable food supply but also supports sustainable agriculture, which is critical for long-term food security by preserving the health of the land and the livelihoods of those who depend on it. Additionally, investments in education and technology that encourage maximum field attention and monitoring can further enhance agricultural productivity and food security (UN, 2019).

DISCUSSIONS

The success of *Pfumvudza* in improving food security in Zimbabwe serves as a powerful example of the way indigenous knowledge and sustainable agricultural practices could make a great difference. The question that arises is, can *Pfumvudza's* ideas be adapted and applied in different countries facing comparable challenges to improve their food security?

In the context of enhancing food security through sustainable agriculture, the adoption of zero-tillage practices, as exemplified by the *Pfumvudza* Programme in Zimbabwe, plays a pivotal role in bolstering crop yield. The success of this Programme can be attributed to the observed benefits of zero-tillage, as highlighted by Mujere (2022). Zero-tillage not only conserves soil moisture more effectively but also contributes to the improvement of soil physical properties and organic matter content. By minimizing soil disturbance, this agricultural practice reduces water evaporation and enhances water retention, crucial factors in arid or semi-arid regions like Zimbabwe. The enhanced soil structure promotes better root development and nutrient uptake by crops, ultimately resulting in increased yields. Therefore, the *Pfumvudza* Programme's emphasis on zero-tillage aligns with sustainable agriculture principles, fostering improved soil health and resilience while concurrently addressing food security challenges in the region.

The findings presented by FAO (2020) underscore the significance of adopting environmentally friendly practices, such as minimum tillage and mulching. The case study of the *Pfumvudza* Programme in Zimbabwe exemplifies the potential benefits of these practices. By embracing minimum tillage and mulching, the Programme has not only realized significant savings in machinery, energy use, and carbon emissions but has also witnessed a notable increase in soil organic matter content and biotic activity. Furthermore, the adoption of these practices has contributed to a reduction in erosion, enhancing the overall resilience of crops to drought conditions. The Programme's success is further manifested in improved aquifer recharge, ensuring sustainable water availability for crop cultivation. Additionally, by mitigating the impact of climate change-related weather variability, the *Pfumvudza* Programme exemplifies a holistic approach to sustainable agriculture that not only bolsters food security but also promotes environmental stewardship and long-term agricultural viability.

The adoption of climate-smart agriculture as one of the pathways to sustainable agriculture through the Climate-proofed Pfumvudza/Intwasa Programme has enabled the country to increase agriculture production and productivity thereby improving overall food security (Sharara, 2023). Government interventions include the National Climate Policy which promotes climate-smart-agriculture through reduced tillage, mulching, crop rotation, water harvesting, efficient irrigation and promotion of drought-tolerant crops, among other practices (Ibid). Through these interventions, Minister Ndhlovu said the Pfumvudza Programme has resulted in a threefold increase in yields.

The Pfumvudza Programme has made a significant contribution to food security in Zimbabwe by implementing sustainable agricultural practices that enhance crop productivity and resilience. The emphasis on minimal soil disturbance, topsoil management, and diversified crop rotation has resulted in improved yields, ensuring a more stable and abundant food supply for local communities. By promoting conservation agriculture, Pfumvudza has helped mitigate the impact of climate change, erratic rainfall patterns, and soil degradation (Penot et. al, 2015), which are critical factors affecting agricultural productivity in the region. The Programme's focus on smallholder farmers has empowered local communities, providing them with the knowledge and tools to enhance their agricultural practices and food production. Through the adoption of Pfumvudza principles, Zimbabwe has seen positive outcomes in terms of increased crop yields, improved food security, and strengthened resilience to environmental challenges, ultimately contributing to the overall well-being of the nation.

The significance of involving farmers with limited land sizes, particularly women, is underscored by the potential demonstrated in cultivating small land pieces. As exemplified in the *Pfumvudza* Programme in Zimbabwe, spearheaded by Mujere (2022), the cultivation of small plots has proven to be an effective strategy for increasing food production and ensuring household food security. By focusing on intensifying cultivation on limited land, farmers, including women, can optimize their resources, leading to higher yields per unit area. This approach aligns with the sustainable agriculture principles advocated by the *Pfumvudza* Programme, emphasizing efficient resource use, conservation practices, and resilience to climate change. Consequently, the cultivation of small land pieces becomes a pivotal component in addressing food security challenges, offering a pragmatic solution to meet the nutritional needs of families and communities, particularly in regions with limited arable land. Zimbabwe has witnessed increased agricultural production and productivity, which has led to the improvement of household incomes as well as food and nutrition security despite not being spared the adverse effects of climate change, thanks to the climate-smart agriculture policies implemented by the government (Sharara, 2023).

The President of Zimbabwe E.D. Mnangagwa in his address at the Climate Smart Agriculture for Food Security at the COP28 in Dubai commended how Zimbabwe's policies on food and nutrition security are based on the ancient wisdom of Zimbabwe's ancestors. This technology in his speech was encouraged for learning adoption by the modern world (Sharara, 2023). The global dissemination of the Pfumvudza Programme represents a transformative initiative in sustainable agriculture with potentially far-reaching impacts. By showcasing the Programme's success and incorporating its principles into broader agricultural practices, countries facing similar challenges in food security, climate change, and resource constraints can draw valuable lessons. The global dissemination of Pfumvudza involves not only sharing its sustainable agricultural techniques but also adapting them to diverse regional contexts. This process can contribute to fostering international collaborations, building resilient agricultural systems, and collectively working towards a more sustainable and food-secure future on a global scale.

The implications of the Pfumvudza Programme extend beyond Zimbabwe, carrying valuable lessons for other nations grappling with food security challenges. In Zimbabwe, Pfumvudza has the potential to revolutionize sustainable agriculture, enhancing food security through its focus on minimal soil disturbance, topsoil management, and diversified crop rotation. The Programme's success could serve as a model for similar agroecological zones globally, especially in regions facing climate change, soil degradation, and resource constraints. The principles of Pfumvudza offer a pathway for other countries to adapt and implement sustainable agricultural practices, fostering resilience against environmental and socio-economic challenges. By sharing knowledge and experiences, Pfumvudza could contribute to a collective effort to address global food security concerns and promote sustainable farming practices on a broader scale.

Challenges Faced in Implementing Pfumvudza/Intwasa

in their study, Mavesere and Dzawanda (2023) found that 64% of respondents mentioned crop diseases and pests, such as maize streak virus, stalk-borers, bacterial stalk rot, and maize lethal necrosis, as significant challenges during Pfumvudza implementation. The identification of crop diseases and pests, such as maize streak virus, stalk-borers, bacterial stalk rot, and maize lethal necrosis, as significant challenges during Pfumvudza implementation underscores the complex interplay between agriculture and conservation practices. These issues not only threaten crop health and productivity but also pose financial burdens on farmers engaged in Pfumvudza Conservation Agriculture. The presence of these challenges demands integrated pest management strategies and a focus on sustainable farming practices. Addressing crop diseases and pests becomes pivotal to mitigating the risk of crop failure, ensuring food security, and sustaining the economic viability of Pfumvudza CA, highlighting the need for educational and extension services to empower farmers with the knowledge and tools for resilient and sustainable agricultural practices.

The impact of climate change, specifically factors like drought and water scarcity, presents formidable challenges to agricultural practices in Zimbabwe and directly influences the effectiveness of Pfumvudza as a resilient strategy (Mpala & Simatele, 2023). The erratic and changing climate patterns exacerbate water shortages, making it difficult for farmers practising Pfumvudza to maintain optimal soil moisture levels. This poses a serious threat to crop growth and productivity, as water is a critical factor in agricultural success. The implications for agriculture and the success of Pfumvudza as a food security initiative are profound. Drought conditions can lead to reduced yields, crop failure, and increased vulnerability to food insecurity. The resilience of Pfumvudza, designed to optimize water usage through its conservation agriculture principles, is compromised in the face of climate-related challenges. To enhance the initiative's effectiveness in the context of climate change, there is a pressing need for adaptive strategies, such as the incorporation of drought-tolerant crop varieties, water management techniques, and sustainable agricultural practices that can withstand the adverse effects of a changing climate.

CONSERVATION FARMING IN OTHER COUNTRIES

Tanzania

CA has been recognized as a potential solution for sustainable development in Africa, as evidenced by fieldwork conducted in the Lindi region of Tanzania (Makumbi et al., 2017). Several studies have been conducted to assess the impacts of CA in Tanzania. For instance, a project initiated to promote CA in Tanzania aimed to improve food security and rural livelihoods of small-scale farmers (Mkonda & Mkonda, 2016). In Arusha, farmers are known for constructing terraces on hilly farms (Mkonda, 2017), by establishing terraces, preventing soil erosion and improving food security.

Dodoma area has seen farmers employing conservational tillage (Ibid). Conservation tillage is a farming practice that aims to reduce soil erosion and improve soil health by minimizing soil disturbance. It involves techniques such as no-till or reduced tillage, where crops are planted directly into the soil with minimal or no prior soil preparation. Research has shown that conservation tillage can lead to decreased soil erosion, reduced leaching of fertilizers and pesticides, and improved water retention in the soil (Subbulakshmi, Saravanan & Subbian, 2009; Busari et al., 2015) The case of the Southern Highlands of Tanzania highlights the potential benefits of CA, including increased crop yields, reduced soil erosion, and improved soil fertility (Shetto & Owenya, 2007). Three case studies conducted in Tanzania demonstrate the effectiveness of CA in promoting sustainable agriculture (Shetto & Owenya, 2007). The studies reveal that CA can lead to increased crop yields, reduced soil erosion, and improved soil fertility. Moreover, CA has been found to enhance the resilience of small-scale farmers to climate change and other environmental challenges (Makumbi et al., 2017). Despite the potential benefits of CA, several challenges remain. For instance, the adoption of CA by small-scale farmers in Tanzania has been limited due to a lack of awareness, limited access to resources, and inadequate support from policymakers (Mkonda & Mkonda, 2016). To address these challenges, it is crucial to invest in capacity building, extension services, and policy reforms that promote the adoption of CA by small-scale farmers (Makumbi et al., 2017).

Conservation agriculture has the potential to promote sustainable agriculture in Tanzania, addressing the challenges of food security and rural livelihoods. However, the adoption of CA by small-scale farmers remains limited due to a lack of awareness, limited access to resources, and inadequate support from policymakers. To address these challenges, it is crucial to invest in capacity building, extension services, and policy reforms that promote the adoption of CA by small-scale farmers.

Zambia

Conservation agriculture (CA) has gained attention in Zambia as a means to transform conventional agricultural systems and ensure food security. CA is based on three main principles: minimal soil disturbance, permanent soil cover, and crop rotations, which are considered universally applicable in all agricultural landscapes and cropping systems. Several studies have been conducted to understand the impact of CA on household welfare and its potential as a climate adaptation strategy among smallholder farmers in Zambia. Baudron and Mwanza (2012) conducted a case study of the Southern Province of Zambia, focusing on the adoption and impact of conservation agriculture. Their research provides insights into the determinants and impact of CA technology adoption on farm household welfare. The study emphasizes the potential of CA to contribute to household welfare in the Zambian context.

Furthermore, Zulu et al. (2019) and Zulu-Mbata and Chapoto (2016) explored the gendered impacts of conservation agriculture on households' livelihoods in Zambia. Their research sheds light on the gender roles and implications of CA adoption, providing a comprehensive understanding of its effects on different household members. In addition, a study by Umar et al. (2011) discussed options for improving smallholder conservation agriculture in Zambia, highlighting the importance of sustainable agricultural practices in the Zambian context. The study provides valuable insights into the challenges and opportunities associated with the adoption of CA by smallholder farmers. According to a study by Baudron and Mwanza, the main barriers to adopting CA practices in Zambia are labour intensity, limited access to farm inputs, limited technical knowledge about CA application, and lack of awareness (Tufa et al., 2023). The study also combined bottom-up and top-down approaches to assess the ex-ante effects of CA-based systems in Zambia (Komarek et al., 2019. Nyanga et al.

Scholarly research on conservation agriculture in Zambia highlights its potential to enhance household welfare, mitigate the impact of climate change, and address gender-specific implications. These studies contribute to a comprehensive understanding of the benefits and challenges of implementing CA in the Zambian agricultural landscape.

Rwanda

Conservation agriculture (CA) is a farming system that is founded around three principles: minimum soil disturbance (reduced or no tillage), permanent soil cover, and crop rotation. Rwanda has been promoting CA as a way to improve soil health, plant biodiversity, and resilience to heat and drought, which helps improve water and nutrient efficiency and increase yields (Murindangacho, Kopecky & Konvalina, 2021; Barnett, 2022). A case study conducted in the Gicumbi district in Rwanda found that farmers' socioeconomic status influenced the adoption of CA practices, access to information, and availability of resources (Murindangacho, Kopecky & Konvalina, 2021). Another study discussed the applicability and potential benefits of CA in Rwanda under the unfolding climate change (Kabirigi et al., 2015). The government of Rwanda has been promoting agricultural intensification focused on the production of food security and livelihood development. However, evidence has shown that many Rwandan farmers have not benefited from this approach, and it is essential to understand the processes through which agricultural transformations unfold for specific types of farmers (Kim, Marshall & Dawson, 2022). The dominant approach to increase the productive capacity of the land, crops, and animal resources has been through large-scale land consolidation, soil fertility management, and the intensive use of biotechnology and external inputs (Ibid). Rwanda's Ministry of

Agriculture and Livestock has implemented the Crop Intensification Programme and the Strategic Plan for the Transformation of Agriculture, which aim to increase agricultural productivity and improve food security (Murindangacho, Kopecky & Konvaliva, 2021; Kim, Marshall & Dawson, 2022). The success of these programmes depends on the adoption of sustainable agricultural practices, including CA, and the availability of resources and information to farmers (Murindangacho, Kopecky & Konvaliva, 2021; Barnett, 2022).

Rwanda has been promoting CA as a way to improve soil health, plant biodiversity, and resilience to heat and drought. Farmers' socioeconomic status has influenced the adoption of CA practices, access to information, and availability of resources. The government of Rwanda has implemented programmes to increase agricultural productivity and improve food security, which depend on the adoption of sustainable agricultural practices, including CA, and the availability of resources and information to farmers.

RESEARCH IMPLICATIONS

The Pfumvudza Programme, a climate-smart agricultural initiative in Zimbabwe, offers promising pathways towards achieving food security and environmental sustainability. However, understanding its implications requires exploration from both theoretical and practical perspectives.

Theoretical Implications

Assessing Long-Term Impact: Developing theoretical frameworks to assess the Programme's long-term impact on food security, environmental sustainability, and socio-economic factors is crucial. This could involve modelling the Programme's influence on food production, access, utilization, and stability, considering climate change and market dynamics (Pretty et al., 2011). Frameworks to analyse the environmental impact on soil health, water use efficiency, carbon sequestration, and biodiversity are crucial to evaluating its long-term sustainability (Godfrey & Garnett, 2014).

Connections to Agro-Ecological Theories: Exploring links between Pfumvudza and existing conservation agriculture frameworks can provide valuable insights. This includes analysing its compatibility with principles like minimum soil disturbance, permanent soil cover, and crop diversification (Mazvimavi & Twomlow, 2017). Additionally, investigating how Pfumvudza aligns with climate-smart agriculture theories can shed light on its potential for adaptation and mitigation strategies in a changing climate.

Gender Dynamics and Sustainable Agriculture: Integrating gender dynamics into theoretical models for understanding Pfumvudza is essential. This involves analysing the specific roles of women in implementation, identifying gender-based constraints to participation, and developing strategies to address them (Mutenje et al., 2019). Further research should explore the Programme's potential for women's empowerment in agriculture.

Policy Influence on Sustainable Agriculture Programmes: Analysing how national agricultural policies, subsidies, and regulations influence Pfumvudza's implementation and scalability is crucial. Frameworks for understanding the complex interplay between policy design and Programme implementation can inform policy decisions and improve Programme effectiveness.

Practical Implications

Implementation Evaluation and Improvement: Regular assessments of the Programme's implementation across different regions are essential. This involves collecting data on farmer adoption rates, yield increases, input utilization, and challenges encountered (Mazvimavi & Twomlow, 2017). Effective feedback mechanisms for farmers to convey their experiences and suggestions are crucial for Programme improvement.

Technological Integration and Innovation: Integrating modern agricultural technologies like precision agriculture tools, climate-smart technologies, and digital information platforms can enhance efficiency, resource management, and decision-making for farmers. Additionally, encouraging research and development of locally appropriate, low-cost technological solutions can address specific challenges faced by smallholder farmers.

Community Participation and Ownership: Establishing community-based committees to oversee implementation, identify local needs, and mobilize resources is crucial. Fostering a culture of knowledge sharing and collaboration through farmer-to-farmer networks and exchange visits can encourage collective problem-solving and adaptation of practices. Ensuring gender-inclusive participation and addressing the specific needs of women farmers is essential for promoting their empowerment.

Policy and Institutional Support: Reviewing and aligning existing agricultural policies and institutional frameworks to support the Programme's scalability is crucial. This may involve addressing issues like land tenure, access to markets, input subsidies, and credit availability. Strengthening the capacity of institutions involved in implementation, such as extension services and research institutions, is essential for effective support to farmers. Public-private partnerships can leverage resources, expertise, and technology for wider adoption and scaling up of Pfumvudza practices.

Monitoring and Evaluation: It is crucial to develop robust monitoring and evaluation frameworks to track the Programme's progress towards its objectives. This should include indicators related to food production, household food security, environmental sustainability, and socio-economic development. Disseminating research findings and best practices widely through publications, workshops, and online platforms can inform stakeholders and support similar Programme development in other contexts.

By understanding both the theoretical and practical implications of the Pfumvudza Programme, scholars and stakeholders can contribute to its success, ultimately leading to a more sustainable and food-secure future for the world.

CONCLUSION

The *Pfumvudza* Programme in Zimbabwe stands as a pivotal pilot initiative in bolstering food security within the smallholder sector through sustainable agricultural practices. This Programme, which can be applied across a diverse range of crops, marks a significant step towards achieving long-term national food security. *Pfumvudza*, with its focus on optimizing resource use and promoting climate-resilient farming techniques, not only aims to increase crop yields but also seeks to ensure the sustainability of agricultural practices. By encouraging smallholder farmers to adopt conservation agriculture principles, such as minimum tillage and mulching (Dick-Sagoe, 2023), *Pfumvudza* enhances soil fertility and water retention, mitigating the impact of climate variability. Furthermore, the Programme's emphasis on crop diversification contributes to a more resilient and balanced food production system. In the broader context of sustainable agriculture, the *Pfumvudza* Programme emerges as a compelling case study, demonstrating the potential for integrated, environmentally conscious approaches to not only address immediate food security concerns but also establish a foundation for enduring agricultural prosperity.

Limitations

The study on Pfumvudza and its contribution to food security in Zimbabwe has several limitations that should be considered. Firstly, the reliance on secondary data may introduce information gaps, and the accuracy and completeness of the available data could be subject to variations. Additionally, there may be potential biases in the sources used, which could influence the study's findings. The research faced limitations in generalizability, as the focus on Zimbabwe's context may not fully capture the diversity of agricultural conditions in other regions or countries. Furthermore, the study's temporal scope limited its ability to capture the long-term sustainability and impact of the Pfumvudza Programme. The research might not comprehensively address socio-economic factors influencing food security, and the assessment of outcomes could be constrained by the availability of relevant data. Lastly, the study lacks an in-depth exploration of the challenges and barriers faced during the implementation of Pfumvudza, which is essential for a more nuanced understanding of the Programme's dynamics. Acknowledging these limitations is crucial for interpreting the study's findings and for guiding future research endeavours.

Key Recommendations

To optimize the impact of Pfumvudza on food security in Zimbabwe, several key recommendations can be considered.

- The Programme should implement a continuous monitoring and evaluation system to assess its ongoing effectiveness and identify areas for improvement.
- Tailoring Pfumvudza practices to suit the diverse agroecological zones within Zimbabwe is essential for ensuring the Programme's adaptability and success across varied environments.
- Investing in comprehensive capacity-building programmes and extension services will empower farmers with the necessary knowledge and skills for sustainable agriculture.
- Exploring the integration of modern agricultural technologies, precision farming techniques, and digital tools into Pfumvudza can enhance overall efficiency and productivity.
- Community engagement and participation should be prioritised to foster local ownership and acceptance.
- Given the challenges of climate change, incorporating climate-resilient agricultural practices is crucial, including drought-resistant crop varieties and water management strategies.
- Gender-inclusive approaches should be integrated to empower women in agriculture, recognizing their specific roles and needs.
- Advocating for supportive agricultural policies at the national level will enhance the scalability and sustainability of Pfumvudza, while ongoing research and innovation should be prioritized to stay abreast of advancements in sustainable agriculture.
- Establishing platforms for knowledge exchange and collaboration will contribute to a global effort to enhance food security, allowing Zimbabwe to share successes and learn from other nations facing similar agricultural challenges.

Suggestions for Future Research

Future research and implementation of the Pfumvudza Programme should prioritize a comprehensive long-term impact assessment to gauge its sustained effects on food security, environmental conservation, and rural livelihoods over an extended period. Additionally, there is a need to delve into the adaptability of Pfumvudza principles to diverse agroecological zones and socio-economic contexts, tailoring the Programme to specific regional needs for broader effectiveness. Understanding the social and economic implications, including gender dynamics and income distribution, will contribute to a more holistic approach to sustainable agricultural development. Exploring the integration of modern technologies, capacity-building strategies, and policy support will further enhance the Programme's efficiency and overall success. It is essential to investigate the resilience of Pfumvudza to climate change and explore climate-smart agriculture approaches for integration. Moreover, fostering global collaboration through knowledge exchange and collaborative research will accelerate the dissemination of Pfumvudza's best practices, contributing to a more sustainable and resilient agricultural future on a global scale.

REFERENCES

AGRITEX. (n.d.). *Adopting the Conservation Agriculture Concept on Maize and Traditional Grains. Pfumvudza/ Intwasa for Sustainable Crop Production.* https://www.zimagrihub.org.zw/sites/default/files/documents/Agritex%20Pfumbvuza%20BROCHURE%20FINAL.pdf

Baudron, F., & Mwanza, R. (2012). *Conservation agriculture in Zambia: A case study of Southern Province*. Academic Press.

Bjornlund, V., Bjornlund, H., & van Rooyen, A. (2022). Why food insecurity persists in sub-Saharan Africa: A review of existing evidence. *Food Security*, *14*(4), 845–864. doi:10.1007/s12571-022-01256-1 PMID:35136455

Bramley, R., & Trengove, S. A. M. (2013). Precision agriculture in Australia: Present status and recent developments. *Engenharia Agrícola*, *33*(3), 575–588. doi:10.1590/S0100-69162013000300014

Braun, V., & Clarke, V. (2006). Using thematic analysis in psychology. *Qualitative Research in Psychology*, *3*(2), 77–101. doi:10.1191/1478088706qp063oa

Busari, M. A., Kukal, S. S., Kaur, A., Bhatt, R., & Dulazi, A. A. (2015). Conservation tillage impacts on soil, crop and the environment. *International Soil and Water Conservation Research*, *3*(2), 119–129. doi:10.1016/j.iswcr.2015.05.002

Dick-Sagoe, C., Hope, K. N., & Asare-Nuamah, P. (2023). Perceived impact of climate variability and change on livelihoods of smallholder farmers in Lesotho. *African Journal of Science, Technology, Innovation and Development*, *15*(2), 175–184. doi:10.1080/20421338.2022.2058339

EEA. (2023). *The Impact of Climate Change on Food Security*. https://emergencyaction.org.au/eaa-2021/posts/the-impact-of-climate-change-on-food-security

FAO. (2011). *Save and Grow: A policymaker's guide to the sustainable intensification of smallholder crop production*.

FAO. (2020). *Crop and food security assessment mission to Zimbabwe*. FAO.

Federal Ministry of Economic Cooperation and Development. (n.d.). *What is sustainable Agriculture?* Deutsche Gesellschaft for International Zuzammenarbeit (GIZ) GmbH.

Frischen, J., Meza, I., Rupp, D., Wietler, K., & Hagenlocher, M. (2020). *Drought Risk to Agricultural Systems in Zimbabwe: A Spatial Analysis of Hazard, Exposure, and Vulnerability*. www.mdpi.com/journal/sustainability

Gerlach, C. (2015). Famine responses in the world food crisis 1972–5 and the World Food Conference of 1974. *European Review of History*, *22*(6), 929–939. doi:10.1080/13507486.2015.1048191

Godfrey, L., & Garnett, T. (2014). Sustainable intensification of agriculture: Narrowing the gap between food security and environmental sustainability. *Food Security*, *6*(2), 135–143.

Guidoboni, M. V., Duparque, A., Boissy, J., Mouny, J.-C., Auberger, J., & van der Werf, H. M. (2023). Conservation agriculture reduces the climate change impact of popcorn and wheat crop rotation. *PLoS One*, *18*(5), e0285586. doi:10.1371/journal.pone.0285586 PMID:37220132

HLPE. (2013). *Investing in Smallholder Agriculture for Food Security*. A Report by the High-Level Panel of Experts on Food Security and Nutrition of the Committee on World Food Security, Committee on World Security, 111. Available at: https://www.fao.org/

Holland, J. E., White, P. J., Glendining, M. J., Goulding, K. W. T., & McGrath, S. P. (2019). Yield responses of arable crops to liming – An evaluation of relationships between yields and soil pH from a long-term liming experiment. *European Journal of Agronomy, 105*, 176–188. doi:10.1016/j.eja.2019.02.016 PMID:31007524

ILO. (2008). *Promotion of rural employment for poverty reduction, international labour conference, 97th Session, 2008*. ILO.

IPCC. (2007). *Fourth Assessment Report.* Cambridge University Press.

Kader, M. A., Senge, M., Mojid, M. A., & Ito, K. (2017). Recent advances in mulching materials and methods for modifying soil environment. *Soil & Tillage Research, 168*, 155–166. doi:10.1016/j.still.2017.01.001

Kapari, M., Hlophe-Ginindza, S., Nhamo, L., & Mpandeli, S. (2023). Contribution of smallholder farmers to food security and opportunities for resilient farming systems. *Frontiers in Sustainable Food Systems, 7*, 1149854. doi:10.3389/fsufs.2023.1149854

Kim, S. K., Marshall, F., & Dawson, N. M. (2022). Revisiting Rwanda's agricultural intensification policy: Benefits of embracing farmer heterogeneity and crop-livestock integration strategies. *Food Security, 14*(3), 637–656. doi:10.1007/s12571-021-01241-0

Komarek, A. M., Hoyoung Kwon, H., Beliyou Haile, B., Christian Thierfelder, C., Mutenje, M. J., & Carlo Azzarri, C. (2015). From plot to scale: Ex-ante assessment of conservation agriculture in Zambia. *Agricultural Systems, 173*, 504–518. doi:10.1016/j.agsy.2019.04.001

Krishnamurthy, P. K., Lewis, K., & Choutartun, R. J. (2012). *Climate Impacts on Food Security and Nutrition. A Review of Existing Knowledge*. Met Office and WFP.

Machingura, G. (2020). *Roundup: Zimbabwe rolls out new farming concept to boost grain production.* Xinhua News. https://www.xinhuanet.com/english/2020-09/11/c_139359235.htm

Makumbi, M., Mugume, M., & Mugisha, A. (2017). Conservation agriculture and sustainable development in Africa: A case study from the Lindi region of Tanzania. *Journal of Agriculture and Rural Development in the Tropics and Subtropics, 121*(2), 121–132.

Makumbi, M., Mugume, M., & Mugisha, A. (2022). *Conservation Agriculture in the Southern Highlands of Tanzania*. CABI Digital Library. doi:10.1079/9781789245745.0006

Manguwo, E. (2023). *Farmers urged to uphold Pfumvudza principles.* https://www.herald.co.zw/farmers-urged-to-uphold-pfumvudza-principles/

Maseya, P. (2020). An analysis of the Pfumvudza/Intwasa Programme in Zimbabwe. *Sustainability, 12*(16), 6390.

Mazvimavi, K., & Twomlow, S. (2017). Conservation agriculture in Zimbabwe: Policy, adoption, and impact. *Journal of International Development, 29*(1), 125–138.

Ministry of Lands, Agriculture, Water Fisheries, Water and Rural Resettlement. (n.d.). http://www.moa.gov.zw/wp-content/uploads/2021/02/pfumvudza-flier.pdf

Mkonda, A. M. (2016). Conservation agriculture (CA) in Tanzania: The case of the Southern Highlands. *International Journal of Agricultural Sustainability, 14*(5), 557–572.

Mkonda, A. M., & Mkonda, A. M. (2016). *Conservation Agriculture in Tanzania.* Semantic Scholar.

Mkonda, M. (2017). *Conservation Agriculture in Tanzania.* . doi:10.1007/978-3-319-48006-0_10

Mujere, N. (2022). An Assessment of the Contribution of The Pfumvudza Concept Towards Climate Smart Agriculture in Zimbabwe: A Review. *Journal of Agriculture and Horticulture Research, 5*(2), 69–76.

Murindangabo, Y. T., Kopecký, M., & Konvalina, P. (2021). Adoption of Conservation Agriculture in Rwanda: A Case Study of Gicumbi District Region. *Agronomy (Basel), 11*(9), 1732. doi:10.3390/agronomy11091732

Mutenje, M. J., Mafongoya, P. L., & Chivenge, P. P. (2019). Climate-smart agriculture in Zimbabwe: Opportunities and challenges. *Climate and Development, 11*(7), 604–610.

National Development Strategy. (2020). Towards a Prosperous & Empowered Upper Middle-Income Society by 2030 January 2021 – December 2025. Author.

Oerke, E. C. (2006). Crop losses to pests. *Journal of Agricultural Science, 144*(1), 31–43. doi:10.1017/S0021859605005708

Penot, E., Fevre, V., Flodrops, P., & Razafimahatratra, H. M. (2015). *Conservation agriculture and agroecology practices to mitigate climatic variations in medium altitude in Madagascar.* CIRAD.

Pepin, I. (2022). *What is Sustainable Agriculture and Why is it Important?* The Humane League. https://thehumanleague.org/article/what-is-sustainable-agriculture-and-why-is-it-important

Pretty, J. N., Toulmin, C., & Williams, S. (2011). Sustainable intensification in African agriculture. *International Journal of Agricultural Sustainability*, *9*(1), 5–24. doi:10.3763/ijas.2010.0583

Scoones, I. (2021). *Can the Pfumvudza Conservation Agriculture Programme deliver food security in Zimbabwe? Article on future agriculture*. https://www.future-agricultures.org/blog/can-the-pfumvudza-conservation-agriculture-Programmeme-deliver-food-security-in-zimbabwe/

Sharara, K. (2023). Pfumvudza mitigates climate change. *The Sunday News*. www.sundaynews.co.zw/pfumvudza-mitigates-climate-change/

Shetto, R. M., & Owenya, M. (2007). *Conservation Agriculture as Practiced in Tanzania: Three Case Studies*. African Conservation Tillage Network.

Subbulakshmi, S., Saravanan, N., & Subbian, P. (2009). Conventional tillage vs conservation tillage - a review. *Agric. Rev., 30*.

Suding, K., Higgs, E., Palmer, M., Callicott, J. B., Anderson, B., Baker, C. B., Gutrich, M., Hondula, J. J., Matthew, K. L., Lafevor, C., Larson, M. C., Randall, B. M. H., Ruhl, A., And, J. B., & Schwartz, K. Z. S. (2015). Committing to ecological restoration. *Environmental Monitoring and Assessment*, *348*(6235), 638–640. doi:10.1126/science.aaa4216 PMID:25953995

UC Sustainable Agriculture Research and Education Programme. (2021). *What is Sustainable Agriculture?* UC Agriculture and Natural Resources. https://sarep.ucdavis.edu/sustainable-ag

Udry, C., Hoddinott, J., Alderman, H., & Haddad, L. (1995). *Gender differentials in farm productivity: Implications for household efficiency and agricultural policy*. Academic Press.

Umar, B. B., Aune, J. B., Johnsen, F. H., & Lungu, O. I. (2011). *Options for improving smallholder conservation agriculture in Zambia*. Academic Press.

Volsi, B., Higashi, G. E., Bordin, I., & Telles, T. S. (2022). The diversification of species in crop rotation increases the profitability of grain production systems. *Scientific Reports*, *12*(1), 19849. doi:10.1038/s41598-022-23718-4 PMID:36400822

Zulu, O., & Chapoto, A. (2016). Conservation agriculture: Gendered impacts on households' livelihoods. *Zambia Social Science Journal*.

ENDNOTES

1 "Conservation agriculture: The 3 principles" Food and Agriculture Organization of the United Nations. Retrieved 9 October 2023.

Compilation of References

Abbasi, J. Y., & Din, M. (2020, June). Fairtrade and Workers' Welfare: A Case Study of the Soccer Industry in Pakistan. *Journal of Asian and African Studies*, *55*(4), 536–551. Advance online publication. doi:10.1177/0021909619885968

Adewale, H. A. (2022). *A Community Knowledge-Based Assessment of Smallholder Farmers' Perception of Cluster Farming and its Potential Role in Sustainable Accessibility to Farmer-Centric Support in Nigeria*. Polytechnic University.

Aditya, S. (2016). *Design and development of single-row power weeder for rice* [Master's thesis].

Adjei, E. (2007). *Impact of mining on livelihoods of rural households. A case study of farmers in the Wassa Mining Region, Ghana* [Master's thesis]. Geografisk institutt.

Adonteng-Kissi, O., & Adonteng-Kissi, B. (2017). Living with conflicts in Ghana's Prestea mining area: Is community engagement the answer? *Journal of Sustainable Mining*, *16*(4), 196–206. doi:10.1016/j.jsm.2017.12.005

Africa, I. S. S. D. (2022). *Policies, laws, and regulations in support of farmer seed systems.* Available: https://issdafrica.org/2022/03/16/policies-laws-and-regulations-in-support-of-farmer-seed-systems/

Agarwal, J., Osiyevskyy, O., & Feldman, P. M. (2015). Corporate Reputation Measurement: Alternative Factor Structures, Nomological Validity, and Organizational Outcomes. *Journal of Business Ethics*, *130*(2), 485–506. doi:10.1007/s10551-014-2232-6

Agrawal, A., & Ostrom, E. (2001). Collective action, property rights, and decentralization in resource use in India and Nepal. *Politics & Society*, *29*(4), 485–514. doi:10.1177/0032329201029004002

AGRITEX. (n.d.). *Adopting the Conservation Agriculture Concept on Maize and Traditional Grains. Pfumvudza/ Intwasa for Sustainable Crop Production.* https://www.zimagrihub.org.zw/sites/default/files/documents/Agritex%20Pfumbvuza%20BROCHURE%20FINAL.pdf

Ahmed, N., Masood, S., Ahmad, S., Bashir, S., Hussain, S., Hassan, W., & Ali, M. A. (2019). *Soil management for better crop production and sustainable agriculture. Agronomic Crops* (Vol. 2). Management Practices.

Ahmed, U. I., Ying, L., Bashir, M. K., Abid, M., & Zulfiqar, F. (2017). Status and determinants of small farming households' food security and role of market access in enhancing food security in rural Pakistan. *PLoS One*, *12*(10), e0185466. doi:10.1371/journal.pone.0185466 PMID:29077719

Akerlof, G. A. (1970). 4. The market for 'lemons': Quality uncertainty and the market mechanism. *The Quarterly Journal of Economics*, *66*(3), 488. Advance online publication. doi:10.2307/1879431

Akinyi, D. P., Karanja Ng'ang'a, S., Ngigi, M., Mathenge, M., & Girvetz, E. (2022). Cost-benefit analysis of prioritized climate-smart agricultural practices among smallholder farmers: Evidence from selected value chains across sub-Saharan Africa. *Heliyon*, *8*(4), e09228. doi:10.1016/j.heliyon.2022.e09228 PMID:35399383

Akmal, N. (2021). *Information Communication Technology (ICT) and Livelihood Improvement in Rural Pakistan: A Comparative Study of Small-and Large-Holder Citrus Farming Households in the Sargodha District*. University of Canberra.

AlBar, A. M., & Hoque, M. R. (2019). Factors affecting the adoption of information and communication technology in small and medium enterprises: A perspective from rural Saudi Arabia. *Information Technology for Development*, *25*(4), 715–738. doi:10.1080/02681102.2017.1390437

Alemu, D., Guinan, J., & Hermanson, J. (2021). Contract farming, cooperatives and challenges of side selling: Malt barley value-chain development in Ethiopia. *Development in Practice*, *31*(4), 496–510. doi:10.1080/09614524.2020.1860194

Alemu, G. T., Nigussie, Z., Haregeweyn, N., Berhanie, Z., Wondimagegnehu, B. A., Ayalew, Z., Molla, D., Okoyo, E. N., & Baributsa, D. (2021). Cost-benefit analysis of on-farm grain storage hermetic bags among small-scale maize growers in northwestern Ethiopia. *Crop Protection (Guildford, Surrey)*, *143*, 105478. doi:10.1016/j.cropro.2020.105478 PMID:33941995

Aliber, M., & Hart, T. G. (2009). Should subsistence agriculture be supported as a strategy to address rural food insecurity? *Agrekon*, *48*(4), 434–458. doi:10.1080/03031853.2009.9523835

Alluvione, F., Moretti, B., Sacco, D., & Grignani, C. (2011). EUE (energy use efficiency) of cropping systems for a sustainable agriculture. *Energy*, *36*(7), 4468–4481. doi:10.1016/j.energy.2011.03.075

Anantha, K., Garg, K. K., Barron, J., Dixit, S., Venkataradha, A., Singh, R., & Whitbread, A. M. (2021). Impact of best management practices on sustainable crop production and climate resilience in smallholder farming systems of South Asia. *Agricultural Systems*, *194*, 103276. doi:10.1016/j.agsy.2021.103276

Andorfer, V. A., & Liebe, U. (2012a). Research on Fair Trade Consumption-A Review. *Journal of Business Ethics*, *106*(4), 415–435. doi:10.1007/s10551-011-1008-5

Andrews, N. (2018). Land versus livelihoods: Community perspectives on dispossession and marginalization in Ghana's mining sector. *Resources Policy*, *58*, 240–249. doi:10.1016/j.resourpol.2018.05.011

Angadi, R. (2019). Introduction to Smart City and Agricultural Revolution: Big Data and Internet of Things (IoT). In Smart Cities and Smart Spaces: Concepts, Methodologies, Tools, and Applications (pp. 135-176). IGI Global.

Archer, M., & Elliott, H. (2021). 'It's up to the market to decide': Revealing and concealing power in the sustainable tea supply chain. *Critique of Anthropology*, *41*(3), 227–246. doi:10.1177/0308275X211038607

Aslam, A. A., Aslam, M. S., & Aslam, A. A. (2022). An Overview on Green synthesis of Nanoparticles and their Advanced Applications in Sustainable Agriculture. *International Journal of Applied Chemical and Biological Sciences*, *3*(2), 70–99.

Ayesiga, C. (2023). *Digitalising African agricultural value chains: the role of ICT platforms in smallholders' uptake of agricultural inputs in Uganda*. Wageningen University.

Bacon, C. (2005). Confronting the coffee crisis: Can Fair Trade, organic, and specialty coffees reduce small-scale farmer vulnerability in Northern Nicaragua? *World Development*, *33*(3), 497–511. doi:10.1016/j.worlddev.2004.10.002

Baiyeri, P. K., Foleng, H. N., Machebe, N. S., & Nwobodo, C. E. (2019). Crop-livestock interaction for sustainable agriculture. *Innovations in Sustainable Agriculture*, 557-582.

Baker, S. E., Edwards, R., Adler, P., Becker, H. S., & Doucet, A. (2012). *How many qualitative interviews is enough? Expert voices and early career reflections on sampling Expert voices*. National Centre for Research Methods Review.

Bala, M. (2021). Role of Information, Communication and Technology (ICT) in Rural Development. *National Research Journal of Social Sciences*, *6*, 57–63.

Balineau, G. (2012). *Disentangling the Effects of Fair Trade on the Quality of Malian Cotton*. http://hdl.handle.net/10419/269320

Ballet, J., & Carimentrand, A. (2010). Fair Trade and the Depersonalization of Ethics. *Journal of Business Ethics*, *92*(S2), 317–330. Advance online publication. doi:10.1007/s10551-010-0576-0

Ba, M. N. (2016). Strategic agricultural commodity value chains in Africa for increased food: The regional approach for food security. *Agricultural Sciences*, *7*(09), 549–585. doi:10.4236/as.2016.79055

Bansal, P., & Song, H. C. (2017). Similar but not the same: Differentiating corporate sustainability from corporate responsibility. *The Academy of Management Annals*, *11*(1), 105–149. doi:10.5465/annals.2015.0095

Barker, R., & Chapman, D. (2020). The economics of sustainable agricultural systems in developing countries. In *Sustainable Agricultural Systems* (pp. 478–494). CRC Press. doi:10.1201/9781003070474-32

Barrientos, S., & Dolan, C. (2012). *Ethical Sourcing in the Global Food System*. doi:10.4324/9781849771269

Bartley, S. J., Blanton, P. W., & Gilliard, J. L. (2005). Husbands and Wives in Dual-Earner Marriages: Decision-Making, Gender Role Attitudes, Division of Household Labor, and Equity. *Marriage & Family Review*, *37*(4), 69–94. doi:10.1300/J002v37n04_05

Bartley, T. (2007). Institutional emergence in an era of globalization: The rise of transnational private regulation of labor and environmental conditions. *American Journal of Sociology*, *113*(2), 297–351. doi:10.1086/518871

Bassett, T. J. (2010). Slim pickings: Fairtrade cotton in West Africa. *Geoforum*, *41*(1), 44–55. doi:10.1016/j.geoforum.2009.03.002

Baudron, F., & Mwanza, R. (2012). *Conservation agriculture in Zambia: A case study of Southern Province*. Academic Press.

Baughman, S. B., & Barclay, S. H. (2011). *Fair Trade and Child Labor*. https://papers.ssrn.com/abstract=1823546

Bebbington, A., Bebbington, D. H., Bury, J., Lingan, J., Muñoz, J. P., & Scurrah, M. (2018). Mining and social movements: Struggles over livelihood and rural territorial development in the Andes. *World Development*, *36*(12), 2888–2905. doi:10.1016/j.worlddev.2007.11.016

Bebbington, A., & Humphreys Bebbington, D. (2018). Mining, movements and sustainable development: Concepts for a framework. *Sustainable Development (Bradford)*, *26*(5), 441–449. doi:10.1002/sd.1888

Behnassi, M., Shahid, S. A., & D'silva, J. (2011). *Sustainable agricultural development*. Springer. doi:10.1007/978-94-007-0519-7

Benson, E., & Connell, K. Y. H. (2014). Fair trade consumption from the perspective of US Baby Boomers. *Social Responsibility Journal*, *10*(2), 364–382. doi:10.1108/SRJ-08-2012-0094

Bessou, C., Ferchaud, F., Gabrielle, B., & Mary, B. (2011). Biofuels, greenhouse gases and climate change. *Sustainable Agriculture*, *2*, 365–468.

Bigirimana, S., & Masengu, R. (2015). The Contribution of the International Coffee Agreement (ICA) to the Development of Export Markets for Small Scale Coffee Farmers in Manicaland, Zimbabwe. *World Journal of Business and Management*, *1*(1), 57. doi:10.5296/wjbm.v1i1.7931

Bissinger, K. (2019). Price Fairness: Two-Stage Comparison of Conventional and Fairtrade Prices. *Journal of International Consumer Marketing*, *31*(2), 86–97. doi:10.1080/08961530.2018.1482525

Bjornlund, V., Bjornlund, H., & van Rooyen, A. (2022). Why food insecurity persists in sub-Saharan Africa: A review of existing evidence. *Food Security*, *14*(4), 845–864. doi:10.1007/s12571-022-01256-1 PMID:35136455

Blanchet, V. (2012). *The two faces of Janus: a postcolonial problematization of the fair trade ambivalence*. https://shs.hal.science/halshs-00676060

Blimpo, M. P., & Cosgrove-Davies, M. (2019). *Electricity access in Sub-Saharan Africa: Uptake, reliability, and complementary factors for economic impact*. World Bank Publications.

Bolarinwa, K. K. L., Irene, S. E., & Apantaku, S. O. (2014). Utilization of information communication technology's components for coordination of marketing, agricultural commodities in oyo State Nigeria. *European Journal of Business and Management*, *6*(6), 12–16.

Boniface, M. (2006). The meaning of adventurous activities for 'women in the outdoors. *Journal of Adventure Education and Outdoor Learning*, *6*(1), 9–24. doi:10.1080/14729670685200711

Bonisoli, L., Galdeano-Gómez, E., Piedra-Muñoz, L., & Pérez-Mesa, J. C. (2019). Benchmarking agri-food sustainability certifications: Evidences from applying SAFA in the Ecuadorian banana agri-system. *Journal of Cleaner Production*, *236*, 117579. Advance online publication. doi:10.1016/j.jclepro.2019.07.054

Borthakur, A., & Singh, P. (2021). Indigenous Agricultural Knowledge Towards Achieving Sustainable Agriculture. In V. Kumar Singh, R. Singh, & E. Lichtfouse (Eds.), *Sustainable Agriculture Reviews 50. Emerging contaminants in agriculture*. Springer. doi:10.1007/978-3-030-63249-6_15

Bramley, R., & Trengove, S. A. M. (2013). Precision agriculture in Australia: Present status and recent developments. *Engenharia Agrícola*, *33*(3), 575–588. doi:10.1590/S0100-69162013000300014

Braun, V., & Clarke, V. (2006). Using thematic analysis in psychology. *Qualitative Research in Psychology*, *3*(2), 77–101. doi:10.1191/1478088706qp063oa

Bremborg, A. D. (2011). Interviewing. In M. Stausberg & S. Engler (Eds.), *The Routledge Handbook of Research Methods in the Study of Religion*. Routledge.

Briggs, J., & Moyo, B. (2012). The Resilience of Indigenous Knowledge in Small-scale African Agriculture: Key Drivers. *Scottish Geographical Journal*, *128*(1), 64–80. doi:10.1080/14702541.2012.694703

Brooks, J., Deconinck, K., & Giner, C. (2019). *Three key challenges facing agriculture and how to start solving them*. Agriculture and Fisheries.

Brunner, T. A. (2014). Applying neutralization theory to fair trade buying behaviour. *International Journal of Consumer Studies*, *38*(2), 200–206. doi:10.1111/ijcs.12081

Brymer, F., & Feletti, F. (2020). Beyond risk: The importance of adventure in the everyday life of young people. *Annals of Leisure Research*, *23*(3), 429–446. doi:10.1080/11745398.2019.1659837

Bucci, G., Bentivoglio, D., & Finco, A. (2018). Precision agriculture as a driver for sustainable farming systems: State of art in literature and research. *Calitatea*, *19*(S1), 114–121.

Bürgin, D., & Wilken, R. L. (2021). Increasing Consumers' Purchase Intentions Toward Fair-Trade Products Through Partitioned Pricing. *Journal of Business Ethics*. Advance online publication. doi:10.1007/s10551-021-04938-6

Busari, M. A., Kukal, S. S., Kaur, A., Bhatt, R., & Dulazi, A. A. (2015). Conservation tillage impacts on soil, crop and the environment. *International Soil and Water Conservation Research, 3*(2), 119–129. doi:10.1016/j.iswcr.2015.05.002

Buthelezi, N. N., & Hughes, J. C. (2014). Indigenous knowledge systems and agricultural rural development in South Africa: Past and present perspectives. *Indilinga, 13*, 231–250.

Cappelli, S. L., Domeignoz-Horta, L. A., Loaiza, V., & Laine, A. L. (2022). Plant biodiversity promotes sustainable agriculture directly and via belowground effects. *Trends in Plant Science, 27*(7), 674–687. doi:10.1016/j.tplants.2022.02.003 PMID:35279365

Carlisle, L., DeLonge, M. S., Iles, A., Calo, A., Getz, C., Ory, J., Galt, R., Melone, B., & Knox, R. (2019). Transitioning to Sustainable Agriculture Requires Growing and Sustaining an Ecologically Skilled Workforce. *Frontiers in Sustainable Food Systems, 3*, 3. doi:10.3389/fsufs.2019.00096

Chandra, A., McNamara, K. E., & Dargusch, P. (2018). Climate-smart agriculture: Perspectives and framings. *Climate Policy, 18*(4), 526–541. doi:10.1080/14693062.2017.1316968

Chari, F., Novukela, C., & Ngcamu, B. S. (2022). Artisanal mining versus sustainability of agricultural food supply chains: Effects of the conflicts in Southern Zimbabwe. *Development in Practice, 32*(3), 349–360. doi:10.1080/09614524.2021.1937543

Chen, J. C., Huang, Y., Lin, K. Y., Chang, Y. S., Lin, H. C., Lin, C. Y., & Hsiao, H. S. (2020). Developing a hands-on activity using virtual reality to help students learn by doing. *Journal of Computer Assisted Learning, 36*(1), 46–60. doi:10.1111/jcal.12389

Chen, J., & Chen, Y. J. (2021). The impact of contract farming on agricultural product supply in developing economies. *Production and Operations Management, 30*(8), 2395–2419. doi:10.1111/poms.13382

Chen, L., Lee, H. L., & Tang, C. S. (2022). Supply Chain Fairness. *Production and Operations Management, 31*(12), 4304–4318. Advance online publication. doi:10.1111/poms.13849

Chihambakwe, M. T. (2018). *Urban and peri-urban agriculture: an analysis of the perceptions and use of food sovereignty among low-income dwellers in Harare.*

Chimhowu, A., & Woodhouse, P. (2006). Customary vs private property rights? Dynamics and trajectories of vernacular land markets in Sub-Saharan Africa. *Journal of Agrarian Change, 6*(3), 346–371. doi:10.1111/j.1471-0366.2006.00125.x

Chineka, J. (2016). Analysis of drought incidence, gendered vulnerability and adaptation in Chivi South. Academic Press.

Chipuriro, R. M. (2021). *Land reform and local economic development: elderly women farmers' narratives in Shamva District, Zimbabwe.* University of Johannesburg.

Chirume, S. (2019). The use of Zimbabwe Open University's MyVista platform in e-communication, accessing and uploading learning materials, and assessing students' work. *American Educational Research Journal, 7*(1), 1–13. doi:10.30918/AERJ.64.18.079

Chisango, T., Mafa, I., & Maunganidze, L. (2022). Investigating the Sexist Implications of Bride Price (Lobola) in Zimbabwe. *Society, 59*(5), 564–575. doi:10.1007/s12115-022-00743-7

Chitungo, S., & Munongo, S. (2013). Extending the Technology Acceptance Model to Mobile Banking Adoption in Rural Zimbabwe. *Journal of Business Administration and Education, 3*(1), 51–79.

Choudhary, M., Ghasal, P. C., Kumar, S., Yadav, R. P., Singh, S., Meena, V. S., & Bisht, J. K. (2016). Conservation agriculture and climate change: an overview. *Conservation Agriculture: An Approach to Combat Climate Change in Indian Himalaya*, 1-37.

CIAT. (2018). *Sloping Agricultural Land Technology (SALT) in the Philippines: A Case Study*. Bogor, Indonesia: CIAT. Available online at https://lib.icimod.org/record/21492/files/Sloping%20 Agricultural%20Land%20Technology%20(SALT).pdf

Clark-Kazak, C. R. (2009). Towards a working definition and application of social age in international development studies. *The Journal of Development Studies, 45*(8), 1307–1324. doi:10.1080/00220380902862952

Coca-Cola. (2023). *Sustainable Agriculture*. https://www.coca-colacompany.com/sustainability/ sustainable-agriculture

Conroy, T., & Weiler, S. (2016). Does gender matter for job creation? Business ownership and employment growth. *Small Business Economics, 47*(2), 397–419. doi:10.1007/s11187-016-9735-8

Corbett, T., O'Faircheallaigh, C., & Regan, A. (2017). 'Designated areas' and the regulation of artisanal and small-scale mining. *Land Use Policy, 68*, 393–401. doi:10.1016/j. landusepol.2017.08.004

Corsi, S., & Muminjanov, H. (2019). *Conservation Agriculture: Training guide for extension agents and farmers in Eastern Europe and Central Asia*. Food and Agriculture Organization.

Crane, A., Matten, D., & Spence, L. (2014). Chapter 4: Responsibilities to stakeholders. *Corporate Social Responsibility: Readings and Cases in a Global Context*, 133–202. https://ebookcentral-proquest-com.ez.library.latrobe.edu.au/lib/latrobe/reader.action?docID=5725906&amp

CraneA.MattenD.SocialC. (2010). https://www.routledge.com/Corporate-Social-Responsibility-Readings-and-Cases-in-a-Global-Context/Crane-Matten-Spence/p/book/9780415683258

Crawford, G., & Botchwey, G. (2017). Conflict, collusion and corruption in small-scale gold mining: Chinese miners and the state in Ghana. *Commonwealth and Comparative Politics, 55*(4), 444–470. doi:10.1080/14662043.2017.1283479

Cruz, J. R. D., Baldovino, R. G., Bandala, A. A., & Dadios, E. P. (2017, October 18). Water usage optimization of Smart Farm Automated Irrigation System using artificial neural network. *2017 5th International Conference on Information and Communication Technology, ICoIC7 2017*. 10.1109/ICoICT.2017.8074668

CTA. (2019, June-August). Digitalising extension: Smart Advisory Services For Farmers. *Spore*, (194), 1–48.

Czyzewski, B., & Kryszak, L. (2022). Definitions and Origins of the Concept of Sustainable Agriculture. In *Sustainable Agriculture Policies for Human Well-Being: Integrated Efficiency Approach* (pp. 1–21). Springer International Publishing. doi:10.1007/978-3-031-09796-6_1

Dai, X., Lin, Y. T., Shi, R., & Xu, D. (2020). A manufacturer's responsible sourcing strategy: Going organic or participating in fair trade? *Annals of Operations Research*, *291*(1–2), 195–218. doi:10.1007/s10479-018-3090-x

Das, B. (2014). Adoption for accessing agricultural information: Evidence from Indian agriculture§. *Agricultural Economics Research Review*, *27*(2), 199–208. doi:10.5958/0974-0279.2014.00024.X

Das, R. (2011). Gender in Practices of Fair Trade and Decent Work. *Gender, Technology and Development*, *15*(1), 101–125. Advance online publication. doi:10.1177/097185241101500105

Davis, F. D. (1989). Perceived usefulness, perceived ease of use, and user acceptance of information technology. *Management Information Systems Quarterly*, *13*(3), 319–340. doi:10.2307/249008

de Groot, R. S., Wilson, M. A., & Boumans, R. M. (2002). A typology for the classification, description and valuation of ecosystem functions, goods and services. *Ecological Economics*, *41*(3), 393–408. doi:10.1016/S0921-8009(02)00089-7

de Sherbinin, A., VanWey, L. K., McSweeney, K., Aggarwal, R., Barbieri, A., Henry, S., Hunter, L. M., Twine, W., & Walker, R. (2020). Rural household demographics, livelihoods and the environment. *Global Environmental Change*, *18*(1), 38–53. doi:10.1016/j.gloenvcha.2007.05.005 PMID:19190718

de Theije, M., & Salman, T. (2018). Conflicts in marginal locations: Small-scale gold-mining in the Amazon. *The Pick*, 261.

del Barrio-García, S., Arquero, J. L., & Romero-Frías, E. (2015). Personal learning environments acceptance model: The role of need for cognition, e-learning satisfaction and students' perceptions. *Journal of Educational Technology & Society*, *18*(3), 129–141.

Denis, N., Dilda, V., Kalouche, R., & Sabah, R. (2020). *Agriculture supply-chain optimization and value creation.*

Dethier, J.-J., & Effenberger, A. (2012). Agriculture and development: A brief review of the literature. *Economic Systems*, *36*(2), 175–205. doi:10.1016/j.ecosys.2011.09.003

Diakosavvas, D., & Frezal, C. (2019). *Bioeconomy and the sustainability of the agriculture and food system: Opportunities and policy challenges*. Academic Press.

Dick-Sagoe, C., Hope, K. N., & Asare-Nuamah, P. (2023). Perceived impact of climate variability and change on livelihoods of smallholder farmers in Lesotho. *African Journal of Science, Technology, Innovation and Development*, *15*(2), 175–184. doi:10.1080/20421338.2022.2058339

Diringer, S. E., Feingold, B. J., Ortiz, E. J., Gallis, J. A., Araújo-Flores, J. M., Berky, A., Pan, W. K. Y., & Hsu-Kim, H. (2015). River transport of mercury from artisanal and small-scale gold mining and risks for dietary mercury exposure in Madre de Dios, Peru. *Environmental Science. Processes & Impacts, 17*(2), 478–487. doi:10.1039/C4EM00567H PMID:25573610

Doherty, B. (2011). Resource advantage theory and fair trade social enterprises. *Journal of Strategic Marketing, 19*(4), 357–380. doi:10.1080/0965254X.2011.581379

Doherty, B., Davies, I. A., & Tranchell, S. (2013). Where now for fair trade? *Business History, 55*(2), 161–189. doi:10.1080/00076791.2012.692083

Dopelt, K., Radon, P., & Davidovitch, N. (2019). Environmental effects of the livestock industry: The relationship between knowledge, attitudes, and behavior among students in Israel. *International Journal of Environmental Research and Public Health, 16*(8), 1359. doi:10.3390/ijerph16081359 PMID:31014019

Dragusanu, R., Montero, E., Nunn, N., Alfaro, L., Atkin, D., Bandiera, O., Donaldson, D., Field, E., Macchiavello, R., Miquel-Florensa, P., Morjaria, A., Muendler, M., Olken, B., Pavcnik, N., Podhorsky, A., & Volpe, C. (2022). *The Effects of Fair Trade Certification: Evidence From Coffee Producers in Costa Rica*. Academic Press.

Dragusanu, R., Giovannucci, D., & Nunn, N. (2014). The economics of fair trade. *The Journal of Economic Perspectives, 28*(3), 217–236. doi:10.1257/jep.28.3.217

Duchene, O., Vian, J., & Celette, F. (2017). Intercropping with legume for agroecological cropping systems: Complementarity and facilitation processes and the importance of soil microorganisms. A review. *Agriculture, Ecosystems & Environment, 240*, 148–161. doi:10.1016/j.agee.2017.02.019

Duncan, E., Abdulai, A.-R., & Fraser, E. D. (2021). Modernizing agriculture through digital technologies: Prospects and challenges. *Handbook on the human impact of agriculture*, 138-161.

Dunjana, N., Zengeni, R., Muchaonyerwa, P., & Wuta, M. (2018). Typological characterisation of farms in a smallholder food-cash crop production system in zimbabwe – opportunities for livelihood sustainability. *Journal of Agriculture and Rural Development in the Tropics and Subtropics, 119*(2), 11–22.

Dupa, H. J. P. (2022). Reasons of Engagement and Impact of Contract Farming to Smallhold Farmers: A Literature Review. *International Journal of Advanced Research in Economics and Finance, 4*(1), 56–60. https://myjms.mohe.gov.my/index.php/ijaref/article/view/17435

Durevall, D. (2020). Fairtrade and Market Efficiency: Fairtrade-Labeled Coffee in the Swedish Coffee Market. *Economies, 8*(2), 30. doi:10.3390/economies8020030

EEA. (2023). *The Impact of Climate Change on Food Security*. https://emergencyaction.org.au/eaa-2021/posts/the-impact-of-climate-change-on-food-security

Egeru, A. (2012). Role of Indigenous Knowledge in Climate Change Adaptation: A Case Study of the Teso Sub-Region, Eastern Uganda. *Indian Journal of Traditional Knowledge, 11*(2), 217–224.

El Bilali, H., & Allahyari, M. S. (2018). Transition towards sustainability in agriculture and food systems: Role of information and communication technologies. *Information Processing in Agriculture*, *5*(4), 456–464. doi:10.1016/j.inpa.2018.06.006

Elizabeth, C., & Baines, S. (1998). Does gender affect business 'performance? A study of microbusinesses in business services in the UK. *Entrepreneurship and Regional Development*, *10*(2), 117–135. doi:10.1080/08985629800000007

Eludoyin, A. O., Ojo, A. T., Ojo, T. O., & Awotoye, O. O. (2017). Effects of artisanal gold mining activities on soil properties in a part of south-western Nigeria. *Cogent Environmental Science*, *3*(1), 1305650. doi:10.1080/23311843.2017.1305650

Emeana, E. M., Trenchard, L., & Dehnen-Schmutz, K. (2020). The revolution of mobile phone-enabled services for agricultural development (m-Agri services) in Africa: The challenges for sustainability. *Sustainability (Basel)*, *12*(2), 485. doi:10.3390/su12020485

Ezedike, E. U. (2013). Globalization, multinationals and business ethics in Nigeria: Ethical imperatives for sustainable development. *Journal of Sustainable Development in Africa*, *15*(6).

Fabregas, R., Kremer, M., & Schilbach, F. (2019). Realizing the potential of digital development: The case of agricultural advice. *Science*, *366*(6471), eaay3038. doi:10.1126/science.aay3038 PMID:31831641

Fairtrade. (2023). *The Fairtrade system*. https://www.fairtrade.net/about/fairtrade-system

FAO & CIAT. (2018). *Climate-Smart Agriculture in the Dry Corridor of Central America: Case Studies*. Available online at https://www.fao.org/3/CA2386EN/ca2386en.pdf

FAO. (2011). *Save and Grow: A policymaker's guide to the sustainable intensification of smallholder crop production*.

FAO. (2020). *Crop and food security assessment mission to Zimbabwe*. FAO.

Faulkner, O. T., & Mackie, J. R. (2013). *West African Agriculture*. Cambridge University Press. doi:10.1017/CBO9781316530122

Federal Ministry of Economic Cooperation and Development. (n.d.). *What is sustainable Agriculture?* Deutsche Gesellschaft for International Zuzammenarbeit (GIZ) GmbH.

Ferreira, M. J., Moreira, F., Pereira, C. S., & Durão, N. (2020). The Digital Transformation at Organizations – The Case of Retail Sector. *Advances in Intelligent Systems and Computing, 1159 AISC*, 560–567. https://doi.org// doi:10.1007/978-3-030-45688-7_56/COVER

Field, A. (2018). *Discovering Statistics Using IBM SPSS Statistics*. SAGE Publications Ltd.

Flick, U. (1998). *An introduction to qualitative research: Theory, method, and applications*. Sage.

Flick, U. (2018). Doing qualitative data collection – charting the routes. In *The SAGE Handbook of Qualitative Data Collection*. SAGE Publications Ltd. doi:10.4135/9781526416070.n1

Food and Agriculture Organisation (FAO). (2017). *Degradation/Restoration.* Available: https://www.fao.org/soils-portal/soil-degradation-restoration/en/

Food and Agriculture Organization of the United Nations. (2008). *Certification in the value chain for fresh fruits: The example of banana industry.* http://www.fao.org

Forte, A., Fagnano, M., & Fierro, A. (2017). Potential role of compost and green manure amendment to mitigate soil GHG emissions in Mediterranean drip-irrigated maize production systems. *Journal of Environmental Management, 192*, 68–78. doi:10.1016/j.jenvman.2017.01.037 PMID:28142125

Frances, R., Coughlan, M., & Cronin, P. (2006). Interviewing in qualitative research. *International Journal of Therapy and Rehabilitation, 16*(16).

Fraser, J. A., Frausin, V., & Jarvis, A. (2015). An intergenerational transmission of sustainability? Ancestral habitus and food production in a traditional agro-ecosystem of the Upper Guinea Forest, West Africa. *Global Environmental Change, 31*, 226–238. doi:10.1016/j.gloenvcha.2015.01.013

Friha, O., Ferrag, M. A., Shu, L., Maglaras, L., & Wang, X. (2021). Internet of Things for the Future of Smart Agriculture: A Comprehensive Survey of Emerging Technologies. *IEEE/CAA Journal of Automatica Sinica, 8*(4), 718–752. doi:10.1109/JAS.2021.1003925

Frischen, J., Meza, I., Rupp, D., Wietler, K., & Hagenlocher, M. (2020). *Drought Risk to Agricultural Systems in Zimbabwe: A Spatial Analysis of Hazard, Exposure, and Vulnerability.* www.mdpi.com/journal/sustainability

Funoh, K. N. (2014). *The impacts of artisanal gold mining on local livelihoods and the environment in the forested areas of Cameroon* (Vol. 150). CIFOR.

Gavai, P., Musungwini, S., & Mugoniwa, B. (2018). A Model for the Adoption and Effective Utilization of ICTs in Commercial Agriculture in Zimbabwe. *Journal of Systems Integration, 9*(4), 40.

Gavali & Kulkarni. (2014). Comparative Analysis of Portable Weeders. *International Journal of Innovative Research in Science, Engineering and Technology, 3*(4), 11004–11013.

Gawande, V., Saikanth, D. R. K., Sumithra, B. S., Aravind, S. A., Swamy, G. N., Chowdhury, M., & Singh, B. V. (2023). Potential of Precision Farming Technologies for Eco-Friendly Agriculture. *International Journal of Plant and Soil Science, 35*(19), 101–112. doi:10.9734/ijpss/2023/v35i193528

Gerlach, C. (2015). Famine responses in the world food crisis 1972–5 and the World Food Conference of 1974. *European Review of History, 22*(6), 929–939. doi:10.1080/13507486.2015.1048191

Geysmans, R., & Hustinx, L. (2016). Placing the Distant Other on the Shelf: An Analysis and Comparison of (Fair Trade) Coffee Packages in Relation to Commodity Fetishism. *Sociological Research Online, 21*(1), 35–40. Advance online publication. doi:10.5153/sro.3854

Gibson, K. (2022). Bridging the digital divide: Reflections on using WhatsApp instant messenger interviews in youth research. *Qualitative Research in Psychology*, *19*(3), 611–631. doi:10.1080/14780887.2020.1751902

Gilbert, D., & Albert, O. B. (2016). Illegal small-scale gold mining in Ghana: A threat to food security. *Journal of Food Security*, *4*(5), 112–119.

Girma, Y., & Abebe, A. (2019). ICTs and Agricultural Marketing in Africa: A Review. *Marketing*, *9*(10).

Giro, A., & Kumar, N. (2022). Climate Smart Livestock System [Review]. *Journal of Agricultural Research Pesticides and Biofertilizers*, *3*(1).

Githui, D. M. (2012). *Responsible Purchasing and Supply Chain Management in Kenya: A Critical Analysis of the Ethical Considerations in Procurement Management*. http://repository.dkut.ac.ke:8080/xmlui/handle/123456789/124

Godfrey, L., & Garnett, T. (2014). Sustainable intensification of agriculture: Narrowing the gap between food security and environmental sustainability. *Food Security*, *6*(2), 135–143.

Goff, S. C. (2018). Fair trade: Global problems and individual responsibilities. *Critical Review of International Social and Political Philosophy*, *21*(4), 521–543. doi:10.1080/13698230.2016.1252993

Gohary, A., Madani, F., Chan, E. Y., & Tavallaei, S. (2023). Political ideology and fair-trade consumption: A social dominance orientation perspective. *Journal of Business Research*, *156*, 113535. doi:10.1016/j.jbusres.2022.113535

Gold, S., Seuring, S., & Beske, P. (2010). Sustainable Supply Chain Management and Inter-Organizational Resources: A Literature Review. *Corporate Social Responsibility and Environmental Management*, *17*(4), 230–245. doi:10.1002/csr.207

Gonalons-Pons, P., & Gangl, M. (2021). Marriage and Masculinity: Male-Breadwinner Culture, Unemployment, and Separation Risk in 29 Countries. *American Sociological Review*, *86*(3), 465–502. doi:10.1177/00031224211012442 PMID:34149053

González-Prieto, A. M. (2018). Conservation of Nearctic Neotropical Migrants: The Coffee Connection Revisited. *Avian Conservation & Ecology*, *13*(1), art19. Advance online publication. doi:10.5751/ACE-01223-130119

Granic, A., & Marangunic, N. (2019). Technology acceptance model in educational context: A systematic literature review. *British Journal of Educational Technology*, *50*(5), 1–40. doi:10.1111/bjet.12864

Grossi, G., Goglio, P., Vitali, A., & Williams, A. G. (2019). Livestock and climate change: Impact of livestock on climate and mitigation strategies. *Animal Frontiers*, *9*(1), 69–76. doi:10.1093/af/vfy034 PMID:32071797

Gruber, V., & Schlegelmilch, B. B. (2014). How Techniques of Neutralization Legitimize Norm- and Attitude-Inconsistent Consumer Behavior. *Journal of Business Ethics*, *121*(1), 29–45. doi:10.1007/s10551-013-1667-5

Guidoboni, M. V., Duparque, A., Boissy, J., Mouny, J.-C., Auberger, J., & van der Werf, H. M. (2023). Conservation agriculture reduces the climate change impact of popcorn and wheat crop rotation. *PLoS One*, *18*(5), e0285586. doi:10.1371/journal.pone.0285586 PMID:37220132

Günther, J., Overbeck, A. K., Muster, S., Tempel, B. J., Schaal, S., Schaal, S., Kühner, E., & Otto, S. (2022). Outcome indicator development: Defining education for sustainable development outcomes for the individual level and connecting them to the SDGs. *Global Environmental Change*, *74*, 102526. doi:10.1016/j.gloenvcha.2022.102526

Gupta, D., Gujre, N., Singha, S., & Mitra, S. (2022). Role of existing and emerging technologies in advancing climate-smart agriculture through modeling: A review. *Ecological Informatics*, *71*, 101805. doi:10.1016/j.ecoinf.2022.101805

Gupta, M., Abdelsalam, M., Khorsandroo, S., & Mittal, S. (2020). Security and privacy in smart farming: Challenges and opportunities. *IEEE Access : Practical Innovations, Open Solutions*, *8*, 34564–34584. doi:10.1109/ACCESS.2020.2975142

Gwaka, L. T. (2022). Computer Supported Livestock Systems: The Potential of Digital Platforms to Revitalize a Livestock System in Rural Zimbabwe. *Proceedings of the ACM on Human-Computer Interaction, 6*(CSCW2), 1-28. 10.1145/3555085

Gwaka, L. T., May, J., & Tucker, W. (2018). Towards low-cost community networks in rural communities: The impact of context using the case study of Beitbridge, Zimbabwe. *The Electronic Journal on Information Systems in Developing Countries*, *84*(3), e12029. doi:10.1002/isd2.12029

Gwanongodza, T. (2020). *Agricultural exports and economic growth in Zimbabwe*. Stellenbosch University.

Hair, J. F., Black, W. C., Babin, B. J., & Anderson, R. E. (2019). *Multivariate data analysis* (8th ed.). Cengage.

Hair, J. F. Jr, Page, M., & Brunsveld, N. (2020). *Essentials of Business Research Methods (4th ed.)*. Routledge.

Hambloch, C. (2022). Contract farming and everyday acts of resistance: Oil palm contract farmers in the Philippines. *Journal of Agrarian Change*, *22*(1), 58–76. doi:10.1111/joac.12462

Hansda, R. (2017). Small-scale farming and gender-friendly agricultural technologies: The interplay between gender, labour, caste, policy and practice. *Gender, Technology and Development*, *21*(3), 189–205. doi:10.1080/09718524.2018.1434990

Haq, S. M., Pieroni, A., Bussmann, R. W., Abd-ElGawad, A. M., & El-Ansary, H. O. (2023). Integrating traditional ecological knowledge into habitat restoration: Implications for meeting forest restoration challenges. *Journal of Ethnobiology and Ethnomedicine*, *19*(1), 33. doi:10.1186/s13002-023-00606-3 PMID:37559120

Harlow, D. E., Hurley, K., Fox, A., Vargas-Guerra, A., & Gibson, J. (2019). *Small-scale & artisanal mining: Impacts on biodiversity in Latin America*. The Cadmus Group and USAID.

Havemann, T., Negra, C., & Werneck, F. (2022). Blended finance for agriculture: exploring the constraints and possibilities of combining fi nancial instruments for sustainable transitions. In *Social innovation and sustainability transition* (pp. 347–358). Springer. doi:10.1007/978-3-031-18560-1_23

HeeksR. (2015). *ICT4D 2016: New priorities for ICT4D policy, practice and WSIS in a Post-2015 World.* doi:10.2139/ssrn.3438433

Heeks, R., Morgan, S., & Nicholson, B. (2014). *IT sourcing and development : new drivers, models and impacts*. Issue October.

Hendriks, S., Soussana, J. F., Cole, M., Kambugu, A., & Zilberman, D. (2023). Ensuring Access to Safe and Nutritious Food for All Through the Transformation of Food Systems. *Science and Innovations for Food Systems Transformation, 31.*

Hennessy, L. (2015). Where there is no company: Indigenous peoples, sustainability, and the challenges of mid-stream mining reforms in Guyana's small-scale gold sector. *New Political Economy*, *20*(1), 126–153. doi:10.1080/13563467.2014.914158

Hentschel, T. (2003). *Artisanal and small-scale mining: challenges and opportunities*. Academic Press.

Hentschel, T., Hruschka, F., & Priester, M. (2002). *Global report on artisanal and small-scale mining*. Report commissioned by the Mining, Minerals and Sustainable Development of the International Institute for Environment and Development. http://www.iied. org/mmsd/mmsd_pdfs/asm_global_report_draft_jan02.pdf

Herrell, K. M., Student, G., Tewari, R., & Mehlhorn, J. (2017). *Honduran Coffee Trade: Economic Effects of Fair Trade Certification On Individual Producers*. Academic Press.

He, W., Zhang, Z., & Li, W. (2020, June). (Justin), & Li, W. (2021). Information technology solutions, challenges, and suggestions for tackling the COVID-19 pandemic. *International Journal of Information Management*, *57*, 102287. Advance online publication. doi:10.1016/j.ijinfomgt.2020.102287 PMID:33318721

Hilson, G. (2002). An overview of land use conflicts in mining communities. *Land Use Policy*, *19*(1), 65–73. doi:10.1016/S0264-8377(01)00043-6

Hilson, G., Gillani, A., & Kutaula, S. (2018). Towards Sustainable Pro-Poor Development? A Critical Assessment of Fair Trade Gold. *Journal of Cleaner Production*, *186*, 894–904. doi:10.1016/j.jclepro.2018.03.123

HLPE. (2013). *Investing in Smallholder Agriculture for Food Security*. A Report by the High-Level Panel of Experts on Food Security and Nutrition of the Committee on World Food Security, Committee on World Security, 111. Available at: https://www.fao.org/

Holland, J. E., White, P. J., Glendining, M. J., Goulding, K. W. T., & McGrath, S. P. (2019). Yield responses of arable crops to liming – An evaluation of relationships between yields and soil pH from a long-term liming experiment. *European Journal of Agronomy*, *105*, 176–188. doi:10.1016/j.eja.2019.02.016 PMID:31007524

Ibrahim, R., Leng, N. S., Yusoff, R. C. M., Samy, G. N., Masrom, S., & Rizman, Z. I. (2017). E-learning acceptance based on technology acceptance model (TAM). *Revue des Sciences Fondamentales et Appliquées*, *9*(4S), 871–889. doi:10.4314/jfas.v9i4S.50

Ikeda, S., & Natawidjaja, R. S. (2022). The sustainability of contract farming with specialised suppliers to modern retailers: Insights from vegetable marketing in Indonesia. *Agriculture*, *12*(3), 380. doi:10.3390/agriculture12030380

ILO. (2008). *Promotion of rural employment for poverty reduction, international labour conference, 97th Session, 2008*. ILO.

Independent. (2022). *Starbucks cuts back sale of Fairtrade coffee in UK*. https://www.independent.co.uk/news/uk/starbucks-fairtrade-europe-middle-east-fairtrade-foundation-b2017451.html

International Labour Organisation. (2021). *Child labour in cotton a briefing*. ILO.

International Trade Centre (ITC). (2011). *Trends in the trade of certified coffees*. www.intracen.org

IPCC. (2007). *Fourth Assessment Report*. Cambridge University Press.

Ireri, D. M., Awuor, M., Ogalo, J., & Nzuki, D. (2021). Role of ICT in the dissemination and access of agricultural information by smallholder farmers in South Eastern Kenya. *Acta Informatica Malaysia*, *5*(1), 31–41. doi:10.26480/aim.01.2021.31.41

Irungu, K. R. G., Mbugua, D., & Muia, J. (2015). Information and Communication Technologies (ICTs) Attract Youth into Profitable Agriculture in Kenya. *East African Agricultural and Forestry Journal*, *81*(1), 24–33. doi:10.1080/00128325.2015.1040645

Issah, M., & Umejesi, I. (2018). Risks and vulnerability in uranium mining: A synthesis of local perspectives in the Great Karoo region of South Africa. *The Extractive Industries and Society*, *5*(3), 284–293. doi:10.1016/j.exis.2018.04.002

Jacob-John, J., D'Souza, C., Marjoribanks, T., & Singaraju, S. (2022). Sustainable Development Practices for SDGs: A Systematic Review of Food Supply Chains in Developing Economies. In N. Nguyen, H. V. Nguyen, C. D'Souza, & C. Strong (Eds.), *Environmental Sustainability in Emerging Markets Consumer, Organisation and Policy Perspective*. Springer. doi:10.1007/978-981-19-2408-8_10

Jaffee, D. (2012). Weak coffee: Certification and co-optation in the fair trade movement. *Social Problems*, *59*(1), 94–116. doi:10.1525/sp.2012.59.1.94

Jaffee, D., & Howard, P. H. (2015). Who's the Fairest of Them All? The Fractured Landscape of U.S. Fair Trade Certification. *Agriculture and Human Values*. Advance online publication. doi:10.1007/s10460-015-9663-2

Janssen, S. J., Porter, C. H., Moore, A. D., Athanasiadis, I. N., Foster, I., Jones, J. W., & Antle, J. M. (2017). Towards a new generation of agricultural system data, models and knowledge products: Information and communication technology. *Agricultural Systems*, *155*, 200–212. doi:10.1016/j.agsy.2016.09.017 PMID:28701813

Jiri, O., Mafongoya, P. L., Mubaya, C., & Mafongoya, O. (2016). Seasonal climate prediction and adaptation using indigenous knowledge systems in agriculture systems in Southern Africa: A review. *Journal of Agricultural Science (Toronto)*, *8*(5), 156–172. doi:10.5539/jas.v8n5p156

Jnr, S. D., Cieem, G., Ayensu-Ntim, A., Twumasi-Ankrah, B., & Barimah, P. T. (2016). Effects of loss of agricultural land due to large-scale gold mining on agriculture in Ghana: The case of the Western Region. *Br. J. Res*, *2*(6).

Kader, M. A., Senge, M., Mojid, M. A., & Ito, K. (2017). Recent advances in mulching materials and methods for modifying soil environment. *Soil & Tillage Research*, *168*, 155–166. doi:10.1016/j.still.2017.01.001

Kamoyo, M., Muranda, Z., & Chikuya, T. (2015). Agricultural export crop participation, contract farming and rural livelihood in Zimbabwe: The case of cotton farming in Rushinga District. IOSR. *Journal of Economics and Finance*, *6*(6), 110–120. doi:10.9790/5933-0661110120

Kamutuezu, E. U., Winschiers-Theophilus, H., & Peters, A. (2021). An exploration of factors influencing the adoption of ICT enabled entrepreneurship applications in Namibian rural communities. *arXiv preprint arXiv:2108.09789.*

Kanyanda, S., Markhof, Y., Wollburg, P., & Zezza, A. (2021). Acceptance of COVID-19 vaccines in sub-Saharan Africa: Evidence from six national phone surveys. *BMJ Open*, *11*(12), e055159. doi:10.1136/bmjopen-2021-055159 PMID:34911723

Kapari, M., Hlophe-Ginindza, S., Nhamo, L., & Mpandeli, S. (2023). Contribution of smallholder farmers to food security and opportunities for resilient farming systems. *Frontiers in Sustainable Food Systems*, *7*, 1149854. doi:10.3389/fsufs.2023.1149854

Kashmanian, R. M. (2017). Building Greater Transparency in Supply Chains to Advance Sustainability. *Environmental Quality Management*, *26*(3), 73–104. doi:10.1002/tqem.21495

Kassegn, A., & Endris, E. (2022). Factors affecting loan repayment rate among smallholder farmers got loans from the Amhara Credit and Saving Institution: In the case of Habru District, Amhara Regional State, Ethiopia. *International Area Studies Review*, *25*(1), 73–96. doi:10.1177/22338659211040993

Kassem, H. S., Alotaibi, B. A., Ghoneim, Y. A., & Diab, A. M. (2021). Mobile-based advisory services for sustainable agriculture: Assessing farmers' information behavior. *Information Development*, *37*(3), 483–495. doi:10.1177/0266666920967979

Kaur, P., Singla, N. & Singh, S. (2021). Role of Contract Farming in Crop Diversification and Employment Generation: Empirical Evidence from Indian Punjab. *Millennial Asia*, *12*(3), 350-366. . doi:10.1177/09763996211051300

Kaur, K., & Kaur, P. (2018). Agricultural extension approaches to enhance the knowledge of farmers. *International Journal of Current Microbiology and Applied Sciences*, *7*(2), 2367–2376. doi:10.20546/ijcmas.2018.702.289

Kazmi, D. H., Afzaal, M., Ali, S., & Nasim, W. (2022). Agro-meteorological Aspect of Climate Change. *Building Climate Resilience in Agriculture: Theory, Practice and Future Perspective*, 17-24.

Kemper, L., & Partzsch, L. (2019). Saving Water While Doing Business: Corporate Agenda-Setting and Water Sustainability. *Water (Basel)*, *11*(2), 297. Advance online publication. doi:10.3390/w11020297

Kennelly, I., Merz, S. N., & Lorber, J. (2001). What Is Gender? *American Sociological Review*, *66*(4), 598–605. doi:10.1177/000312240106600407

Keyhanpour, M. J., Jahromi, S. H. M., & Ebrahimi, H. (2021). System dynamics model of sustainable water resources management using the Nexus Water-Food-Energy approach. *Ain Shams Engineering Journal*, *12*(2), 1267–1281. doi:10.1016/j.asej.2020.07.029

Khan, N. A., Qijie, G., Ali, S., Shahbaz, B., & Shah, A. A. (2019). Farmers' use of mobile phone for accessing agricultural information in Pakistan: A case of Punjab province. *Ciência Rural*, *49*(10), 1–12. doi:10.1590/0103-8478cr20181016

Kiel, D., Arnold, C., & Voigt, K.-I. (2017). The influence of the Industrial Internet of Things on business models of established manufacturing companies–A business level perspective. *Technovation*, *68*, 4–19. doi:10.1016/j.technovation.2017.09.003

Kikuchi, M., Mano, Y., Njagi, T. N., Merrey, D., & Otsuka, K. (2021). Economic viability of large-scale irrigation construction in sub-Saharan Africa: What if mwea irrigation scheme were constructed as a brand-new scheme? *The Journal of Development Studies*, *57*(5), 772–789. doi:10.1080/00220388.2020.1826443

Kimathi, L. M. (2023). *Impact of Conservation Agriculture on Smallholder Farmer Livelihoods in Kenya and Zimbabwe*. University of Pretoria.

Kim, J., Shah, P., Gaskell, J. C., & Prasann, A. (2020). *Scaling up disruptive agricultural technologies in Africa*. World Bank Publications. doi:10.1596/978-1-4648-1522-5

Kim, S. K., Marshall, F., & Dawson, N. M. (2022). Revisiting Rwanda's agricultural intensification policy: Benefits of embracing farmer heterogeneity and crop-livestock integration strategies. *Food Security*, *14*(3), 637–656. doi:10.1007/s12571-021-01241-0

Kim, S., & Chae, S. (2022). Shareholder Value Effects of Ethical Sourcing: Comparing Reactive and Proactive Initiatives. *Journal of Business Ethics*, *179*(3), 887–906. doi:10.1007/s10551-021-04841-0

Kim, S., Colicchia, C., & Menachof, D. (2018). Ethical Sourcing: An Analysis of the Literature and Implications for Future Research. *Journal of Business Ethics*, *152*(4), 1033–1052. doi:10.1007/s10551-016-3266-8

Kitula, A. G. N. (2006). The environmental and socio-economic impacts of mining on local livelihoods in Tanzania: A case study of Geita District. *Journal of Cleaner Production, 14*(3-4), 405–414. doi:10.1016/j.jclepro.2004.01.012

Komarek, A. M., Hoyoung Kwon, H., Beliyou Haile, B., Christian Thierfelder, C., Mutenje, M. J., & Carlo Azzarri, C. (2015). From plot to scale: Ex-ante assessment of conservation agriculture in Zambia. *Agricultural Systems, 173*, 504–518. doi:10.1016/j.agsy.2019.04.001

Koomson, J. E., Donkor, E., & Owusu, V. (2022). Contract scheme for rubber production in Western region of Ghana: Why do farmers side sell? *Forests, Trees and Livelihoods, 31*(3), 139–152. doi:10.1080/14728028.2022.2079007

Koos, S. (2021). Moralising Markets, Marketizing Morality. The Fair Trade Movement, Product Labeling and the Emergence of Ethical Consumerism in Europe. *Journal of Nonprofit & Public Sector Marketing, 33*(2), 168–192. Advance online publication. doi:10.1080/10495142.2020.1865235

Kremen, C., & Miles, A. (2012). Ecosystem services in biologically diversified versus conventional farming systems: Benefits, externalities, and trade-offs. *Ecology and Society, 17*(4), art40. doi:10.5751/ES-05035-170440

Krishnamurthy, P. K., Lewis, K., & Choutartun, R. J. (2012). *Climate Impacts on Food Security and Nutrition. A Review of Existing Knowledge.* Met Office and WFP.

Kristjanson, P., Bryan, E., Bernier, Q., Twyman, J., Meinzen-Dick, R., Kieran, C., Ringler, C., Jost, C., & Doss, C. (2017). Addressing gender in agricultural research for development in the face of a changing climate: Where are we and where should we be going? *International Journal of Agricultural Sustainability, 15*(5), 482–500. doi:10.1080/14735903.2017.1336411

Kubzansky, M. (2012). *Old Problems, New Solutions : Harnessing Technology and Innovation in the Fight Against Global Poverty.* Global Economy and Development.

Kufeoglu, S. (2022). SDG-15: Life on Land. In *Emerging Technologies: Value Creation for Sustainable Development* (pp. 469–486). Springer International Publishing. doi:10.1007/978-3-031-07127-0_17

Kuipa, P. J. T. (2019). *The agricultural financing gap in Zimbabwe: rationing, sustainable credit access and participation in rural financial markets.* University of the Free State.

Kumar Sharma, N. (2020). *Ethical Dilemma and Marketing Decisions: A Case Analysis of Food Products.* doi:10.26643/think-india.v16i3.7818

Kwapong, N. A., Ankrah, D. A., Boateng-Gyambiby, D., Asenso-Agyemang, J., & Fening, L. O. (2020). Assessment of agricultural advisory messages from farmer-To-farmer in making a case for scaling up production: A qualitative study. *The Qualitative Report, 25*(8), 2011–2025. doi:10.46743/2160-3715/2020.4241

Ladha, J. K., Jat, M. L., Stirling, C. M., Chakraborty, D., Pradhan, P., Krupnik, T. J., Sapkota, T. B., Pathak, H., Rana, D. S., Tesfaye, K., & Gerard, B. (2019). Achieving the Sustainable Development Goals in Agriculture: The Crucial Role of Nitrogen in Cereal-based Systems. *Advances in Agronomy*, *163*, 39–116. doi:10.1016/bs.agron.2020.05.006

Ladhari, R., & Tchetgna, N. M. (2017). Values, Socially Conscious Behaviour and Consumption Emotions as Predictors of Canadians' Intent to Buy Fair Trade Products. *International Journal of Consumer Studies*, *41*(6), 696–705. Advance online publication. doi:10.1111/ijcs.12382

Lajoie-O'Malley, A., Bronson, K., van der Burg, S., & Klerkx, L. (2020). The future (s) of digital agriculture and sustainable food systems: An analysis of high-level policy documents. *Ecosystem Services*, *45*, 101183. doi:10.1016/j.ecoser.2020.101183

Lamsal, R. R., Karthikeyan, P., Otero, P., & Ariza, A. (2023). Design and Implementation of Internet of Things (IoT) Platform Targeted for Smallholder Farmers: From Nepal Perspective. *Agriculture*, *13*(10), 1900. doi:10.3390/agriculture13101900

Lasco, R. D., Delfino, R. J. P., Catacutan, D. C., Simelton, E. S., & Wilson, D. M. (2014). Climate risk adaptation by smallholder farmers: The roles of trees and agroforestry. *Current Opinion in Environmental Sustainability*, *6*, 83–88. doi:10.1016/j.cosust.2013.11.013

Lashgarara, F., Mohammadi, R., & Najafabadi, M. O. (2017b). Identifying appropriate information and communication technology (ICT) in improving marketing of agricultural products in Garmsar City, Iran. *African Journal of Biotechnology*, *10*(5), 11537–11540.

Lawrence, X., Mendoza, D., Tadeo, J. B., Dacanay, J. M., Marla, A. N., Aszhnie, C., & Vergara, B. (2023). Enterprise management strategies in agricultural fairtrade products. *Journal of Enterprise and Development*, *5*(2), 2023.

Lee, C.-L., Strong, R., & Dooley, K. E. (2021). Analyzing precision agriculture adoption across the globe: A systematic review of scholarship from 1999–2020. *Sustainability (Basel)*, *13*(18), 10295. doi:10.3390/su131810295

Lee, T. H., & Tseng, C. H. (2015). How personality and risk-taking attitude affect the behavior of adventure recreationists. *Tourism Geographies*, *17*(3), 307–331. doi:10.1080/14616688.2014.1000955

Lele, U., & Goswami, S. (2017). The fourth industrial revolution, agricultural and rural innovation, and implications for public policy and investments: A case of India. *Agricultural Economics*, *48*(S1), 87–100. doi:10.1111/agec.12388

Le, M. H., & Nguyen, P. M. (2022). Integrating the Theory of Planned Behavior and the Norm Activation Model to Investigate Organic Food Purchase Intention: Evidence from Vietnam. *Sustainability (Basel)*, *14*(2), 816. Advance online publication. doi:10.3390/su14020816

Lemaire, G., Franzluebbers, A., de Faccio Carvalho, P. C., & Dedieu, B. (2014). Integrated crop–livestock systems: Strategies to achieve synergy between agricultural production and environmental quality. *Agriculture, Ecosystems & Environment*, *190*, 4–8. doi:10.1016/j.agee.2013.08.009

Lemaire, G., Gastal, F., Franzluebbers, A., & Chabbi, A. (2015). Grassland–cropping rotations: An avenue for agricultural diversification to reconcile high production with environmental quality. *Environmental Management, 56*(5), 1065–1077. doi:10.1007/s00267-015-0561-6 PMID:26070897

Lezoche, M., Hernandez, J. E., Díaz, M. M. E. A., Panetto, H., & Kacprzyk, J. (2020). Agri-food 4.0: A survey of the supply chains and technologies for the future agriculture. *Computers in Industry, 117*, 103187. doi:10.1016/j.compind.2020.103187

Li, J., Qing, P., Hu, W., & Li, M. (2022). Contract farming, community effect, and farmer valuation of biofortified crop varieties in China: The case of high-zinc wheat. *Review of Development Economics, 26*(2), 1035–1055. doi:10.1111/rode.12847

Lim, X. (2022). *How technology can help address challenges in agriculture. World Economic Forum.*

Lindqvist, A., Sendén, M. G., & Renström, E. A. (2021). What is gender, anyway: A review of the options for operationalising gender. *Psychology and Sexuality, 12*(4), 332–344. doi:10.1080/19419899.2020.1729844

Lipinski, B., Hanson, C., Waite, R., Searchinger, T., & Lomax, J. (2013). *Reducing food loss and waste.* Academic Press.

Lipper, L., Thornton, P., Campbell, B. M., Baedeker, T., Braimoh, A., Bwalya, M., Caron, P., Cattaneo, A., Garrity, D., Henry, K., Hottle, R., Jackson, L., Jarvis, A., Kossam, F., Mann, W., McCarthy, N., Meybeck, A., Neufeldt, H., Remington, T., ... Torquebiau, E. F. (2014). Climate-smart agriculture for food security. *Nature Climate Change, 4*(12), 1068–1072. doi:10.1038/nclimate2437

Liu, W., Shao, X.-F., Wu, C.-H., & Qiao, P. (2021). A systematic literature review on applications of information and communication technologies and blockchain technologies for precision agriculture development. *Journal of Cleaner Production, 298*, 126763. doi:10.1016/j.jclepro.2021.126763

Liu, Y., Sheng, H., Mundorf, N., Redding, C., & Ye, Y. (2017). Integrating norm activation model and theory of planned behavior to understand sustainable transport behavior: Evidence from China. *International Journal of Environmental Research and Public Health, 14*(12), 1593. Advance online publication. doi:10.3390/ijerph14121593 PMID:29258245

Loures, L., Chamizo, A., Ferreira, P., Loures, A., Castanho, R., & Panagopoulos, T. (2020). Assessing the effectiveness of precision agriculture management systems in mediterranean small farms. *Sustainability (Basel), 12*(9), 1–15. doi:10.3390/su12093765

Luna, J. K., Chalit Hernandez, B., & Sawadogo, A. (2021). The paradoxes of purity in organic agriculture in Burkina Faso. *Geoforum, 127*, 46–56. doi:10.1016/j.geoforum.2021.09.014

Lyon, S., Bezaury, J. A., & Mutersbaugh, T. (2010). Gender equity in fairtrade-organic coffee producer organizations: Cases from Mesoamerica. *Geoforum, 41*(1), 93–103. doi:10.1016/j.geoforum.2009.04.006

Lysons, K. & Farrington, B. 2020. *Procurement and supply chain management.* Pearson UK.

Ma, M. (2005). Chapter 11: Biodiversity Regulation of Ecosystem Services. Ecosystems and human well-being: Current State and Trends.

Machingura, G. (2020). *Roundup: Zimbabwe rolls out new farming concept to boost grain production.* Xinhua News. https://www.xinhuanet.com/english/2020-09/11/c_139359235.htm

MacPherson, J., Voglhuber-Slavinsky, A., Olbrisch, M., Schöbel, P., Dönitz, E., Mouratiadou, I., & Helming, K. (2022). Future agricultural systems and the role of digitalization for achieving sustainability goals. A review. *Agronomy for Sustainable Development*, *42*(4), 70. doi:10.1007/s13593-022-00792-6 PMID:35818482

Maertens, M., & Velde, K. V. (2017). Contract-farming in Staple Food Chains: The Case of Rice in Benin. *World Development*, *95*, 73–87. doi:10.1016/j.worlddev.2017.02.011

Maher, T. C. (2011). Defining Education. *The Journal of Continuing Higher Education*, *59*(2), 108–109. doi:10.1080/07377363.2011.583184

Majumdar, K., & Singh, R. K. (2019). Impact of information and communication technology on marketing of rice: A study of Uttar Pradesh. *International Journal of Social Economics*, *46*(9), 1061–1080. doi:10.1108/IJSE-02-2019-0105

Makate, C. (2019). Effective scaling of climate smart agriculture innovations in African smallholder agriculture: A review of approaches, policy and institutional strategy needs. *Environmental Science & Policy*, *96*, 37–51. doi:10.1016/j.envsci.2019.01.014

Makita, R. (2011). Fair Trade Certification: The Case of Tea Plantation Workers in India. *Development Policy Review*. Advance online publication. doi:10.1111/j.1467-7679.2012.00561.x

Makumbi, M., Mugume, M., & Mugisha, A. (2017). Conservation agriculture and sustainable development in Africa: A case study from the Lindi region of Tanzania. *Journal of Agriculture and Rural Development in the Tropics and Subtropics*, *121*(2), 121–132.

Makumbi, M., Mugume, M., & Mugisha, A. (2022). *Conservation Agriculture in the Southern Highlands of Tanzania.* CABI Digital Library. doi:10.1079/9781789245745.0006

Malone, A., Smith, N. M., & Zeballos, E. Z. (2021). Coexistence and conflict between artisanal mining, fishing, and farming in a Peruvian boomtown. *Geoforum*, *120*, 142–154. doi:10.1016/j.geoforum.2021.01.012

Mandipaza, B. F. (2022). *Socio-Cultural Dimensions of Farming, Small Farm Households and Conservation Agriculture in Nyanga District, Zimbabwe.* University of Pretoria.

Mango, L., & Kugedera, A. T. (2022). Pitfalls and prospects of contract farming to smallholder tobacco farmers in Shamva District, Zimbabwe. *International Journal of Agriculture Environment and Food Sciences*, *6*(1), 189–196. doi:10.31015/jaefs.2022.1.24

Manguwo, E. (2023). *Farmers urged to uphold Pfumvudza principles.* https://www.herald.co.zw/farmers-urged-to-uphold-pfumvudza-principles/

Mapfumo, P., Mtambanengwe, F., & Chikowo, R. (2015). Building on indigenous knowledge to strengthen the capacity of smallholder farming communities to adapt to climate change and variability in southern Africa. *Climate and Development*.

Mapiye, O., Makombe, G., Molotsi, A., Dzama, K., & Mapiye, C. (2021). Towards a revolutionized agricultural extension system for the sustainability of smallholder livestock production in developing countries: The potential role of icts. *Sustainability (Basel)*, *13*(11), 5868. doi:10.3390/su13115868

Mapiye, O., Makombe, G., Molotsi, A., Dzama, K., & Mapiye, C. (2023). Information and communication technologies (ICTs): The potential for enhancing the dissemination of agricultural information and services to smallholder farmers in sub-Saharan Africa. *Information Development*, *39*(3), 638–658. doi:10.1177/02666669211064847

Marambi, S., Musungwini, S., & Mzikamwi, T. (2018). Strategy for the adoption and effective utilization of mobile phone technologies in smallholder agriculture in Zimbabwe. In *Proceedings of 6th International Conference on M4D Mobile Communication for Development* (pp. 15–16). Universitetstryckeriet, Karlstad. www.researchjournal.co.in

Mark, B., Wendela, H., Elmy, S.-F., & Terya, S. (2018). *Fairtrade Standard for Small Producer Organizations*. www.fairtrade.net/standards.html

Marshall, B., Cardon, P., Poddar, A., Fontenot, R., Marshall, B., Cardon, P., Poddar, A., & Does, R. F. (2015). *Does Sample Size Matter in Qualitative Research ? A Review of Qualitative Interviews in is*. doi:10.1080/08874417.2013.11645667

Maseya, P. (2020). An analysis of the Pfumvudza/Intwasa Programme in Zimbabwe. *Sustainability*, *12*(16), 6390.

Masimba, F., Appiah, M., & Zuva, T. (2019). Influencing factors of mobile technology adoption by small holder farmers in Zimbabwe. *Advances in Intelligent Systems and Computing*, *1046*, 125–134. doi:10.1007/978-3-030-30329-7_12

Matema, T. (2021). *State-led Agricultural Reform and Food Security in Zimbabwe: Command Agriculture and Food Production in Makoni Rural District, Manicaland*. University of Witwatersrand Johannesburg.

Maulu, S., Hasimuna, O. J., Mutale, B., Mphande, J., & Siankwilimba, E. (2021). Enhancing the role of rural agricultural extension programs in poverty alleviation: A review. *Cogent Food & Agriculture*, *7*(1), 1886663. doi:10.1080/23311932.2021.1886663

Maurial, M. (1999). Indigenous knowledge and schooling: A continuum between conflict and dialogue. In L. M. Semali & J. L. Kincheloe (Eds.), *What is indigenous knowledge system: Voices from the academy* (pp. 59–77). Falmer.

Mayoyo, A., Chapoto, A., Matchaya, G., Aheeyar, M., Chiwunze, G., Ebrahim, G., Ajayi, O. C., Afun-Ogidan, K., Fakudze, B., & Kasoma-Pele, W. (2023). *Digital climate adaptation in agriculture profile for Zimbabwe*. Academic Press.

Mazvimavi, K., & Twomlow, S. (2017). Conservation agriculture in Zimbabwe: Policy, adoption, and impact. *Journal of International Development*, *29*(1), 125–138.

Mazwi, F. (2020). Sugar production dynamics in Zimbabwe: An analysis of contract farming at Hippo Valley. *Review of African Political Economy*, *47*(166), 568–584. doi:10.1080/03056244.2020.1832022

Mazwi, F., Chambati, W., & Mudimu, G. T. (2020). Tobacco Contract Farming in Zimbabwe: Power Dynamics, Accumulation Trajectories, Land Use Patterns and Livelihoods. *Journal of Contemporary African Studies*, *38*(1), 55–71. doi:10.1080/02589001.2020.1746752

Mazwi, F., Chambati, W., & Mutodi, K. (2018). *Contract farming arrangement and poor resourced farmers in Zimbabwe. Policy brief. Sam Moyo African Institute for agrarian studies*. SMAIAS.

Mazwi, F., Chemura, A., Mudimu, G. T., & Chambati, W. (2019). Political Economy of Command Agriculture in Zimbabwe: A State-led Contract Farming Model. *Agrarian South - the Journal of Political Economy*, *8*(1–2), 232–257. doi:10.1177/2277976019856742

Mboga, J. (2017). *Ethical sourcing to ensure sustainability in the food and beverage industry and eliciting millennial perspectives.* Advance online publication. doi:10.5281/zenodo.1009174

Mbow, C., Van Noordwijk, M., Luedeling, E., Neufeldt, H., Minang, P. A., & Kowero, G. (2014). Agroforestry solutions to address food security and climate change challenges in Africa. *Current Opinion in Environmental Sustainability*, *6*, 61–67. doi:10.1016/j.cosust.2013.10.014

McCarthy, N., Lipper, L., & Zilberman, D. (2018). Economics of climate smart agriculture: An overview. *Climate smart agriculture: Building resilience to climate change*, 31-47.

Meemken, E. M., & Bellemare, M. F. (2020). Smallholder farmers and contract farming in developing countries. *Proceedings of the National Academy of Sciences of the United States of America*, *117*(1), 259–264. doi:10.1073/pnas.1909501116 PMID:31836695

Ministry of Lands, Agriculture, Water Fisheries, Water and Rural Resettlement. (n.d.). http://www.moa.gov.zw/wp-content/uploads/2021/02/pfumvudza-flier.pdf

Misaki, E., Apiola, M., Gaiani, S., & Tedre, M. (2018). Challenges facing sub-Saharan small-scale farmers in accessing farming information through mobile phones: A systematic literature review. *The Electronic Journal on Information Systems in Developing Countries*, *84*(4), e12034. doi:10.1002/isd2.12034

Mishra, P. K., & Rai, S. C. (2013). Use of Indigenous Soil and Water Conservation Practices among Farmers in Sikkim Himalaya. *Indian Journal of Traditional Knowledge*, *12*(3), 554–564.

Mishra, S. K., & Mishra, P. (2017). Determinants of households' resistance against land acquisition for mining: Experiences at Talcher coalfields in India. *Land Use Policy*, *66*, 10–17. doi:10.1016/j.landusepol.2017.04.024

Mkodzongi, G. (2013). New people, new land and new livelihoods: A micro-study of Zimbabwe's fast-track land reform. *Agrarian South - the Journal of Political Economy*, *2*(3), 345–366. doi:10.1177/2277976013517320

Mkodzongi, G., & Spiegel, S. (2019). Artisanal gold mining and farming: Livelihood linkages and labour dynamics after land reforms in Zimbabwe. *The Journal of Development Studies*, *55*(10), 2145–2161. doi:10.1080/00220388.2018.1516867

Mkonda, M. (2017). *Conservation Agriculture in Tanzania.*. doi:10.1007/978-3-319-48006-0_10

Mkonda, A. M. (2016). Conservation agriculture (CA) in Tanzania: The case of the Southern Highlands. *International Journal of Agricultural Sustainability*, *14*(5), 557–572.

Mkonda, A. M., & Mkonda, A. M. (2016). *Conservation Agriculture in Tanzania*. Semantic Scholar.

Mnwana, S., & Bowman, A. (2018). Mine mechanisation and distributional conflict in rural South Africa. *Resources Policy*, *59*, 227–237. doi:10.1016/j.resourpol.2018.07.008

Modi, R. (2019). *The Role of Agriculture for Food Security and Poverty Reduction in Sub-Saharan Africa*. The Palgrave Handbook of Contemporary International Political Economy. doi:10.1057/978-1-137-45443-0_25

Molieleng, L., Fourie, P., & Nwafor, I. (2021). Adoption of climate smart agriculture by communal livestock farmers in South Africa. *Sustainability (Basel)*, *13*(18), 10468. doi:10.3390/su131810468

Morse, J. M. (1995). *The significance of saturation.pdf.*

Moyo, S. (2016). *Family farming in sub-Saharan Africa: its contribution to agriculture, food security and rural development*. Academic Press.

Moyo, F., Ncube, M., & Ndlovu, T. (2022). *The Competing Nature-Based Livelihood-Strategies: Artisanal Small-Scale Mining (ASM)*. Perspectives in Agricultural-Communities in Umzingwane District.

Moyo, F., Ndlovu, T., Francis, B., & Ncube, M. (2018). The effects of artisanal mining on irrigation farming-the case of Umzinyathini Irrigation Scheme in Umzingwane District, Southern Matabeleland, Zimbabwe. *African Journal of Public Affairs*, *10*(2), 139–162.

Mtero, F. (2017). Rural livelihoods, large-scale mining and agrarian change in Mapela, Limpopo, South Africa. *Resources Policy*, *53*, 190–200. doi:10.1016/j.resourpol.2017.06.015

Mugweni, E., Pearson, S., & Omar, M. (2012). Traditional gender roles, forced sex and HIV in Zimbabwean marriages. *Culture, Health & Sexuality*, *14*(5), 577–590. doi:10.1080/13691058.2012.671962 PMID:22472019

Mugwisi, T. (2015). Communicating agricultural information for development: The role of the media in Zimbabwe. *Libri*, *65*(4), 281–299. doi:10.1515/libri-2015-0094

Mugwisi, T., Mostert, J., & Ocholla, D. N. (2015). Access to and Utilization of Information and Communication Technologies by Agricultural Researchers and Extension Workers in Zimbabwe. *Information Technology for Development*, *21*(1), 67–84. Advance online publication. doi:10.1080/02681102.2013.874317

Mujere, N. (2022). An Assessment of the Contribution of The Pfumvudza Concept Towards Climate Smart Agriculture in Zimbabwe: A Review. *Journal of Agriculture and Horticulture Research*, *5*(2), 69–76.

Mujuzi, J. D. (2023). Marriage by Cohabitation (Common Law Marriage) in Seychelles: Emerging Issues. In Cohabitation and the Evolving Nature of Intimate and Family Relationships. Emerald Publishing Limited. doi:10.1108/S1530-35352023009

Mujuzi, J. D. (2023). Distribution of property at the termination of *de facto* unions (marriages by cohabitation/repute) in some African countries. *International Journal of Law, Policy and the Family*, *37*(1), ebad015. Advance online publication. doi:10.1093/lawfam/ebad015

Mukucha, P. & Chari, F. (2021). The influence of supplier development, in the form of contract farming, on performance in Zimbabwean tobacco industry. *Cogent Business & Management*, *8*(1). . doi:10.1080/23311975.2021.1968732

Mukucha, P., Tsekea, S., Jaravaza, D. C. & Jaravaza, N. (2024). Contract farming: a comparison of production and marketing contracts as sourcing strategies in the fast food restaurant industry. *Journal of Foodservice Business Research*, 1-18. https://doi.org/. doi:10.1080/15378020.2024.2303540

Mukucha, P., & Chari, F. (2022). Supply chain resilience: The role of supplier development in the form of contract farming in fast-food outlets in Zimbabwe. *Continuity & Resilience Review*, *4*(3), 280–299. doi:10.1108/CRR-03-2022-0006

Mukucha, P., Chari, F., & Manzini, S. (2023). An assessment of differential supplier category performance in the agro processing industry: A case for supplier development in the Zimbabwean tobacco industry. *Journal of Future Sustainability*, *3*(4), 223–232. doi:10.5267/j.jfs.2023.2.003

Müller-Lindenlauf, M. (2009). *Organic agriculture and carbon sequestration. Possibilities and constrains for the consideration of organic agriculture within carbon accounting systems*. Natural Resources Management and Environment Department, Food and Agriculture Organization of the United Nations.

Murindangabo, Y. T., Kopecký, M., & Konvalina, P. (2021). Adoption of Conservation Agriculture in Rwanda: A Case Study of Gicumbi District Region. *Agronomy (Basel)*, *11*(9), 1732. doi:10.3390/agronomy11091732

Musungwini, S., van Zyl, I., & Kroeze, J. H. (2022). The perceptions of smallholder farmers on the use of mobile technology: A naturalistic inquiry in Zimbabwe. *Lecture Notes in Networks and Systems, 439 LNNS*, 530–544. doi:10.1007/978-3-030-98015-3_37

Musungwini, S. (2016). A model for harnessing the power of the mobile phone technology to improve smallholder agriculture in Zimbabwe. *Proceedings of the 5th International Conference on M4D Mobile Communication Technology for Development: M4D 2016, General Tracks, December*, 237–252.

Musungwini, S. (2018). Mobile phone use by Zimbabwean smallholder farmers : A baseline study. *The African Journal of Information and Communication*, *22*(22), 29–52. doi:10.23962/10539/26171

Musungwini, S., Gavai, P. V., Munyoro, B., & Chare, A. (2023). Emerging ICT technologies for agriculture, training, and capacity building for farmers in developing countries: A case study in Zimbabwe. *Advances in Environmental Engineering and Green Technologies*, (January), 12–30. doi:10.4018/978-1-6684-6413-7.ch002

Mutambara, J., & Mujeyi, K. (2021). Enhancing competitiveness of Zimbabwe's cotton production under contract farming. *African Journal of Science, Technology, Innovation and Development*, *13*(3), 283–291. doi:10.1080/20421338.2020.1744270

Mutenje, M. J., Farnworth, C. R., Stirling, C., Thierfelder, C., Mupangwa, W., & Nyagumbo, I. (2019). A cost-benefit analysis of climate-smart agriculture options in Southern Africa: Balancing gender and technology. *Ecological Economics*, *163*, 126–137. doi:10.1016/j.ecolecon.2019.05.013

Mutenje, M. J., Mafongoya, P. L., & Chivenge, P. P. (2019). Climate-smart agriculture in Zimbabwe: Opportunities and challenges. *Climate and Development*, *11*(7), 604–610.

Muthee, D. W., Gwademba, G. K., & Masinde, J. M. (2019). The role of indigenous knowledge systems in enhancing agricultural productivity in Kenya. *Eastern Africa Journal of Contemporary Research, 1*(1).

Mwambi, M. M., Oduol, J., Mshenga, P., & Saidi, M. (2016). Does contract farming improve smallholder income? The case of avocado farmers in Kenya. *Journal of Agribusiness in Developing and Emerging Economies*, *6*(1), 2–20. doi:10.1108/JADEE-05-2013-0019

Naegele, H. (2020). Where Does the Fair Trade Money Go? How Much Consumers Pay Extra for Fair Trade Coffee and How This Value Is Split Along the Value Chain. *World Development*, *133*, 105006. Advance online publication. doi:10.1016/j.worlddev.2020.105006

National Development Strategy. (2020). Towards a Prosperous & Empowered Upper Middle-Income Society by 2030 January 2021 – December 2025. Author.

Ncube, M., Moyo, F., & Mamhute, S. T. (2021). Unpacking Negative Externalities of Social Capital in the Sustainability of Smallholder Rural Irrigation Farming: The Case of Rozva Irrigation Scheme in Bikita District, Zimbabwe. Academic Press.

Ndiaye, P. G., Diouf, B. L., Tindall, C., Kébé, M., & Holvoet, K. (2009). Evaluating the opportunities, constraints and implications of eco and ethical fish labelling on the octopus value chain in Senegal. *FAO Fisheries and Aquaculture Report*, (904), 165–174.

Ndlovu, I., & Ndlovu, M. (2013). Mobile banking the future to rural financial inclusion: Case study of Zimbabwe. *IOSR Journal of Humanities And Social ….* http://iosrjournals.org/iosr-jhss/papers/Vol9-issue4/K0947075.pdf

Neeliah, H., Rajkomar, B., Dookun-Saumtully, A., & Ramkissoon, J. (2006). *Food Agriculture and Natural Resources Policy Analysis Network*. FANRPAN.

Neergaard, H., Shaw, E., & Carter, S. (2005). The impact of gender, social capital and networks on business ownership: A research agenda. *International Journal of Entrepreneurial Behaviour & Research, 11*(5), 338–357. doi:10.1108/13552550510614999

Negi, M. S., & Verma, H. (2015). Prospects of Conservation Agriculture in Hill Situation". Training Course Series 32, Compendium of lectures. *Tillage and Nutrient Dynamics for Better Crop Production (October 10-30, 2015), organized by CAFT, Department of Agronomy, GBPUA&T, Pantnagar, Sponsored by ICAR, New Delhi*, 66-72.

Neilsno, J., & Pritchard, B. (2010). Fairness and Ethicality in Their Place: The Regional Dynamics of Fair Trade and Ethical Sourcing Agendas in the Plantation Districts of South India. *Environment and Planning a Economy and Space*. doi:10.1068/a4260

New Business Models in the Accommodation Industry - Benchmarking of Rules and Regulations in the Short-term Rental Market. (2019). In New Business Models in the Accommodation Industry - Benchmarking of Rules and Regulations in the Short-term Rental Market. World Tourism Organization (UNWTO). doi:10.18111/9789284421084

Ngarava, S. (2020). Impact of the fast-track land reform programme (FTLRP) on agricultural production: A tobacco success story in Zimbabwe? *Land Use Policy, 99*, 105000. doi:10.1016/j.landusepol.2020.105000

Ngundu, O. A. (2008). *Mission churches and African marriage in Southern Rhodesia 1890-1970: contemporary African Christian dilemmas over marriage in historical and theological perspective* [Doctoral dissertation]. University of Cambridge. doi:10.17863/CAM.100699

Nicholls, A., & Opal, C. (2006). *Fair Trade: Market-Driven Ethical Consumption*. Choice Reviews Online. doi:10.5860/CHOICE.43-2905

Ning, C., Qu, J., He, L., Yang, R., Chen, Q., Luo, S., & Cai, K. (2017). Improvement of yield, pest control and Si nutrition of rice by rice-water spinach intercropping. *Field Crops Research, 208*, 34–43. doi:10.1016/j.fcr.2017.04.005

Njoh, A. J. (2018). The relationship between modern information and communications technologies (ICTs) and development in Africa. *Utilities Policy, 50*, 83–90. doi:10.1016/j.jup.2017.10.005

Norton, G. W., & Alwang, J. (2020). Changes in Agricultural Extension and Implications for Farmer Adoption of New Practices. *Applied Economic Perspectives and Policy, 42*(1), 8–20. doi:10.1002/aepp.13008

Norton, G. W., Alwang, J., & Masters, W. A. (2021). *Economics of agricultural development: world food systems and resource use*. Routledge. doi:10.4324/9780429316999

Nyagadza, B., & Chuchu, T. (2023). *Mobile Fin-Tech Ecosystem Shaping Financial Inclusion in Zimbabwean Banking and Financial Services Markets Metadata of the chapter that will be visualized online.* doi:10.1007/978-3-031-23863-5

O'Donnell, M. (2013). Using ICT to enhance marketing for small agricultural producers. *USAID Briefing Paper,* 1–6.

Obi-Egbedi, O., Osanyinlusi, O. I., & Adekunle, O. B. (2022). Does contract farming really matter in cassava farms productivity in Iseyin Local Government Area, Oyo State, Nigeria? *Agricultura Tropica et Subtropica*, *55*(1), 65–73. doi:10.2478/ats-2022-0008

Odunze, D. (2015). *An Analysis of the Impact of the Use of Mobile Communica tion Technologies by Farmers in Zimbabwe. A Case Study of Esoko and EcoFarmer Platforms.* https://www.researchgate.net/publication/330347260

Odunze, D., & Hove, M. T. (2022). *An Analysis of the Impact of the Use of Mobile Communica tion Technologies by Farmers in Zimbabwe. A Case Study of Esoko and EcoFarmer Platforms.* Academic Press.

Oerke, E. C. (2006). Crop losses to pests. *Journal of Agricultural Science*, *144*(1), 31–43. doi:10.1017/S0021859605005708

Okediran, O. O. (2019). An e-agriculture framework for inclusive agricultural value chains in Nigeria. Annals Computer Science Series, Tibiscus University of Timisoara, Romania, 17(2), 209-219.

Omilola, B., & Robele, S. (2017). *The Central position of agriculture within the 2030 Agenda for sustainable development* (Vol. 1683). Intl Food Policy Res Inst.

Osman, R. S. (2019). *Enhanced biogas production by anaerobic co-digestion from food waste and agriculture byproducts* [Master's thesis]. PAUWES.

Otsuka, K., Nakano, Y., & Takahashi, K. (2016). Contract farming in developed and developing countries. *Annual Review of Resource Economics*, *8*(1), 353–376. doi:10.1146/annurev-resource-100815-095459

Owolabi, O., Oku, A.-R. O., Alejo, A., Ogunbiyi, T., & Ubah, J. I. (2021). Access to electricity, information and communications technology (ICT), and financial development: Evidence from West Africa. *International Journal of Energy Economics and Policy*, *11*(2), 247–259. doi:10.32479/ijeep.10344

Oya, C., Schaefer, F., Skalidou, D., McCosker, C., & Langer, L. (2017). Effects of Certification Schemes for Agricultural Production on Socio-Economic Outcomes in Low- And Middle-Income Countries. *Systematic Reviews*. Advance online publication. doi:10.23846/SR0034

Palinkas, L. A., Horwitz, S. M., Green, C. A., Wisdom, J. P., Duan, N., & Hoagwood, K. (2015). Purposeful Sampling for Qualitative Data Collection and Analysis in Mixed Method Implementation Research. *Administration and Policy in Mental Health*, *42*(5), 2–4. doi:10.1007/s10488-013-0528-y PMID:24193818

Pangapanga-Phiri, I., Mungatana, E., & Mhondoro, G. (2024). Does contract farming arrangement improve smallholder tobacco productivity? Evidence from Zimbabwe. *Heliyon*, *10*(1), e23862. Advance online publication. doi:10.1016/j.heliyon.2023.e23862 PMID:38205304

Paredes, M. (2016). The glocalization of mining conflict: Cases from Peru. *The Extractive Industries and Society*, *3*(4), 1046–1057. doi:10.1016/j.exis.2016.08.007

Paroda, R. S., & Joshi, P. K. (2019). Sustainable development goals: role of agriculture. *2030 Agenda and India: Moving from Quantity to Quality: Exploring Convergence and Transcendence*, 17-40.

Partovifar, P. (2016). *Role of ICT in marketing of garden products* [MSc thesis]. Islamic Azad University.

Partzsch, L., & Kemper, L. (2019). Cotton certification in Ethiopia: Can an increasing demand for certified textiles create a 'fashion revolution'? *Geoforum*, *99*, 111–119. doi:10.1016/j.geoforum.2018.11.017

Parvathi, P., Grote, U., & Waibel, H. (2017). *Fair Trade and Organic Agriculture: A Winning Combination?* doi:10.1079/9781786393050.0000

Paul, C., Weber, M., & Knoke, T. (2017). Agroforestry versus farm mosaic systems–Comparing land-use efficiency, economic returns and risks under climate change effects. *The Science of the Total Environment*, *587*, 22–35. doi:10.1016/j.scitotenv.2017.02.037 PMID:28189309

Pazderka, C. (2010). *Procurement for Development: Aligning Social with Commercial Considerations in the Supply Chain*. Academic Press.

Pedzisai, C. (2020). *An analysis of indigenous agricultural knowledge systems in the Zimbabwe secondary school agriculture curriculum: Prospects and opportunities*. Academic Press.

Pedzisa, T. (2016). *Determinants of yield impact and adoption of conservation agriculture among smallholder farmers in Zimbabwe*. University of Pretoria.

Penot, E., Fevre, V., Flodrops, P., & Razafimahatratra, H. M. (2015). *Conservation agriculture and agroecology practices to mitigate climatic variations in medium altitude in Madagascar*. CIRAD.

Pepin, I. (2022). *What is Sustainable Agriculture and Why is it Important?* The Humane League. https://thehumanleague.org/article/what-is-sustainable-agriculture-and-why-is-it-important

Pepsico. (2023). *Sustainable sourcing*. https://www.pepsico.com/our-impact/esg-topics-a-z/sustainable-sourcing

Peters, K. A. (2010). Creating a sustainable urban agriculture revolution. *J. Envtl. L. & Litig.*, *25*, 203.

Piñeiro, V., Arias, J., Dürr, J., Elverdin, P., Ibáñez, A. M., Kinengyere, A., Opazo, C. M., Owoo, N., Page, J. R., Prager, S. D., & Torero, M. (2020). A scoping review on incentives for adoption of sustainable agricultural practices and their outcomes. *Nature Sustainability*, *3*(10), 809–820. doi:10.1038/s41893-020-00617-y

Poku, K. O. (2016). *Effects of Illegal Mining on the Agriculture Sector in Ghana*. Policy Hub. https://blogs.ubc.ca/kop7865/2016/11/30/effects-of-illegal-mining-on-the-agriculture-sector-in-ghana

Ponelis, S. R. (2015). Using interpretive qualitative case studies for exploratory research in Doctoral studies : A case of information systems research in small and medium enterprises. *International Journal of Doctoral Studies*, *10*, 535–550. doi:10.28945/2339

Posey, D. (2002). Upsetting the sacred balance: Can the study of indigenous knowledge reflect cosmic connectedness? In P. Sillitoe, A. Bicker, & J. Pottier (Eds.), *Participating in development: Approaches to indigenous knowledge* (Vol. 39, pp. 24–42). Routledge. doi:10.4324/9780203428603_chapter_2

Pratama, I. P., Winarso, H., Hudalah, D., & Syabri, I. (2021). Extended urbanization through capital centralization: Contract farming in palm oil-based agroindustrialization. *Sustainability (Basel)*, *13*(18), 10044. doi:10.3390/su131810044

Pretty, J. N., Toulmin, C., & Williams, S. (2011). Sustainable intensification in African agriculture. *International Journal of Agricultural Sustainability*, *9*(1), 5–24. doi:10.3763/ijas.2010.0583

Pryzgoda, J., & Chrisler, J. C. (2000). Definitions of gender and sex: The subtleties of meaning. *Sex Roles*, *43*(7/8), 553–569. doi:10.1023/A:1007123617636

Quandt, A., Salerno, J. D., Neff, J. C., Baird, T. D., Herrick, J. E., Terrence McCabe, J., Xu, E., & Hartter, J. (2020). Mobile phone use is associated with higher smallholder agricultural productivity in Tanzania, East Africa. *PLoS One*, *15*(8), e0237337. doi:10.1371/journal.pone.0237337 PMID:32760125

Quayson, M., Bai, C., & Osei, V. (2020). Digital inclusion for resilient post-COVID-19 supply chains: Smallholder farmer perspectives. *IEEE Engineering Management Review*, *48*(3), 104–110. doi:10.1109/EMR.2020.3006259

Rabo Usha, M. (2022). *Impact of tillage practices on runoff, soil and nutrient loss and wheat yield at kulumsa, central Ethiopia* [Doctoral dissertation]. Hawassa University.

Ratinho, T., Amezcua, A., Honig, B., & Zeng, Z. (2020). Supporting entrepreneurs: A systematic review of literature and an agenda for research. *Technological Forecasting and Social Change*, *154*, 119956. doi:10.1016/j.techfore.2020.119956

Rehman, A., Farooq, M., Lee, D. J., & Siddique, K. H. (2022). Sustainable agricultural practices for food security and ecosystem services. *Environmental Science and Pollution Research International*, *29*(56), 84076–84095. doi:10.1007/s11356-022-23635-z PMID:36258111

Reinecke, M., & Prinsloo, T. (2017). The influence of drone monitoring on crop health and harvest size. *2017 1st International Conference on Next Generation Computing Applications, NextComp 2017*, 5–10. 10.1109/NEXTCOMP.2017.8016168

Renn, B. N., Hoeft, T. J., Lee, H. S., Bauer, A. M., & Areán, P. A. (2019). Preference for in-person psychotherapy versus digital psychotherapy options for depression: Survey of adults in the US. *NPJ Digital Medicine*, *2*(1), 6. doi:10.1038/s41746-019-0077-1 PMID:31304356

Reynar, R. (1999). Indigenous people's knowledge and education: A tool for development? In L. M. Semali & J. L. Kicheloe (Eds.), *What is indigenous knowledge system: Voices from the academy* (pp. 285–304). Falmer.

Ribeiro-Duthie, A. C., Gale, F., & Murphy-Gregory, H. (2021a). Fair trade and staple foods: A systematic review. In *Journal of Cleaner Production* (Vol. 279). Elsevier Ltd. doi:10.1016/j.jclepro.2020.123586

Ribeiro-Duthie, A. C., Gale, F., & Murphy-Gregory, H. (2021b). Fair trade governance: Revisiting a framework to analyse challenges and opportunities for sustainable development towards a green economy. *Discover Sustainability*, *2*(1), 58. Advance online publication. doi:10.1007/s43621-021-00063-6 PMID:35425912

Richards, L., & Morse, J. M. (2006). Qualitative research design. Read Me First for a User's Guide to Qualitative Methods, 73–104. doi:10.1186/1472-6963-11-23

Rivera-Ferre, M. G., Ortega-Cerdà, M., & Baumgärtner, J. (2013). Rethinking study and management of agricultural systems for policy design. *Sustainability (Basel)*, *5*(9), 3858–3875. doi:10.3390/su5093858

Rizqiyanto, S. (2017). Starbucks's Fair Trade in The Edge of Globalization. *ETIKONOMI*, *16*(2), 231–248. doi:10.15408/etk.v16i2.5546

Robinson, O. C. (2014). Sampling in Interview-Based Qualitative Research: A Theoretical and Practical Guide. *Qualitative Research in Psychology*, *11*(1), 25–41. doi:10.1080/14780887.2013.801543

Roebuck, J. (1979). When does "old age begin? The evolution of the English definition. *Journal of Social History*, *12*(3), 416–428. doi:10.1353/jsh/12.3.416

Rohde, R. F., Moleele, N. M., Mphale, M., Allsopp, N., Chanda, R., Hoffman, M. T., Magole, L., & Young, E. (2006). Dynamics of grazing policy and practice: Environmental and social impacts in three communal areas of southern Africa. *Environmental Science & Policy*, *9*(3), 302–316. doi:10.1016/j.envsci.2005.11.009

Rola-Rubzen, M. F., Paris, T., Hawkins, J., & Sapkota, B. (2020). Improving gender participation in agricultural technology adoption in Asia: From rhetoric to practical action. *Applied Economic Perspectives and Policy*, *42*(1), 113–125. doi:10.1002/aepp.13011

Rose, D. C., Sutherland, W. J., Barnes, A. P., Borthwick, F., Ffoulkes, C., Hall, C., Moorby, J. M., Nicholas-Davies, P., Twining, S., & Dicks, L. V. (2019). Integrated farm management for sustainable agriculture: Lessons for knowledge exchange and policy. *Land Use Policy, 81*, 834–842. doi:10.1016/j.landusepol.2018.11.001

Rose, M. R., Flatt, T., Graves, J. L., Greer, L. F., Martínez, D. E., Matos, M., Mueller, L. D., Shmookler Reis, R. J., & Shahrestani, P. (2012). What is aging? *Frontiers in Genetics, 3*, 134. doi:10.3389/fgene.2012.00134 PMID:22833755

Roulston, K., & Choi, M. (2018). Qualitative Interviews. In U. Flick (Ed.), *The SAGE Handbook of Qualitative Data Collection*. SAGE Publications. doi:10.4135/9781526416070.n15

Roy, R., Phau, I., & Lee, T. (2012). Quality and Ethical Sourcing among Smallholder Coffee Producers in Papua New Guinea. *Proceedings of the 12th International Conference of the Society for Global Business & Economic Development*, 1488–1496. https://espace.curtin.edu.au/bitstream/20.500.11937/25122/2/162893_39843_21.2.pdf

Ruckert, A., Ciurlia, D., Labonte, R., Lencucha, R., Drope, J., Nhamo, N., Kadungure, A., & Mlambo, Z. (2022). *The Political Economy of Tobacco Production and Control in Zimbabwe: A Document Analysis*. https://tobacconomics.org/files/research/763/political-economy-of-tobacco-production-and-control-in-zimbabwe-draft-6-final.pdf

Ruml, A., & Qaim, M. (2020). Effects of marketing contracts and resource providing contracts in the African small farm sector: Insights from oil palm production in Ghana. *World Development, 136*, 105110. doi:10.1016/j.worlddev.2020.105110

Ruml, A., & Qaim, M. (2021). New evidence regarding the effects of contract farming on agricultural labor use. *Agricultural Economics, 52*(1), 51–66. doi:10.1111/agec.12606

Sánchez Prieto, J. C., Olmos Migueláñez, S., & García-Peñalvo, F. J. (2016). Informal toolsin formal contexts: Development of a model to assess the acceptance of mobile technologies among teachers. *Computers in Human Behavior, 55A*, 519–528. doi:10.1016/j.chb.2015.07.002

Sandelowski, M. (1995). Sample size in qualitative research. *Research in Nursing & Health, 18*(2), 179–183. doi:10.1002/nur.4770180211 PMID:7899572

Sardar, A., Kiani, A. K., & Kuslu, Y. (2021). Does adoption of climate-smart agriculture (CSA) practices improve farmers' crop income? Assessing the determinants and its impacts in Punjab province, Pakistan. *Environment, Development and Sustainability, 23*(7), 10119–10140. doi:10.1007/s10668-020-01049-6

Sarker, M. N. I., Wu, M., Alam, G. M., & Islam, M. S. (2019). Role of climate smart agriculture in promoting sustainable agriculture: A systematic literature review. *International Journal of Agricultural Resources, Governance and Ecology, 15*(4), 323–337. doi:10.1504/IJARGE.2019.104199

Saunders, B., Sim, J., Kingstone, T., Baker, S., Waterfield, J., Bartlam, B., Burroughs, H., & Jinks, C. (2018). Saturation in qualitative research: Exploring its conceptualization and operationalization. *Quality & Quantity*, *52*(4), 1893–1907. doi:10.1007/s11135-017-0574-8 PMID:29937585

Saunders, M., Lewis, P., & Thornhill, A. (2007). *Research methods for business students*. Pearson Education.

Saunders, M., Lewis, P., & Thornhill, A. (2018). *Research Methods for Business Students* (8th ed.). Pearson Education Limited.

Scalise, A., Pappa, V. A., Gelsomino, A., & Rees, R. M. (2017). Pea cultivar and wheat residues affect carbon/nitrogen dynamics in pea-triticale intercropping: A microcosms approach. *The Science of the Total Environment*, *592*, 436–450. doi:10.1016/j.scitotenv.2017.03.012 PMID:28340454

Schenk, P. (2019). A Matter of Principle: Comparing Norm-Based Explanations for Fair Trade Consumption. *Journal of Consumer Policy*, *42*(3), 397–423. doi:10.1007/s10603-018-9401-4

Schenk, P., Rössel, J., & Weingartner, S. (2021). It's all about distinction: The lifestyle embeddedness of fair trade consumption. *Sustainability (Basel)*, *13*(19), 10997. Advance online publication. doi:10.3390/su131910997

Scherr, S. J., Shames, S., & Friedman, R. (2012). From climate-smart agriculture to climate-smart landscapes. *Agriculture & Food Security*, *1*(1), 1–15. doi:10.1186/2048-7010-1-12

Schlegelmilch, F., & Ollenburg, C. (2013). Marketing the adventure: Utilizing the aspects of risk/fear/thrill to target the youth traveller segment. *Tourism Review*, *68*(3), 44–54. doi:10.1108/TR-03-2013-0010

Schmidt, E. (1990). Negotiated spaces and contested terrain: Men, women, and the law in colonial Zimbabwe, 1890–1939. *Journal of Southern African Studies*, *16*(4), 622–648. doi:10.1080/03057079008708254

Scoones, I. (1998). *Sustainable rural livelihoods: a framework for analysis*. Academic Press.

Scoones, I. (2021). *Can the Pfumvudza Conservation Agriculture Programme deliver food security in Zimbabwe? Article on future agriculture*. https://www.future-agricultures.org/blog/can-the-pfumvudza-conservation-agriculture-Programmeme-deliver-food-security-in-zimbabwe/

Scoones, I., Mavedzenge, B., Murimbarimba, F., & Sukume, C. (2018). Tobacco, contract farming, and agrarian change in Zimbabwe. *Journal of Agrarian Change*, *18*(1), 22–42. doi:10.1111/joac.12210

Semali, M. J., & Kincheloe, J. L. (1999). What is indigenous knowledge, and why we should study it? In M. L. Semali & J. L. Kincheloe (Eds.), *What is indigenous knowledge system: Voices from the academy* (pp. 3–57). Falmer.

Sen, D. (2018). Fempreneurs or Organic Tea Farmers? Entrepreneurialism, Resilience and Alternative Agriculture in Darjeeling, India. *Journal of Political Ecology*, *25*(1). Advance online publication. doi:10.2458/v25i1.22386

Shackleton, R. T. (2020). Loss of land and livelihoods from mining operations: A case in the Limpopo Province, South Africa. *Land Use Policy*, *99*, 104825. doi:10.1016/j.landusepol.2020.104825

Shackleton, S., & Luckert, M. (2015). Changing livelihoods and landscapes in the rural Eastern Cape, South Africa: Past influences and future trajectories. *Land (Basel)*, *4*(4), 1060–1089. doi:10.3390/land4041060

Shafiee-Jood, M., & Cai, X. (2016). Reducing food loss and waste to enhance food security and environmental sustainability. *Environmental Science & Technology*, *50*(16), 8432–8443. doi:10.1021/acs.est.6b01993 PMID:27428555

Shah, D. (2022). Role of Information and Communication Technology in Agricultural Development of India. *Review of Market Integration*, *14*(2-3), 113–132. doi:10.1177/09749292221119286

Sharara, K. (2023). Pfumvudza mitigates climate change. *The Sunday News*. www.sundaynews.co.zw/pfumvudza-mitigates-climate-change/

Sharma, R. K., Abidi, N., & Khan, K. M. (2018). Comparison of conventional and fair trade systems on dimensions of sustainability: A study of basmati rice procurement in India. *International Journal of Innovation and Sustainable Development*, *12*(4), 446. doi:10.1504/IJISD.2018.095066

Shetto, R. M., & Owenya, M. (2007). *Conservation Agriculture as Practiced in Tanzania: Three Case Studies*. African Conservation Tillage Network.

Shih-Tse Wang, E., & Chen, Y.-C. (2019). Effects of perceived justice of fair trade organizations on consumers' purchase intention toward fair trade products. *Journal of Retailing and Consumer Services*, *50*, 66–72. https://doi.org/https://doi.org/10.1016/j.jretconser.2019.05.004. doi:10.1016/j.jretconser.2019.05.004

Shiringo, L., Chundu, M., & Sithole, P. (2022). Contributing of smallholder tobacco contract farming to economic and human development of rural communities in Zimbabwe: A case of Igava community, Marondera District. *Open Journal of Social Sciences*, *10*(9), 36–57. doi:10.4236/jss.2022.109004

Shonhe, T., & Scoones, I. (2021). Private and state-led contract farming in Zimbabwe: Accumulation, social differentiation and rural politics. *Journal of Agrarian Change*, *22*(1), 118–138. doi:10.1111/joac.12473

Shorette, K. (2014). *Nongovernmental Regulation and Construction of Value in Global Markets*. doi:10.1177/0731121414543027

Siankwilimba, E., Hiddlestone-Mumford, J., Hoque, M. E., Hang'ombe, B. M., Mumba, C., Hasimuna, O. J., Maulu, S., Mphande, J., Chibesa, M., Moono, M. B., Muhala, V., Cavaliere, L. P. L., Faccia, A., & Prayitno, G. (2023). Sustainability of agriculture extension services in the face of COVID-19: A study on gender-specific market systems. *Cogent Food & Agriculture*, *9*(2), 2284231. doi:10.1080/23311932.2023.2284231

Sillars, A. L., & Kalbflesch, P. J. (1989). Implicit and Explicit Decision-Making Styles in Couples. In D. Brinberg & J. Jaccard (Eds.), *Dyadic Decision Making*. Springer. doi:10.1007/978-1-4612-3516-3_8

Simangunsong, E., Hendry, L., & Stevenson, M. (2016). Managing Supply Chain Uncertainty With Emerging Ethical Issues. *International Journal of Operations & Production Management*, *36*(10), 1272–1307. Advance online publication. doi:10.1108/IJOPM-12-2014-0599

Singh, A. (2023). *Shweta Sharma*. Academic Press.

Singh, A. K. (2013). Income and livelihood issues of farmers: A field study in Uttar Pradesh. *Agricultural Economics Research Review Conference No*, *26*, 89–96.

Singh, R. K., & Sureja, A. K. (2008). Indigenous Knowledge and Sustainable Agricultural Resources Management under Rainfed Agro-Ecosystems. *Indian Journal of Traditional Knowledge*, *7*(4), 642–654.

Singh, R., & Singh, G. S. (2017). Traditional agriculture: A climate-smart approach for sustainable food production. *Energy, Ecology & Environment*, *2*(5), 296–316. doi:10.1007/s40974-017-0074-7

Si, R., Zhang, X., Yao, Y., Liu, L., & Lu, Q. (2021). Influence of contract commitment system in reducing information asymmetry, and prevention and control of livestock epidemics: Evidence from pig farmers in China. *One Health*, *13*, 100302. doi:10.1016/j.onehlt.2021.100302 PMID:34430696

Slack, K. (2013). *The growing battle between mining and agriculture*. Oxfam Politics of Poverty.

Smidt, H. J., & Jokonya, O. (2022). Factors affecting digital technology adoption by small-scale farmers in agriculture value chains (AVCs) in South Africa. *Information Technology for Development*, *28*(3), 558–584. doi:10.1080/02681102.2021.1975256

Snyder, H. (2019). Literature review as a research methodology: An overview and guidelines. *Journal of Business Research*, *104*, 333–339. doi:10.1016/j.jbusres.2019.07.039

Starbucks. (2020). *Starbucks: Ethical Sourcing - Coffee: Starbucks Coffee Company*. https://www.starbucks.com/responsibility/sourcing/coffee/

Steenwerth, K. L., Hodson, A. K., Bloom, A. J., Carter, M. R., Cattaneo, A., Chartres, C. J., Hatfield, J. L., Henry, K., Hopmans, J. W., Horwath, W. R., Jenkins, B. M., Kebreab, E., Leemans, R., Lipper, L., Lubell, M. N., Msangi, S., Prabhu, R., Reynolds, M. P., Sandoval Solis, S., ... Jackson, L. E. (2014). Climate-smart agriculture global research agenda: Scientific basis for action. *Agriculture & Food Security*, *3*(1), 1–39. doi:10.1186/2048-7010-3-11

Subbulakshmi, S., Saravanan, N., & Subbian, P. (2009). Conventional tillage vs conservation tillage - a review. *Agric. Rev., 30*.

Suding, K., Higgs, E., Palmer, M., Callicott, J. B., Anderson, B., Baker, C. B., Gutrich, M., Hondula, J. J., Matthew, K. L., Lafevor, C., Larson, M. C., Randall, B. M. H., Ruhl, A., And, J. B., & Schwartz, K. Z. S. (2015). Committing to ecological restoration. *Environmental Monitoring and Assessment*, *348*(6235), 638–640. doi:10.1126/science.aaa4216 PMID:25953995

Talukder, B., vanLoon, G. W., & Hipel, K. W. (2019). Energy efficiency of agricultural systems in the southwest coastal zone of Bangladesh. *Ecological Indicators*, *98*, 641–648. doi:10.1016/j.ecolind.2018.11.030

Tang, C., Zhang, H., & Xie, J. (2022). Optimal contract design in contract farming under asymmetric effort information. *Sustainability (Basel)*, *14*(22), 15000. doi:10.3390/su142215000

Tanveer, A., Ikram, R. M., & Ali, H. H. (2019). Crop Rotation: Principles and Practices. In M. Hasanuzzaman (Ed.), *Agronomic Crops*. Springer. doi:10.1007/978-981-32-9783-8_1

Tenkorang, E. Y., & Osei-Kufuor, P. (2013). The impact of gold mining on local Farming Communities in Ghana. Journal of Global Initiatives: Policy, Pedagogy. *Perspective*, *8*(1), 3.

Terstappen, V., Hanson, L., & McLaughlin, D. (2012). Gender, Health, Labor, and Inequities: A Review of the Fair and Alternative Trade Literature. *Agriculture and Human Values*. Advance online publication. doi:10.1007/s10460-012-9377-7

Thao Tran, P., Vettersand Tutor, E., & Blombäck, A. (2012). *Fairtrade-a Competitive Imperative? Title: Fairtrade-a Competitive Imperative?* Author.

Tiwari, S. P. (2022). Information and communication technology initiatives for knowledge sharing in agriculture. *arXiv preprint arXiv:2202.08649.*

Ton, G., Vellema, W., Desiere, S., Weituschat, S., & D'Haese, M. (2018). Contract farming for improving smallholder incomes: What can we learn from effectiveness studies? *World Development*, *104*, 46–64. doi:10.1016/j.worlddev.2017.11.015

Toor, M. D. (2021). Nutrients and Their Importance in Agriculture Crop Production: A Review. *Indian Journal of Pure & Applied Biosciences*, *9*(1), 1–6. doi:10.18782/2582-2845.8527

Triberti, L., Nastri, A., & Baldoni, G. (2016). Long-term effects of crop rotation, manure and mineral fertilisation on carbon sequestration and soil fertility. *European Journal of Agronomy*, *74*, 47–55. doi:10.1016/j.eja.2015.11.024

U.S. Agency for International Development's Support. (2022). *Adapting food systems in Africa*. Author.

UC Sustainable Agriculture Research and Education Programme. (2021). *What is Sustainable Agriculture?* UC Agriculture and Natural Resources. https://sarep.ucdavis.edu/sustainable-ag

Udry, C., Hoddinott, J., Alderman, H., & Haddad, L. (1995). *Gender differentials in farm productivity: Implications for household efficiency and agricultural policy*. Academic Press.

Umar, B. B., Aune, J. B., Johnsen, F. H., & Lungu, O. I. (2011). *Options for improving smallholder conservation agriculture in Zambia*. Academic Press.

Umesha, S., Manukumar, H. M., & Chandrasekhar, B. (2018). Sustainable agriculture and food security. In *Biotechnology for sustainable agriculture* (pp. 67–92). Woodhead Publishing. doi:10.1016/B978-0-12-812160-3.00003-9

Umrani, R. (2010). *Agroforestry systems and practices*. Academic Press.

Utting, K. (2008). Assessing the Impact of Fair Trade Coffee: Towards an Integrative Framework. *Journal of Business Ethics*. Advance online publication. doi:10.1007/s10551-008-9761-9

Vamuloh, V. V., Kozak, R. A., & Panwar, R. (2020). Voices unheard: Barriers to and opportunities for small farmers' participation in oil palm contract farming. *Journal of Cleaner Production*, *275*, 121955. doi:10.1016/j.jclepro.2020.121955

Van Baar, A. (n.d.). *A Fair Price for Human Rights Due Diligence*. https://ecco-plan.com/

Van Dam, P. (2015). The limits of a success story: Fair trade and the history of postcolonial globalization. *Comparativ (Leipzig)*, *25*(1). Advance online publication. doi:10.26014/j.comp.2015.01.05

van der Werff, E., & Steg, L. (2016). The psychology of participation and interest in smart energy systems: Comparing the value-belief-norm theory and the value-identity-personal norm model. *Energy Research & Social Science*, *22*, 107–114. doi:10.1016/j.erss.2016.08.022

van Noordwijk, M., Minang, P. A., Dewi, S., Hall, J., & Rantala, S. (2009). *Reducing Emissions from All Land Uses: The case for a whole landscape approach*. Academic Press.

van Wijk, M. T., Merbold, L., Hammond, J., & Butterbach-Bahl, K. (2020). Improving assessments of the three pillars of climate smart agriculture: Current achievements and ideas for the future. *Frontiers in Sustainable Food Systems*, *4*, 558483. doi:10.3389/fsufs.2020.558483

Vasileiou, K., Barnett, J., Thorpe, S., & Young, T. (2018). *Characterising and justifying sample size sufficiency in interview-based studies: Systematic analysis of qualitative health research over a 15-year period*. Academic Press.

Velten, S., Leventon, J., Jager, N., & Newig, J. (2015). What is sustainable agriculture? A systematic review. *Sustainability (Basel)*, *7*(6), 7833–7865. doi:10.3390/su7067833

Vicol, M. (2018). Potatoes, petty commodity producers and livelihoods: Contract farming and agrarian change in Maharashtra, India. *Journal of Agrarian Change*, *19*(1), 135–161. doi:10.1111/joac.12273

Volsi, B., Higashi, G. E., Bordin, I., & Telles, T. S. (2022). The diversification of species in crop rotation increases the profitability of grain production systems. *Scientific Reports*, *12*(1), 19849. doi:10.1038/s41598-022-23718-4 PMID:36400822

WFTO. (2023). *Our 10 Fair Trade Principles - Word Fair Trade Organisation*. https://wfto.com/our-fair-trade-system/our-10-principles-of-fair-trade/

White, B., & Wijaya, H. (2022). What kind of labour regime is contract farming? Contracting and sharecropping in Java compared. *Journal of Agrarian Change*, *22*(1), 19–35. doi:10.1111/joac.12459

White, G., & Samuel, A. (2016). Fairtrade and Halal Food Certification and Labeling. *Journal of Macromarketing, 36*(4), 388–399. Advance online publication. doi:10.1177/0276146715620236

Wolfert, S., Verdouw, C., van Wassenaer, L., Dolfsma, W., & Klerkx, L. (2023). Digital innovation ecosystems in agri-food: Design principles and organizational framework. *Agricultural Systems, 204*, 103558. doi:10.1016/j.agsy.2022.103558

Wong, S., & Guggenheim, S. (2018). Community-driven development: myths and realities. *World Bank Policy Research Working Paper* (8435).

World Economic Forum. (2021). *What effect do insects and pests have on crop production.* Available at https://www.weforum.org./agenda/2021/06/

Xiao, Y., & Watson, M. (2019). Guidance on Conducting a Systematic Literature Review. *Journal of Planning Education and Research, 39*(1), 93–112. doi:10.1177/0739456X17723971

Yadav, P., Davies, P. J., & Abdullah, S. (2018). Reforming capital subsidy scheme to finance energy transition for the below poverty line communities in rural India. *Energy for Sustainable Development, 45*, 11–27. doi:10.1016/j.esd.2018.04.001

Zhang, K., Liu, Z., & Wang, P. (2020). Exploring the supply chain management of fair trade business: Case study of a fair trade craft company in China. *Cogent Business and Management, 7*(1), 1734345. Advance online publication. doi:10.1080/23311975.2020.1734345

Zhang, Q. F. (2012). The Political Economy of Contract Farming in China's Agrarian Transition. *Journal of Agrarian Change, 12*(4), 460–483. doi:10.1111/j.1471-0366.2012.00352.x

Zimbalist, Z. (2018). "Fear-of-the-State Bias" in Survey Data. *International Journal of Public Opinion Research, 30*(4), 631–651. doi:10.1093/ijpor/edx020

Zomer, R. J., Neufeldt, H., Xu, J., Ahrends, A., Bossio, D., Trabucco, A., van Noordwijk, M., & Wang, M. (2016). Global Tree Cover and Biomass Carbon on Agricultural Land: The contribution of agroforestry to global and national carbon budgets. *Scientific Reports, 6*(1), 29987. doi:10.1038/srep29987 PMID:27435095

Zulu, O., & Chapoto, A. (2016). Conservation agriculture: Gendered impacts on households' livelihoods. *Zambia Social Science Journal.*

Zysk, W. (2020). Fair trade phenomenon and its evolution in Visegrad countries. *International Entrepreneurship Review, 6*(4), 81–98. doi:10.15678/IER.2020.0604.06

About the Contributors

Jabulani Garwi is an academic researcher, diplomat, and development practitioner specialising in sustainable development and African rural studies. With a strong interdisciplinary background, he holds a Bachelor of Arts Honours degree in Economic History from the University of Zimbabwe, a Master of Arts degree in Development Studies from the Midlands State University, and a Doctor of Philosophy degree in Sustainable Agriculture from the University of the Free State, South Africa. Dr. Garwi's research spans various critical areas including rural development, livelihoods, poverty dynamics, agrarian systems, food security, and the interplay between climate change and agriculture. His expertise also extends to contract farming, a vital mechanism for connecting smallholder farmers to agribusinesses and improving their market access and resource availability. Through his academic work, Dr. Garwi has made contributions to the field, publishing scholarly articles, participating in international conferences, and engaging in policy dialogues with governmental and non-governmental stakeholders. His research offers valuable insights and evidence-based recommendations, informing decision-making, shaping policy frameworks, and facilitating effective interventions for inclusive and sustainable development in rural Africa. Dr. Garwi's commitment to bridging the gap between academia and practice is evident through his active involvement in diplomatic and development initiatives.

Reason Masengu is currently a Senior Lecturer of Marketing in the Department of Business Management at Middle East College, Muscat, Oman. He is a lecturer in courses such as Global Marketing, Marketing Management, Digital Marketing, and other courses that fall within the business management profile. His duties range from student advisory, curriculum development, quality control, and any other activities related to teaching and learning.

Option Takunda Chiwaridzo is currently a 4th year PhD student at the University of Science and Technology in Beijing, China. He holds a Masters degree in Business Management and Corporate Governance from Midlands State University in Zimbabwe. For the past 8 years, Mr. Chiwaridzo has worked at the Ministry on SMEs in Zimbabwe. He has also been a part-time lecturer for 4 years at the Izeikel Guti University in Zimbabwe. Originally from Zimbabwe, Mr. Chiwaridzo moved to China to pursue his doctoral studies and research in Business Administration. His work focuses on tourism marketing, nation branding, energy sustainability and efficiency. In his free time, Mr. Chiwaridzo enjoys reading, writing, playing and watching soccer. With his advanced education, teaching experience, and background working in the public sector, Mr. Chiwaridzo is well-positioned to make valuable contributions in his field.

* * *

Mohd Aarif has been employed with the Global Research Network as a researcher. He graduated from Aligarh Muslim University in India with an MTA and JJT University in India with a PhD in Commerce and Management. He has worked in the educational field, industry, research, teaching, and consulting for more than ten years. He has produced a number of exceptional and inventive contributions to scholarly inquiry. He has written many books, several patents, and more than 45 papers in worldwide peer-reviewed impact journals, Elsevier, WoS, Scopus, and SCIE.

Annabel Banda is a lecturer and researcher within the department of Horticulture and Crop Production at Gwanda State University.

Trymore Chingwaru (PhD) is a lecturer in the department of marketing at Bindura University of Science Education. He has published on entrepreneurship, business management and marketing.

Phillip Dangaiso is a lecturer at Zimbabwe Ezekiel Guti University. He has published on marketing management, social marketing and consumer behaviour.

Blessing Hodzi is a Public Administration-specialising in peace studies, PhD student at the Durban University of Technology. She holds a Master of Science degree in Peace, Leadership, and Conflict Resolution from Zimbabwe Open University.

Hope Hogo is a lecturer at the Midlands State University. She holds a Master of Science Degree in Marketing and a Bachelor of Commerce Honours degree in Marketing obtained from the National University of Science and Technology. She has research interests in agribusiness marketing and marketing technology.

Divaries Cosmas Jaravaza is a Witswatersrand business school doctoral graduate, with a vast experience in business training, business subjects pedagogics and research. His research interests are in transformative consumer research, emerging and subsistence marketplaces, social, healthcare and sustainable marketing, entrepreneurship and consumer behavior.

Samuel M. (PhD) is a lecturer at Midlands State University, in the Department of Information and Marketing Sciences. He holds a PhD in Information Systems from University of South Africa (UNISA), a Post Graduate Diploma in Tertiary education from Midlands State University, a Bachelor of Science Honours Degree in Information Systems from Midlands State University, a Master Science Degree in Information Systems Management from Midlands State University. He is a founding member of the Zimbabwe Conference on Information Technology (ZCICT). He is a member of the ZIMCHE physical assessment for facilities and resources checklist for academic programs. He has contributed to the 2019 Technology Transfer Workshop (Intellectual property (IP)) at Gweru Polytechnic. He is a published author who has published research articles in Scopus indexed journals and non-Scopus indexed journals. He has also attended and presented at a number of international conferences (ICT4D, Springer and Elsevier organised).

Yvonne Madongonda is a Lecturer at Midlands State University, in the Department of Information and Marketing Sciences. She holds a Masters Degree in Marketing Strategy from Midlands State University, a Bachelor's Honours Degree in Marketing Management from Great Zimbabwe University and a Higher National Diploma In Accounting from Gweru Polytechnic College.

Neil B. Maheve is a PhD student at Rhodes University in the Political Science Department. He holds a Master's degree in African History from the University of Zimbabwe.

Reason Masengu holds the position of Senior Lecturer of Marketing in the Department of Business Management at Middle East College, located in Muscat, Oman. In this role, he imparts knowledge in courses such as Global Marketing, Marketing Management, Digital Marketing, and other subjects pertaining to the business management domain. He undertakes various responsibilities, including student counseling, developing curricula, ensuring quality standards, and facilitating teaching and learning activities. Reason holds the distinction of being a Fellow of the Chartered Institute of Marketing, a Senior Member of the Marketers Association of Zimbabwe, and a Professional Member of the

Australian Marketing Institute. As an ardent researcher in the fields of marketing, logistics, and supply management, he has acquired international exposure in administering research grants.

Fungai Mauchi (PhD) is the Deputy Dean of the Faculty of Commerce at Bindura University of Science Education.

Paul Mukucha has a PhD from Nelson Mandela Metropolitan University, South Africa. He has earned book prize and distinction in Supply chain management degree. He is the chairman of the department of marketing with vast experience in university teaching. His research interests are in supply chain management, services marketing, social and healthcare marketing.

Mthuthukisi Ncube is a Research Lecturer at the Institute of Development Studies, National University of Science and Technology with interests in balanced socio-ecological systems, ecosystem services, quality assurance in higher education, rights-based approaches, conflict and sustainability of natured livelihoods.

Joshua Risiro has a PhD from Kwazulu-Natal University, South Africa. He has vast experience in geography education, teacher education and environmental management. His research interests are in indigenous knowledge systems, curriculum studies, geography education and environmental management.

Index

R

S

T

V

W

Z

www.ingramcontent.com/pod-product-compliance
Lightning Source LLC
LaVergne TN
LVHW081340300926
849161LV00008B/2033